Luis Martinez-Sobrido (Ed.)

Replication-Competent Reporter-Expressing Viruses

MDPI

This book is a reprint of the Special Issue that appeared in the online, open access journal, *Viruses* (ISSN 1999-4915) from 2015–2016, available at:

http://www.mdpi.com/journal/viruses/special_issues/reporter_expressing_viruses

Guest Editor
Luis Martinez-Sobrido
University of Rochester
School of Medicine and Dentistry
USA

Editorial Office
MDPI AG
St. Alban-Anlage 66
Basel, Switzerland

Publisher
Shu-Kun Lin

Managing Editor
Delphine Guerin

1. Edition 2016

MDPI • Basel • Beijing • Wuhan • Barcelona • Belgrade

ISBN 978-3-03842-258-7 (Hbk)
ISBN 978-3-03842-259-4 (PDF)

Table of Contents

List of Contributors

Sally Al Ali Department of Preventive Medicine, Public Health and Microbiology, Universidad Autónoma, E-28029 Madrid, Spain.

Razim Ali University of Karachi, Karachi 75270, Pakistan.

Tong-Qing An The Key Laboratory of Veterinary Public Health, Ministry of Agriculture, State Key Laboratory of Veterinary Biotechnology, Harbin Veterinary Research Institute of the Chinese Academy of Agricultural Sciences, Harbin 150001, China.

Weiya Bai Key Laboratory of Medical Molecular Virology (MOH & MOE) and Institutes of Biomedical Sciences, School of Basic Medical Sciences, Shanghai Medical College, Fudan University, Shanghai 200032, China.

Steven F. Baker Department of Microbiology and Immunology, University of Rochester School of Medicine and Dentistry, 601 Elmwood Avenue, Rochester, NY 14642, USA.

Timothy S. Baker Department of Chemistry and Biochemistry and Division of Biological Sciences, University of California, San Diego, La Jolla, CA 92093, USA.

Sara Baldanta Department of Preventive Medicine, Public Health and Microbiology, Universidad Autónoma, E-28029 Madrid, Spain.

Stéphane Biacchesi VIM, INRA, Université Paris-Saclay, Jouy-en-Josas 78350, France.

Michael Breen Department of Microbiology and Immunology, University of Rochester School of Medicine and Dentistry, 601 Elmwood Avenue, Rochester, NY 14642, USA.

Michel Brémont VIM, INRA, Université Paris-Saclay, Jouy-en-Josas 78350, France.

Xue-Hui Cai The Key Laboratory of Veterinary Public Health, Ministry of Agriculture, State Key Laboratory of Veterinary Biotechnology, Harbin Veterinary Research Institute of the Chinese Academy of Agricultural Sciences, Harbin 150001, China.

Xiaoxian Cui Key Laboratory of Medical Molecular Virology (MOH & MOE) and Institutes of Biomedical Sciences, School of Basic Medical Sciences, Shanghai Medical College, Fudan University, Shanghai 200032, China.

Michael C. Currier Department of Pediatrics, Emory University School of Medicine, Atlanta, GA 30307, USA / Children's Healthcare of Atlanta, 1405 Clifton Road, Atlanta, GA 30322, USA.

Juan Carlos de la Torre Department of Immunology and Microbial Science, The Scripps Research Institute, La Jolla, CA 92037, USA.

Cheng-Lin Deng Key Laboratory of Special Pathogens and Biosafety, Center for Emerging Infectious Diseases, Wuhan Institute of Virology, Chinese Academy of Sciences, Wuhan 430071, China.

Qiong-Qiong Fang The Key Laboratory of Veterinary Public Health, Ministry of Agriculture, State Key Laboratory of Veterinary Biotechnology, Harbin Veterinary Research Institute of the Chinese Academy of Agricultural Sciences, Harbin 150001, China.

Mercedes Fernández-Escobar Department of Preventive Medicine, Public Health and Microbiology, Universidad Autónoma, E-28029 Madrid, Spain.

Susana Guerra Department of Preventive Medicine, Public Health and Microbiology, Universidad Autónoma, E-28029 Madrid, Spain.

Takayuki Hishiki Viral Infectious Diseases Project, Tokyo Metropolitan Institute of Medical Science, Tokyo 156-8506, Japan.

Weijin Huang Division of HIV/AIDS and Sexually Transmitted Virus Vaccines, National Institutes for Food and Drug Control (NIFDC), No. 2 Tiantanxili, Beijing 100050, China.

James K. Jancovich Department of Biological Sciences, California State University San Marcos, 333 S. Twin Oaks Valley Rd., San Marcos, CA 92096, USA.

Joyce Jose Department of Biological Sciences, Purdue University, West Lafayette, IN 47907, USA.

Fumihiro Kato Department of Virology 1, National Institute of Infectious Diseases, Tokyo 162-8640, Japan.

Shin-Hee Kim Virginia-Maryland Regional College of Veterinary Medicine, University of Maryland, College Park, MD 20742, USA.

Richard J. Kuhn Department of Bindley Bioscience Center; Department of Biological Sciences, Purdue University, West Lafayette, IN 47907, USA.

Lian-Feng Li State Key Laboratory of Veterinary Biotechnology, Harbin Veterinary Research Institute, Chinese Academy of Agricultural Sciences, 427 Maduan Street, Harbin 150001, China.

Peng-Hui Li Key Laboratory of Special Pathogens and Biosafety, Center for Emerging Infectious Diseases, Wuhan Institute of Virology, Chinese Academy of Sciences, Wuhan 430071, China.

Weike Li Department of Chemistry, College of Arts and Sciences, Georgia State University, Atlanta, GA 30302, USA.

Xiao-Dan Li School of Medicine, Hunan Normal University, Changsha 410000, China.

Yongfeng Li State Key Laboratory of Veterinary Biotechnology, Harbin Veterinary Research Institute, Chinese Academy of Agricultural Sciences, 427 Maduan Street, Harbin 150001, China.

Ji-Ting Liu College of Animal Science and Technology, Jilin Agriculture University, Changchun 130018, China / The Key Laboratory of Veterinary Public Health, Ministry of Agriculture, State Key Laboratory of Veterinary Biotechnology, Harbin Veterinary Research Institute of the Chinese Academy of Agricultural Sciences, Harbin 150001, China.

Jing Liu Key Laboratory of Medical Molecular Virology (MOH & MOE) and Institutes of Biomedical Sciences, School of Basic Medical Sciences, Shanghai Medical College, Fudan University, Shanghai 200032, China.

Si-Qing Liu Key Laboratory of Special Pathogens and Biosafety, Center for Emerging Infectious Diseases, Wuhan Institute of Virology, Chinese Academy of Sciences, Wuhan 430071, China.

Yangyang Liu Division of HIV/AIDS and Sexually Transmitted Virus Vaccines, National Institutes for Food and Drug Control (NIFDC), No. 2 Tiantanxili, Beijing 100050, China.

Luis Martínez-Sobrido Department of Microbiology and Immunology, University of Rochester School of Medicine and Dentistry, 601 Elmwood Avenue, Rochester, NY 14642, USA.

Emilie Mérour VIM, INRA, Université Paris-Saclay, Jouy-en-Josas 78350, France.

Martin L. Moore Department of Pediatrics, Emory University School of Medicine, Atlanta, GA 30307, USA; Children's Healthcare of Atlanta, 1405 Clifton Road, Atlanta, GA 30322, USA.

Elke Mühlberger National Emerging Infectious Diseases Laboratories (NEIDL); Department of Microbiology, School of Medicine, Boston University, 620 Albany Street, Boston, MA 02118, USA.

Jianhui Nie Division of HIV/AIDS and Sexually Transmitted Virus Vaccines, National Institutes for Food and Drug Control (NIFDC), No. 2 Tiantanxili, Beijing 100050, China.

Aitor Nogales Department of Microbiology and Immunology, University of Rochester School of Medicine and Dentistry, 601 Elmwood Avenue, Rochester, NY 14642, USA.

Cheng-Feng Qin State Key Laboratory of Pathogen and Biosecurity, Beijing Institute of Microbiology and Epidemiology, Beijing 100071, China.

Hua-Ji Qiu State Key Laboratory of Veterinary Biotechnology, Harbin Veterinary Research Institute, Chinese Academy of Agricultural Sciences, 427 Maduan Street, Harbin 150001, China.

Jacques Robert Department of Microbiology and Immunology, University of Rochester Medical Center, Rochester, NY 14642, USA.

Christina A. Rostad Department of Pediatrics, Emory University School of Medicine, Atlanta, GA 30307, USA, Children's Healthcare of Atlanta, 1405 Clifton Road, Atlanta, GA 30322, USA.

Ronan N. Rouxel VIM, INRA, Université Paris-Saclay, Jouy-en-Josas 78350, France.

Siba K. Samal Virginia-Maryland Regional College of Veterinary Medicine, University of Maryland, College Park, MD 20742, USA.

Kristina Maria Schmidt Friedrich-Loeffler-Institut, Federal Research Institute for Animal Health, Institute of Novel and Emerging Infectious Diseases, Greifswald-Insel Riems 17493, Germany.

Ming-Xia Sun The Key Laboratory of Veterinary Public Health, Ministry of Agriculture, State Key Laboratory of Veterinary Biotechnology, Harbin Veterinary Research Institute of the Chinese Academy of Agricultural Sciences, Harbin 150001, China.

Jinghua Tang Department of Chemistry and Biochemistry and Division of Biological Sciences, University of California, San Diego, La Jolla, CA 92093, USA.

Yan-Dong Tang The Key Laboratory of Veterinary Public Health, Ministry of Agriculture, State Key Laboratory of Veterinary Biotechnology, Harbin Veterinary Research Institute of the Chinese Academy of Agricultural Sciences, Harbin 150001, China.

Aaron B. Taylor Department of Bindley Bioscience Center, Purdue University, West Lafayette, IN 47907, USA.

Zhi-Jun Tian The Key Laboratory of Veterinary Public Health, Ministry of Agriculture, State Key Laboratory of Veterinary Biotechnology, Harbin Veterinary Research Institute of the Chinese Academy of Agricultural Sciences, Harbin 150001, China.

Tong-Yun Wang The Key Laboratory of Veterinary Public Health, Ministry of Agriculture, State Key Laboratory of Veterinary Biotechnology, Harbin Veterinary Research Institute of the Chinese Academy of Agricultural Sciences, Harbin 150001, China.

Xiao Wang State Key Laboratory of Veterinary Biotechnology, Harbin Veterinary Research Institute, Chinese Academy of Agricultural Sciences, 427 Maduan Street, Harbin 150001, China.

Youchun Wang Division of HIV/AIDS and Sexually Transmitted Virus Vaccines, National Institutes for Food and Drug Control (NIFDC), No. 2 Tiantanxili, Beijing 100050, China.

Hong-Ping Wei Key Laboratory of Special Pathogens and Biosafety, Center for Emerging Infectious Diseases, Wuhan Institute of Virology, Chinese Academy of Sciences, Wuhan 430071, China.

Libao Xie State Key Laboratory of Veterinary Biotechnology, Harbin Veterinary Research Institute, Chinese Academy of Agricultural Sciences, 427 Maduan Street, Harbin 150001, China.

Youhua Xie Key Laboratory of Medical Molecular Virology (MOH & MOE) and Institutes of Biomedical Sciences, School of Basic Medical Sciences, Shanghai Medical College, Fudan University, Shanghai 200032, China.

Lin-Lin Xu Key Laboratory of Special Pathogens and Biosafety, Center for Emerging Infectious Diseases, Wuhan Institute of Virology, Chinese Academy of Sciences, Wuhan 430071, China.

Han-Qing Ye Key Laboratory of Special Pathogens and Biosafety, Center for Emerging Infectious Diseases, Wuhan Institute of Virology, Chinese Academy of Sciences, Wuhan 430071, China.

Shaoxiong Yu State Key Laboratory of Veterinary Biotechnology, Harbin Veterinary Research Institute, Chinese Academy of Agricultural Sciences, 427 Maduan Street, Harbin 150001, China.

Jiahui Yu State Key Laboratory of Veterinary Biotechnology, Harbin Veterinary Research Institute, Chinese Academy of Agricultural Sciences, 427 Maduan Street, Harbin 150001, China.

Zhi-Ming Yuan Key Laboratory of Special Pathogens and Biosafety, Center for Emerging Infectious Diseases, Wuhan Institute of Virology, Chinese Academy of Sciences, Wuhan 430071, China.

Bo Zhang Key Laboratory of Special Pathogens and Biosafety, Center for Emerging Infectious Diseases, Wuhan Institute of Virology, Chinese Academy of Sciences, Wuhan 430071, China.

Lingkai Zhang State Key Laboratory of Veterinary Biotechnology, Harbin Veterinary Research Institute, Chinese Academy of Agricultural Sciences, 427 Maduan Street, Harbin 150001, China.

Pan-Tao Zhang Key Laboratory of Special Pathogens and Biosafety, Center for Emerging Infectious Diseases, Wuhan Institute of Virology, Chinese Academy of Sciences, Wuhan 430071, China.

Dong-Gen Zhou Ningbo International Travel Healthcare Center, Ningbo 315012, China.

About the Guest Editor

Luis Martinez-Sobrido, Ph.D., is currently an Associate Professor in the Department of Microbiology and Immunology at University of Rochester. His Ph.D. research focused on the study of viral replication and transcription of the respiratory syncytial virus under the guidance of Dr. Jose Antonio Melero at the Instituto de Salud Carlos III in Madrid, Spain. He also conducted post-doctoral research on the molecular biology of influenza viruses under the supervision of Dr. Adolfo Garcia-Sastre at the Icahn School of Medicine at Mount Sinai in New York, USA. His research interest has previously focused on the molecular biology, immunology and pathogenesis of negative-stranded influenza viruses (respiratory syncytial virus, human metapneumovirus, arenavirus, thogoto virus, ebola virus, Crimean Congo hemorrhagic fever virus) and positive-stranded (dengue virus, SARS coronavirus, mouse hepatitis virus) RNA and DNA (human cytomegalovirus and vaccinia) viruses. His current research interest focuses on the molecular biology of RNA viruses, mainly arenaviruses and influenza.

Preface to "Replication-Competent Reporter-Expressing Viruses"

With the development of reverse genetics systems, recombinant viruses expressing reporter fluorescent or bioluminescent genes represent an excellent option to evaluate the dynamics of viral infection progression in both cultured cells and/or validated animal models of infection. Expression of reporter proteins allows for direct viral detection in vitro and in vivo, without the use of secondary methodologies to identify infected cells. By eliminating the need of secondary labeling, fluorescent or bioluminescence tractable replicating-compatible viruses provide an ideal tool to monitor viral infections in real time, representing a significant advance in the study of the biology of viruses, to evaluate vaccination approaches, and to identify new therapeutics against viral infections using high-throughput screening settings. In this Special Issue, we aim to review replication-competent, reporter-expressing viruses belonging to different families, methods of characterization, and applications to facilitate the study of in vitro and in vivo viral infections. Contrasting advantages, we also seek to discuss disadvantages associated with these reporter-expressing viruses. Finally, we will provide rational future perspectives and additional avenues for the development, characterization, and application of recombinant, reporter-expressing, competent viruses.

Luis Martinez-Sobrido
Guest Editor

Fluorescent Protein-Tagged Sindbis Virus E2 Glycoprotein Allows Single Particle Analysis of Virus Budding from Live Cells

Joyce Jose, Jinghua Tang, Aaron B. Taylor, Timothy S. Baker and Richard J. Kuhn

Abstract: Sindbis virus (SINV) is an enveloped, mosquito-borne alphavirus. Here we generated and characterized a fluorescent protein-tagged (FP-tagged) SINV and found that the presence of the FP-tag (mCherry) affected glycoprotein transport to the plasma membrane whereas the specific infectivity of the virus was not affected. We examined the virions by transmission electron cryo-microscopy and determined the arrangement of the FP-tag on the surface of the virion. The fluorescent proteins are arranged icosahedrally on the virus surface in a stable manner that did not adversely affect receptor binding or fusion functions of E2 and E1, respectively. The delay in surface expression of the viral glycoproteins, as demonstrated by flow cytometry analysis, contributed to a 10-fold reduction in mCherry-E2 virus titer. There is a 1:1 ratio of mCherry to E2 incorporated into the virion, which leads to a strong fluorescence signal and thus facilitates single-particle tracking experiments. We used the FP-tagged virus for high-resolution live-cell imaging to study the spatial and temporal aspects of alphavirus assembly and budding from mammalian cells. These processes were further analyzed by thin section microscopy. The results demonstrate that SINV buds from the plasma membrane of infected cells and is dispersed into the surrounding media or spread to neighboring cells facilitated by its close association with filopodial extensions.

Reprinted from *Viruses*. Cite as: Jose, J.; Tang, J.; Taylor, A.B.; Baker, T.S.; Kuhn, R.J. Fluorescent Protein-Tagged Sindbis Virus E2 Glycoprotein Allows Single Particle Analysis of Virus Budding from Live Cells. *Viruses* **2015**, *7*, 6182–6199.

1. Introduction

Alphaviruses are arthropod-borne viruses that cause frequent epidemics in humans and other vertebrates. Sindbis virus (SINV) is the type member of the genus *Alphavirus* that replicates in mammalian host and mosquito vector cells. It has a positive-sense, single-stranded RNA genome of 11,703 nucleotides with a cap at the 5′ end and a 3′ poly(A) tail. Nonstructural proteins (nsP1-nsP4) are translated from the 49S genomic RNA, whereas structural proteins capsid (CP), E3, E2, 6K, and E1 are translated as a polyprotein from a 26S subgenomic RNA [1]. From the structural polyprotein precursor, CP is autoproteolytically cleaved, exposing an N-terminal signal sequence on E3 that translocates the glycoprotein precursor into the

1

endoplasmic reticulum (ER). In the ER lumen, signalase cleavage removes 6K from pE2 (E3-E2) and E1 envelope proteins that are subsequently glycosylated and form heterodimers. These glycoprotein heterodimers trimerize to form glycoprotein spikes that are transported to the plasma membrane (PM) via the secretory pathway [2,3]. Furin cleavage followed by the release of E3 in the late Golgi primes the glycoprotein spikes for subsequent fusogenic activation during cell entry [4]. CP binds genomic RNA in the cytoplasm to form nucleocapsid cores (NCs). Subsequently, virus particles bud from the plasma membrane (PM) where specific interactions between CP and the cytoplasmic domain of E2 (cdE2) drive envelopment and budding of virions [5].

SINV virions are spherical (~70 nm diameter) and contain 240 copies each of CP, E1, and E2 arranged with icosahedral symmetry in a T = 4 lattice [6]. A host-derived lipid bilayer membrane lies sandwiched between the outer glycoprotein shell and the inner nucleocapsid core (NC) that encapsidates the genomic RNA. Virions also contain sub-stoichiometric amounts of the small "6K" and "TF" proteins [7]. There are two types of virus-induced membranous structures found in the infected cells: type I and type II cytopathic vacuoles (CPV-I and CPV-II) [8,9]. CPV-I (0.6 to 2.0 μm diameter) originates from endosomes and lysosomes and contains replication spherules that are the sites of viral RNA synthesis [10]. CPV-II [11] originates from the *trans*-Golgi network ~4 h post-infection (p.i.) [12,13] and contains the E1/E2 glycoproteins with numerous NCs attached to its cytoplasmic face [11,12,14]. Electron tomography studies have revealed that the E1/E2 glycoproteins are arranged in a helical array within CPV-II in a manner that resembles their organization on the viral envelope [2]. CPV-IIs have been proposed earlier to be caused by over-loading of the secretory pathway by the highly expressed viral glycoproteins [12]. Later it was also suggested that CPV-IIs promote the intracellular transport of the glycoproteins from the *trans*-Golgi network to the PM and also the transport of NCs to the site of virus budding at the PM [2]. Subsequently, NC buds through the PM by forming specific interactions with cdE2 [5,15]. The curvature of the preassembled NC, coupled with regularly spaced, strong interactions between the NC and the cytoplasmic domains of the E2 molecules, allows the membrane and embedded glycoprotein spikes to encircle the NC to form enveloped, fully mature virus particles [6].

We and others have described fluorescent fusion proteins including CP [16], E2 [17–20], and tetra cysteine-labeled structural proteins for virus entry and budding studies [21]. Furthermore, generation of CPV-I in cells infected with Semliki Forest virus has been demonstrated by live-cell imaging coupled with transmission electron microscopy (TEM) [22]. Several imaging studies have utilized fluorescent protein-tagged viruses and subviral particles in single-particle tracking to probe virus entry and assembly. Fluorescently tagged derivatives of Gag-containing human immunodeficiency virus (HIV)-1 virus-like particles were employed to demonstrate assembly, budding, and release of particles from live cells [23]. Similar studies in hepatitis B virus (HBV) have found that the incorporation of only a few fluorescent protein-tagged envelope proteins is sufficient to generate functional, fluorescent virions and subviral particles that enter HBV receptor-positive cells [24]. Furthermore, cryo-electron microscopy (cryoEM) reconstructions have been utilized to determine the organization of fluorescent proteins on purified virus particles. Using cryoEM methods it has been previously shown that green fluorescent protein (GFP)-tagged HBV core particles purified from a bacterial expression system retained icosahedral structure and displayed GFP on its surface [25]. Likewise, a Herpes Simplex Virus 1 GFP-tagged UL17 minor capsid protein was used to determine its location in the capsid vertex-specific component using cryoEM studies [26]. SINV with fluorescent protein labels on the E2 envelope protein has been employed to study virus assembly and budding in living cells. Previous correlative light and electron microscopy studies using fluorescent SINV have provided information about alphavirus budding. Such studies established that glycoprotein E2 is enriched on the PM in localized patches that also contain other viral structural proteins, from which capsid protein interacts with E2 protein for virus budding. This study also suggested that SINV induces reorganization of the PM and cytoskeleton, leading to virus budding from specialized sites [18].

In the current study we characterized the structural stability of an FP-tagged virus and determined the arrangement of mCherry on the virus surface. We provide evidence for the structural stability of the FP-tagged virus and demonstrate that single-particle tracking can be employed to visualize SINV budding from live cells. By employing FP-tagged virus to study virus spread in mammalian cells, we observed that SINV buds from the PM and is associated with filopodial extensions that assist in the dispersal of virions. Comparison of wild-type and budding negative mutant viruses confirmed that fluorescent specks budding from filopodial extensions of mCherry-E2 virus-infected cells are individual virions. By treating infected cells with fusogenic low-pH media, we show that the nascent virions were able to fuse to the PM of filopodial extensions of the infected cells, and we provide evidence for the presence of virions on the outside of these filopodia. This FP-tagged virus can be employed as a tool in high-resolution live and fixed cell imaging coupled with other labeled host proteins and other components to study various aspects of the alphavirus lifecycle.

2. Materials and Methods

2.1. Cells and Viruses

Baby hamster kidney fibroblast cells (BHK-15) obtained from the American Type Culture Collection (ATCC) were maintained in minimal essential medium [27] supplemented with 10% fetal bovine serum (FBS). All SINV cDNA clones were constructed using standard overlapping PCR mutagenesis from pToto64, a full-length cDNA clone of SINV, as previously described [28]. Viruses were propagated in BHK-15 cells at 37 °C in Minimum Essential Medium (MEM) supplemented with 5% FBS in the presence of 5% CO_2 unless otherwise noted.

2.2. Construction and Characterization of FP-Tagged Virus and Mutants

Sequences that encode mCherry, with additional Ser residues at the N- and C-termini, were cloned after Ser_1 of E2 replacing E2 Val_2. Previously characterized cdE2 mutations ($_{400}YAL_{402}$/A3 and $_{416}CC_{417}$/A2) [5] and an E1 (G91D) fusion loop mutation [29] were generated by overlapping PCR and were cloned into the mCherry-E2 cDNA plasmid using *BssHII-BsiWI* restriction sites. The full-length WT and tagged cDNA clones were linearized with *SacI*, *in vitro* transcribed with SP6 RNA polymerase, and transfected into BHK-15 cells as previously described [5]. Infectious virus produced from the transfected cells was quantified by standard plaque assay using medium over cells collected at 24 h post-electroporation. Plaque phenotypes and virus titers were determined by comparing the mutant with WT Toto64 plaques.

2.3. One-Step Growth Curve Analysis

One-step growth analyses were performed as described previously to measure growth kinetics of the mCherry-E2-tagged virus [5]. BHK-15 cells in 35-mm culture dishes were infected with virus at a multiplicity of infection (MOI) of 5 for 1 h at room temperature. Infected cells were washed extensively with MEM and incubated further and culture media were harvested at every hour for 12 h. The amount of infectious virus in the virus supernatant was quantified by titration on BHK cells. All experiments were conducted in triplicate.

2.4. Quantitative Real Time RT-PCR

The number of virus particles released at different time points and total RNA molecules in the media over infected cells were determined by qRT-PCR as previously described [30]. RNA was extracted from virus supernatants using the RNeasy kit (Quiagen, Valencia, CA, USA) according to the manufacturer's instructions. qRT-PCR was performed using the SuperScript III Platinum SYBR Green One-Step qRT-PCR Kit (Invitrogen, Grand Island, NY, USA) with primers 5′-TTCCCTgTgTgCACgTACAT-3′ and 5′- TgAgCCCAACCAgAAgTTTT-3′, which bind to nucleotides 1044–1063 and nucleotides 1130–1149 of the SINV genome, respectively. Amplification reactions were carried out in triplicate in 25 μL sample volumes that contained a 5 μL aliquot of purified viral RNA [5]. Cycling conditions were 4 min at 50 °C and 5 min at 95 °C, followed by 40 cycles of 5 s at 95 °C and 1 min at 60 °C. The number of molecules of viral RNA was determined using a standard curve of the cycle threshold values (CT) determined by qRT-PCR *versus* the number of molecules of *in vitro* transcribed genomic RNA using primers.

2.5. Flow Cytometry (FC)

Transport and cell surface expression of E2 in infected cells were assayed using FC and anti-E2 antibody. BHK-15 cells were infected with an MOI of 5. The cells were trypsinized at 6 h, 8 h, and 12 h post-transfection and resuspended in MEM supplemented with 10% FBS. Cells were washed two times with PBS supplemented with 1% FBS and incubated on ice for 1 h with a 1:50 dilution of anti-E2 127 monoclonal antibody. The cells were washed subsequently three times with PBS (1% FBS) and then incubated on ice in the dark for 30 min with a fluorescein-conjugated goat anti-mouse secondary antibody. The cells were washed thrice with PBS (1% FBS) and suspended in 500 µL of PBS and were analyzed on an FC500 flow cytometer (Beckman Coulter, Indianapolis, IN, USA) with the FlowJo software package. Control staining was performed with mock-transfected cells.

2.6. Virus Purification, Cryo-Electron Microscopy (cryoEM), and 3D Image Reconstruction

WT and mCherry-E2 viruses were purified according to standard virus purification protocols. Briefly, cell culture supernatants from SINV or mCherry-E2-tagged virus-infected BHK were collected at 12 h p.i. and the media were harvested and clarified by centrifugation for 15 min at $9000 \times g$. Virus particles were pelleted through a 27% sucrose cushion in a Beckman Ti-50.2 rotor at 38,000 rpm for 2 h. The virus pellets were resuspended and loaded onto a 0 to 30% continuous iodixanol gradient, in TNE (50 mM Tris, pH 7.4, 100 mM NaCl, 1 mM EDTA), and centrifuged at 38,000 rpm in a Beckman SW-41 rotor for 2 h. The virus band was extracted by syringe and buffer exchanged using TNE buffer and the presence of the mCherry-E2 tag was confirmed by SDS PAGE analysis.

Small (3 µL) aliquots of the purified mCherry-E2-tagged virus were vitrified for cryoEM via standard, rapid freeze-plunging procedures [31] on Quantifoil holey grids (Quantifoil, Electron Microscopy Sciences, Hatfield, Pennsylvania, USA). Grids were then loaded into a multi-specimen holder and inserted into an FEI Polara microscope and maintained at liquid-nitrogen temperature. Micrographs were recorded on a $4K^2$ Ultrascan CCD (Gatan, Inc., Pleasanton, CA, USA) at a nominal magnification of 51,000× under low-dose conditions (≈15 e/$Å^2$) with the microscope operated at 200 keV and the objective lens defocused between 0.9 and 4.7 µm underfocus. Micrographs that exhibited some astigmatism or specimen drift were eliminated from the data set. Individual virus particles were boxed from the remaining 103 micrographs with the program RobEM [32].The Random model computation method [33] was employed to generate an initial 3D map at ~25 Å resolution for the mCherry-E2 insertion mutant. This map was then used as the starting model to initiate orientation and origin determinations for the full set of 9235 particle images using the AUTO3DEM program suite [33] to yield a final 3D map at 11 Å resolution. Graphical representations were generated with RobEM and Chimera [34]. A SINV pseudo-atomic model [6] was used to fit and interpret the mCherry-E2 virus reconstruction. The crystal structure of red fluorescent protein [35] was used to model the densities not accounted for by the virus itself in the mutant. Optimal fitting of the red fluorescent protein model was achieved by rigid body refinement with the *Fit in Map* module of Chimera [34].

2.7. Live Cell Imaging

BHK-15 cells were seeded onto a four-chambered borosilicate cover glass (Fischer Scientific, Pittsburgh, PA, USA) and infected with fluorescent virus at an MOI of 50 at 25% confluence. Infected cells were imaged after media were replaced with Opti-MEM I Reduced-Serum Medium (Invitrogen) at specified time points. Live imaging-compatible stains were obtained from Invitrogen/Molecular Probes. These included Hoechst stain (nucleus) and BODIPY FL C5 ceramide (Golgi stain) and were used according to the manufacturer's instructions in conjunction with mCherry-E2 virus. Fluorescent images were acquired at indicated temperatures using Nikon A1R confocal microscope (Nikon, Melville, NY, USA) with $60\times$, 1.4 numerical aperture (NA) lens) using NIS Elements software (Nikon, Melville, NY, USA). Live imaging for 10–30 min periods was conducted using a heated $60\times$ oil immersion objective (1.4 NA) in a live imaging chamber (Tokai Hit, Fujinomiya, Shizuoka Prefecture, Japan) supplied with 5% CO_2 at 37 °C. The lasers and emission band passes used for imaging were as follows: blue, excitation: 405 nm, emission: 425–475 nm; green, excitation: 488 nm, emission: 500–550 nm; red, excitation: 561 nm, emission: 570–620 nm. Differential interference contrast images were collected from transmitted light along with fluorescent images for colocalization of viral proteins in the cellular organelles. NIS-Elements software was used for image acquisition and analysis. For generating videos, live images were collected at frame rates ranging from 0.8 to 1 frames per second (fps) for a time scale of 1–30 min, and time-lapse videos were generated from the acquired images at a frame rate of 5–7 fps using ImageJ (NIH Bethesda, Maryland, USA). To compare the size and fluorescent properties of purified mCherry-E2 virus, purified virus was mixed with 0.1 μm diameter fluorescent microspheres (TetraSpeck Beads, Invitrogen) and imaged on a cover glass using a Nikon A1R system with a 60x oil immersion objective (1.4 NA).

2.8. Immunofluorescence (IF) Analysis

IF analyses were performed on BHK-15 cells grown on glass coverslips. Primary antibodies used in the experiments were Golgi-specific rabbit polyclonal anti-Giantin (Abcam, Cambridge, MA, USA), SINV-specific rabbit polyclonal anti-E1, anti-CP and mouse monoclonal anti-E2. Cells were fixed using 3.7% paraformaldehyde for 15 min at room temperature and permeabilized using 0.1% Triton $\times$ 100 in phosphate-buffered saline (PBS) for 5 min. The secondary antibodies used were fluorescein isothiocyanate (FITC) or tetramethyl rhodamine (TRITC)-conjugated goat anti-rabbit and goat anti-mouse antibodies [36] in PBS with 10 mg/mL bovine serum albumin. Nuclei were stained using Hoechst stain (Invitrogen) according to the manufacturer's instructions. Images were acquired using a Nikon A1R-MP confocal microscope at room temperature with a 60× oil objective and 1.4 NA. Images were processed using the NIS Elements software (Nikon) and the brightness and contrast were adjusted using nonlinear lookup tables.

2.9. Thin-Section Transmission Electron Microscopy (TEM)

BHK-15 cells infected with wild-type or mCherry-E2-tagged SINV at an MOI 5 were fixed at 6 or 12 h p.i. Cells were fixed for three days in 2.5% glutaraldehyde in 0.1 M sodium cacodylate buffer, embedded in 2% agarose, post-fixed for 90 min in buffered 1% osmium tetroxide containing 0.8% potassium ferricyanide, and stained for 45 min in 2% uranyl acetate. They were then dehydrated with a graded series of ethanol, transferred into propylene oxide and embedded in EMbed-812 resin. Thin sections were cut on a Reichert-Jung Ultracut E ultramicrotome and stained with 2% uranyl acetate and lead citrate [37]. Images were acquired in an FEI Tecnai G^2 20 electron microscope equipped with a LaB$_6$ source and operated at 100 keV (Life Science Microscopy Facility, Purdue University, West Lafayette, IN, USA).

3. Results

3.1. Construction and Characterization of mCherry-E2 SINV

FP-tagged SINV constructs have proved to be useful tools to detect replication complexes (RCs) from infected cells and to study virus entry and budding [18,38]. The schematic of the mCherry-E2 construct generated for this study is shown in Figure 1A.

The N-terminus of E2 is known to tolerate insertions [39]. Hence, we cloned the sequence that encodes mCherry into the second residue following the furin cleavage site between E3 and E2 as previously described by us [20] and others [18–20]. We extensively characterized this mCherry-E2 virus and compared it with wild-type (WT) SINV. One-step growth kinetic analyses of the mCherry-E2 virus were performed and replication was found to be reduced by a one log equivalent in virus yield compared to the WT virus (Figure 1B). To determine whether the lower number of virus plaque-forming unit (pfu) observed in the growth kinetic analysis was caused by a reduced specific infectivity, quantitative real-time reverse transcription PCR (qRT-PCR) analysis of the number of RNA molecules released into the media was performed (Figure 1C). This analysis showed a consistent reduction in the number of RNA molecules released into the media compared to the WT virus. Based on the calculated particle-to-pfu ratio at each time point (data not shown), the specific infectivity of the mCherry-E2 virus was found to be comparable to that of WT SINV. To further characterize the defect(s) of the mCherry-E2 virus, we analyzed infected BHK cells by flow cytometry (FC) using monoclonal anti-E2 antibody (Figure 1D) and determined the surface expression of glycoprotein spikes at 6, 8, and 12 h p.i. This revealed that the viral glycoproteins were transported to the PM more slowly for the mCherry-E2 virus compared to the WT virus. Also, fewer glycoproteins accumulated at the PM in the FP-tagged virus compared to the WT virus. Indeed, when glycoprotein transport to the PM reached a maximum, the level of mCherry-E2 only reached 65% of the WT E2. This defect in surface expression is likely a consequence of slower folding of the glycoproteins caused by the mCherry insertion, thus resulting in reduced virus production but no assembly defects.

As a component of the spike, the mCherry-E2 tag was packaged successfully into particles. To confirm that mCherry was incorporated into particles, we purified the mCherry-E2 virus and compared it with the WT virus by sodium dodecyl sulfate polyacrylamide gel electrophoresis (SDS-PAGE) analysis (Figure 2A). As predicted, the mCherry-E2 protein was larger than the WT E2. Even though the mCherry represents a large insertion in the E2 protein, the tagged virus particles proved to be comparable in size and shape to WT SINV in micrographs of vitrified (Figure 2B) as well as negatively stained samples (data not shown). Next we tested the fluorescence of virus particles after red, green, and blue 100 nm TetraSpeck fluorescent microspheres were mixed with purified mCherry-E2 virus and imaged by confocal microscopy. This test confirmed that the mCherry-E2 virus was suitable for single-particle tracking experiments (Figure 2C).

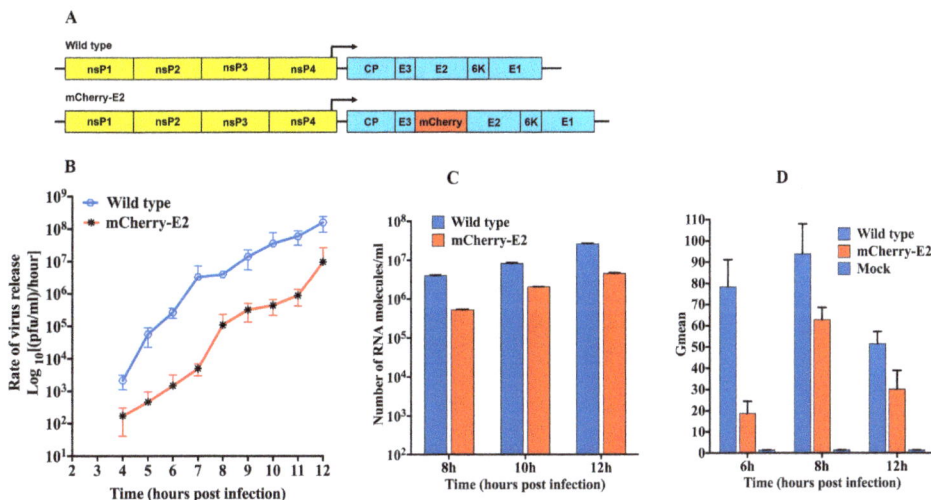

Figure 1. Construction and characterization of mCherry-tagged virus. (**A**) Schematic of wild-type (WT) and mCherry-E2 Sindbis virus (SINV) complementary DNA (cDNA) clones; (**B**) One-step growth curve analysis of WT and mCherry-E2 virus released from Baby Hamster Kidney fibroblast (BHK) cells. BHK cells were infected with WT or mCherry-E2 viruses at a multiplicity of infection (MOI) of 5 and media were changed every hour for 12 h and the rate of virus release (plaque-forming unit (pfu) per ml per hour) was determined using standard plaque assays; (**C**) Quantitation of the number of viral RNA molecules (corresponding to virus particles released into the media) for WT and mCherry-E2 mutant viruses at 8, 10, and 12 h. Total number of genome RNA molecules was determined by quantitative real-time reverse transcription PCR (qRT-PCR) using a standard curve of known amount of *in vitro* transcribed SINV RNA molecules; (**D**) Determination of E2 surface expression in BHK cells infected with WT or mCherry-E2 virus by flow cytometry using anti-E2 monoclonal antibody at 6, 8, and 12 h post-infection (p.i). Y-axis is represented by Gmean, which corresponds to the geometric mean of the fluorescence data calculated by averaging the log of the fluorescence and the scale value of that average in fluorescence units.

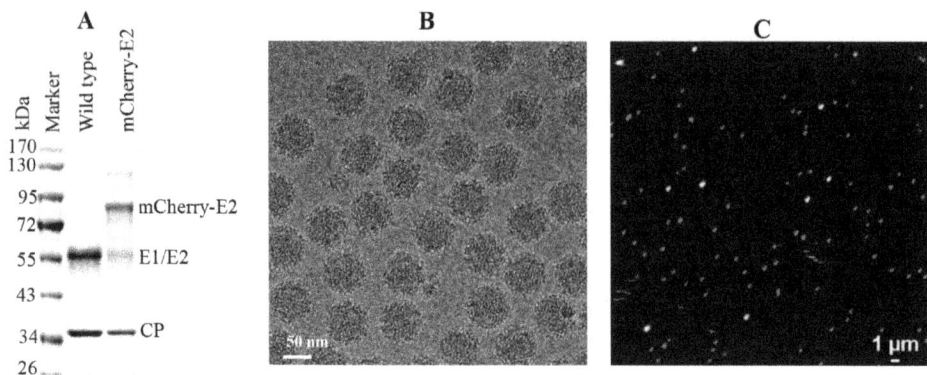

Figure 2. (**A**) Sodium dodecyl sulfate polyacrylamide gel electrophoresis (SDS-PAGE) analysis of purified WT and mCherry-E2 virus showing the size difference of mCherry protein tagged to the E2 and WT E2 protein; (**B**) Cryo-electron microscopy (CryoEM) of purified mCherry-E2 virus exhibiting uniform spherical morphology; (**C**) Confocal merged image of purified mCherry-E2 virus mixed with tetra speck beads (100 nm size fluorescent microspheres fluoresce in green, red, and blue channels, seen as white dots when merged) demonstrating that the individual virions can be observed by confocal live imaging.

3.2. mCherry-E2 SINV Assembles in a Manner Similar to that of WT Virus

The addition of mCherry (236 residues) at the N-terminus of E2 produced virus particles of similar size and shape to native virions. Images of vitrified, FP-tagged particles (Figure 2B) that revealed additional density features at the peripheries of cryoEM reconstructions compared with native virions correlated directly with the presence of mCherry (Figure 3).

FP-tagged particles exhibited additional morphological features on the outer surface compared with the native virion, which we attribute to the mCherry moiety (Figure 3A,C,E). The observation that moderate resolution three-dimensional (3D) cryo-reconstructions could be computed from images of mCherry-E2-tagged particles proved that they were uniform and stable analogous to native SINV (Figures 3E,F and 4A,C). In native SINV, after pE2 is cleaved by furin, the N-terminus of E2 is located at the surface of E2 [40]. Because the mCherry tag occurs at the N-terminus of E2 and coincides with the location of the uncleaved E3, we generated an E3 model for comparison (Figure 4B). The E3 model was built by fitting the E3 structure of Chikungunya virus (CHIKV) [40] into the cryoEM reconstruction of an uncleaved E3 mutant reconstruction. The location of mCherry density near the tip of the glycoprotein spike (Figure 4C) was consistent with the position of the E2 N-terminus and with E3. The mCherry tag is oriented such that it splays away from the E2 receptor binding domain and the E1 fusion peptide. Hence, its presence does not

10

interfere with the infectivity of the mutant virus. However, given that the mCherry tag wedges between adjacent spikes (Figures 3E and 4D), some small conformational changes occur in the spikes and in the NC protein shell (Figure 3A,B).

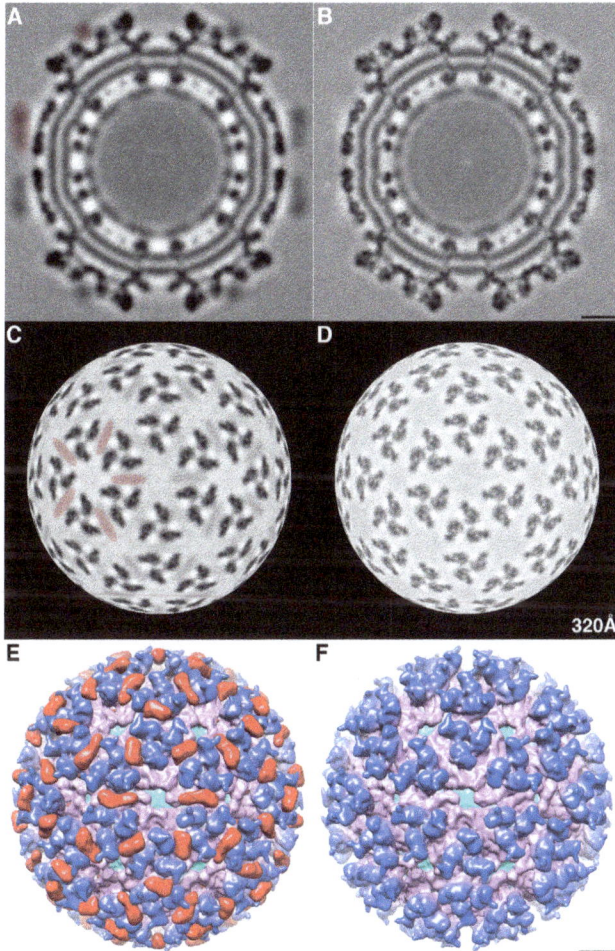

Figure 3. Comparison of mCherry-E2 and WT virus cryoEM structures: (**A**) Central section of mCherry-E2 virus showing extra densities marked in red compared to WT virus; (**B**) Radial projection views at radius of 320 Å show extra densities on the surface of the mCherry-E2 virus (Panel **C**, highlighted in red) compared to WT virus (**D**). Surface view, color-cued by radius (from cyan to pink to blue with increasing radius) of the mCherry-E2 virus (**E**) reveals extra densities (colored in red) compared to the WT virus (**F**). The scale bars represent 100 Å.

11

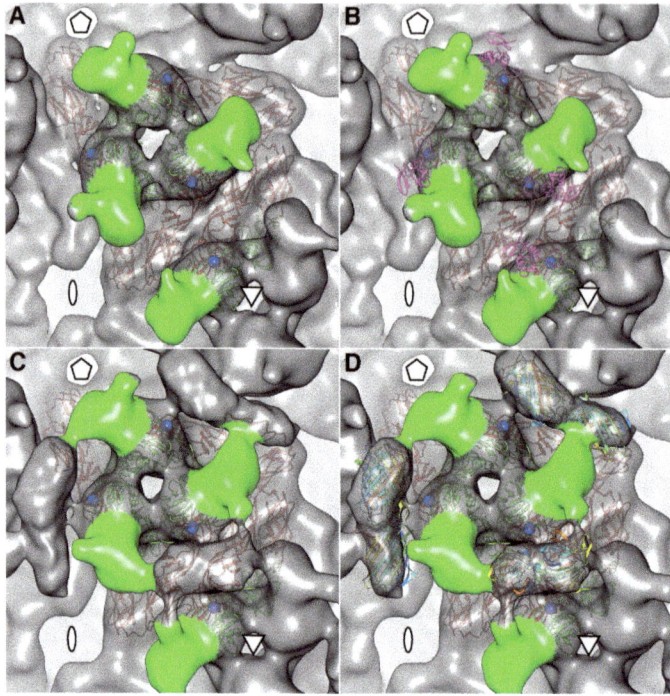

Figure 4. Comparison of mCherry-E2 and WT SINV spike structures. The asymmetric unit of the icosahedral reconstruction is shown with five-, two-, and three-fold axes labeled with a pentagon, oval, and triangle, respectively. The E1 and E2 glycoprotein models in ribbon structures are colored red and green, respectively. The N-terminal residue in each E2 is marked with a blue sphere and the E2 receptor-binding domain represented by the surface structure is colored green. (**A**) WT virus spike; (**B**) WT spike showing the location of E3 on the glycoprotein spike by adding a ribbon structure of E3 (magenta) to the glycoprotein model to show its potential location on the virus surface where the mCherry tag is cloned; (**C**) mCherry-E2 virus spike with extra mCherry densities between the spikes next to the N-termini of E2; (**D**) mCherry-E2 virus spike fitted with the red fluorescent protein (RFP) dimer (rainbow-colored ribbon structure from N- to C-termini) between two neighboring E2 subunits across the spikes.

3.3. SINV Budding Observed in Live Cell Imaging

BHK cells infected with WT or FP-tagged SINV were subjected to immunofluorescence (IF) analyses using antibodies against giantin, E2, E1, and CP (Figure 5). All structural proteins were located on the PM (Figure 5B,C,F) and glycoproteins E1 and E2 were detected on filopodial extensions in both WT and FP-tagged, virus-infected cells (Figure S1). The association of glycoproteins with the Golgi complex in virus-infected cells was studied using anti-Giantin antibody (Figure 5A,D). In BHK cells infected with mCherry-E2 virus, glycoprotein-containing vesicles (Figure 6-1) were transported to the PM (Videos S1A,B and S2A,B). Time-lapse images of mCherry-E2 virus-infected BHK cells showed virus budding in close association with filopodial extensions (Figure 6-2, Video S2A,B). Anterograde trafficking of glycoprotein-containing vesicles to the PM was observed in BHK cells where nascent virions were budding (Videos S1 and S2). Budded viruses were released from filopodial extensions to the surrounding media (Figure 6-2). BHK cells transfected with non-budding cdE2 mutant $_{416}CC_{417}$/AA (Video S3) show the lack of particle budding from filopodial extensions despite glycoprotein transport to the PM and filopodial extensions (Video S3).

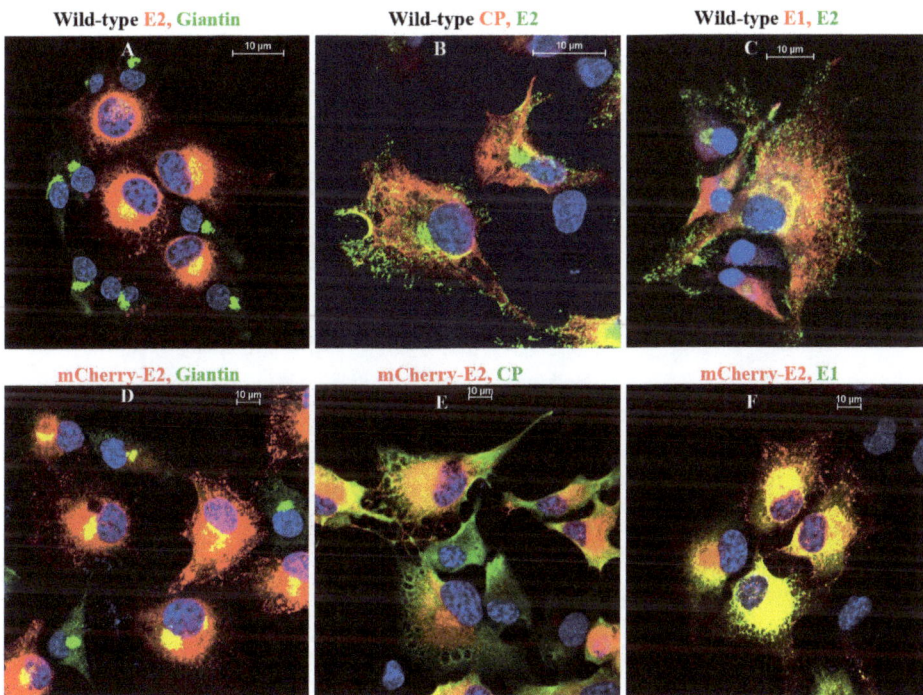

Figure 5. Immunofluorescence (IF) analysis of WT (**A–C**) and mCherry-E2 virus-infected BHK cells (**D–F**) showing the distribution of viral proteins. Cells infected with WT or mCherry-E2 viruses were subjected to IF analysis at 6 h p.i. with antibodies against rabbit polyclonal Giantin (Golgi), rabbit polyclonal anti-CP, rabbit polyclonal anti-E1, and mouse monoclonal anti-E2 primary antibodies and labeled with fluorescein isothiocyanate (FITC)- or tetramethylrhodamine (TRITC)-labeled goat anti-rabbit and goat anti-mouse secondary antibodies.

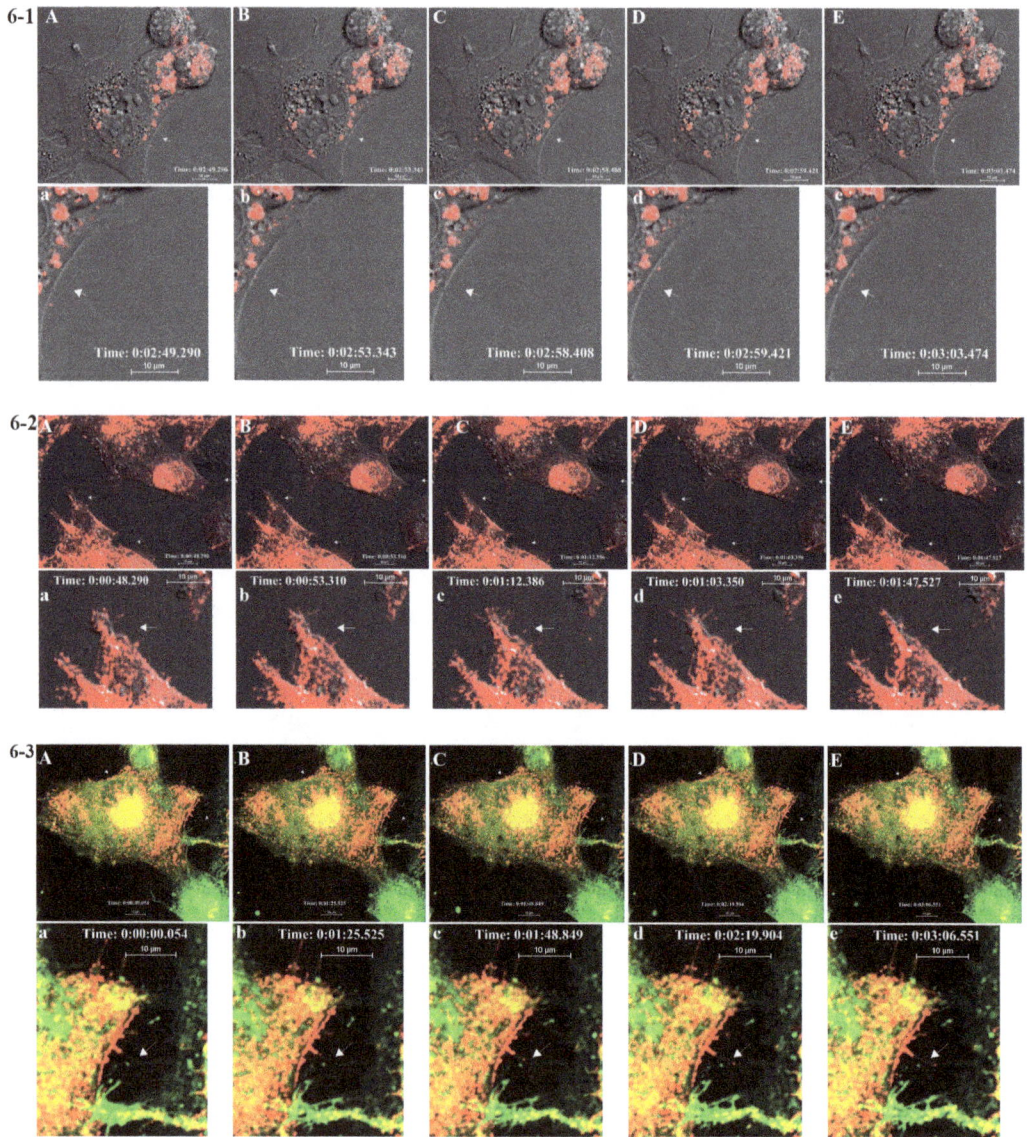

6-1

Time: 0:02:49.290   Time: 0:02:53.343   Time: 0:02:58.408   Time: 0:02:59.421   Time: 0:03:03.474

6-2

Time: 0:00:48.290   Time: 0:00:53.310   Time: 0:01:12.386   Time: 0:01:03.350   Time: 0:01:47.527

6-3

Time: 0:00:00.054   Time: 0:01:25.525   Time: 0:01:48.849   Time: 0:02:19.904   Time: 0:03:06.551

Figure 6. *Cont.*

14

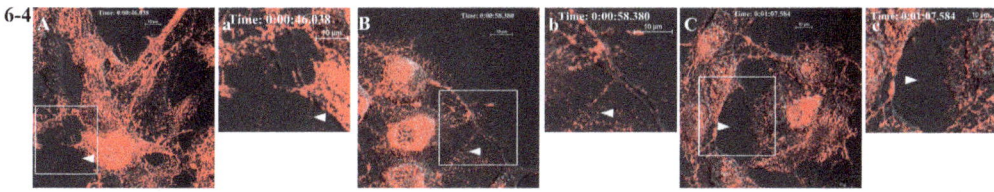

Figure 6. Live imaging of mCherry-E2 virus. (**6-1**) Time-lapse images of mCherry-E2 virus (panels **A–E**)-infected BHK cells exhibiting glycoprotein transport to the PM and virus budding at 3 h p.i. mCherry-E2 virus particles bud from the PM (white arrow) of infected cells. Selected images are shown from supplementary Video S1A (http://dx.doi.org/10.5281/zenodo.34119). Panels (**a–e**) indicate enlarged areas from panels (**A–E**) near the white arrow. Selected images are shown from supplementary Video S1B (http://dx.doi.org/10.5281/zenodo.34119); (**6-2**) Time-lapse images of mCherry-E2 virus budding from infected BHK cells (panels **A–E**) at 6 h p.i. Glycoprotein-containing vesicles traffic to the PM and virus buds from the PM. Budded viruses (white arrow) disperse from filopodial extensions to the surrounding media. Images correspond to Video S2A. Corresponding time-lapse images of enlarged area near the arrow are shown in panels (**a–e**) below, corresponding to supplementary Video S2B (http://dx.doi.org/10.5281/zenodo.34119). Arrow points to virus particle budding from infected cells traveling along the filopodial extension (Figure 6-2 panels **a–e**); (**6-3**) E2 glycoprotein (mCherry-E2; red) colocalizing with Golgi stain (panels **A–E**) observed in live imaging with images from supplementary Video S5A (http://dx.doi.org/10.5281/zenodo.34119). BHK cells were infected with mCherry-E2 virus and stained with BODIPY FL C5 ceramide at 3 h p.i. and imaged at 4 h p.i. Yellow color represents the colocalization of Golgi and mCherry-E2. Glycoprotein-containing (panel **B**) vesicles originate from Golgi and are transported to the PM. Enlarged areas of panels 6-3 **A–E** are shown in panels (**a–e**), and selected images are from Video S5B; (**6-4**) Virus particles budding predominantly from filopodial extensions present at the PM were dispersed into the media after treating cells with pH 7 (panels **A,B**) for 15 min at 6 h p.i. When cells were imaged after low pH (pH 5) treatment, budded virions that were outside the cell and on the filopodial extensions stayed attached to the filopodia outside the cells and fused to the PM of the filopodial extensions (panels **C** and **D**) while retaining red fluorescence (white arrow). When cells were treated with pH 4 (panels **E** and **F**), the fluorescence was lost from the virus particles that were fusing to the PM. However, the fluorescent signal from the mCherry-E2 protein molecules present within the cell was not lost after low pH treatment. Insets labeled **B**, **D**, and **F** represent enlarged rectangle areas from representative panels **A**, **C**, and **E** on the left.

BHK cells transfected with an mCherry-E2 tagged E1 fusion loop mutant (G91D) showed that particles released from transfected cells can enter uninfected neighboring cells, but they appear unable to fuse within the cell to initiate a productive infection (Video S4). The association of mCherry-E2 with Golgi was analyzed in live imaging using BODIPY FL C5 ceramide (Figure 6-3) which stains Golgi (Videos S5A and S5B). Additionally, virus budding in close association with filopodial extensions and fluorescent single-particle trafficking between two cells was observed (Figure 6-3 panels a–e and Video S5B) in the stained cells (green). Furthermore, low pH treatment of FP-tagged, virus-infected BHK cells confirmed that the budded virions were present outside the filopodial extensions. At neutral pH, viruses were present outside the PM of filopodial extensions and were released into the media (Figure 6-4 panels A and B). Low pH treatment of BHK cells infected with FP-tagged virus showed that the budded virions retained the fluorescence and stayed attached to the filopodia after fusing to the PM at pH 5 (Figure 6-4 panels C and D). When cells were treated at pH 4, the fluorescent signals were lost from the fused virions on the filopodia as well as the glycoprotein spikes present on the PM while the fluorescence was retained for the mCherry-E2 present within the cells (Figure 6-4 panels E and F).

3.4. TEM Analysis of BHK Cells Infected with WT-SINV

Thin sections of BHK cells infected with WT SINV at 6 h p.i. (Figure 7A) and 12 h p.i. (Figure 7B) and mCherry-E2 SINV at 6 h p.i. (Figure 7C) and 12 h p.i. (Figure 7D) showed the presence of NCs, budding viruses, and released virions. CPV-IIs with NCs attached to the outer lipid bilayers were seen in the cytoplasm (Figure 7B,D). Virus budding occurred predominantly from the PM. When compared to 6 h p.i., CPV-IIs were abundant at 12 h p.i. (Figure 7B,D). Virus budding is associated with filopodial extensions and filopodia were observed in mCherry-E2 virus-infected BHK cells at 12 h p.i. (Figure S2). Overall, mCherry-E2 virus-infected cells were indistinguishable from WT SINV-infected cells. CPV-II and virus budding from the PM were observed from both types of virus-infected cells, suggesting that these processes were not affected by the presence of the fluorescent protein tag on the mCherry-E2 virus.

16

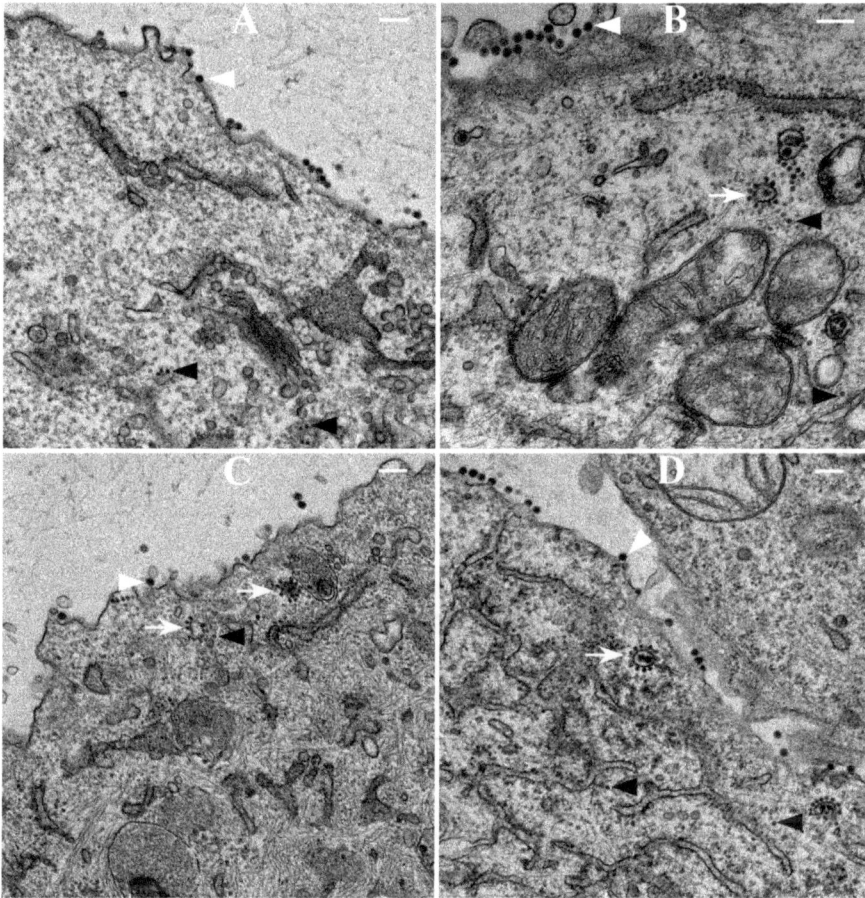

Figure 7. Transmission electron microscopy (TEM) analysis of SINV-infected BHK cells. BHK cells were infected with WT or mCherry-E2 virus at an MOI 5 and fixed for TEM analysis at 6 h and 12 h. p.i. Cells infected with WT (panel **A** 6 h; panel **B** 12 h) and cells infected with mCherry-E2 (Panel **C** 6 h; panel **D** 12 h) viruses are shown. Budding viruses (white arrowhead), and nucleocapsid cores (black arrows) are marked. White arrows indicate type II cytopathic vacuoles (CPV-II). Scale bars represent 200 nm.

4. Discussion

Single-particle tracking and real-time live imaging provide powerful tools for obtaining spatial and temporal resolution information. This contrasts with traditional modes of TEM and super resolution light microscopy that provide high spatial resolution but lack temporal resolution. In this study, we used live imaging coupled with an FP-tagged viral protein to analyze temporal aspects of alphavirus assembly in mammalian cells. We generated an FP-tagged virus with mCherry fused to the N-terminus of the E2 glycoprotein, which is known to tolerate insertions of the immunoglobulin-binding domains of protein L [39] and fluorescent proteins [17–20]. In this study, mCherry was deemed to be an ideal tag based on its monomeric nature, photostability, fast maturation, and resistance to low pH [41]. The mCherry tag has a low pK_a value of 4.5, and hence retains its fluorescence when it encounters the cellular secretory pathway [42]. During maturation, E3 packs against the acid-sensitive region of E2, which maintains the A and B domains of E2 and the B domain to cover the E1 fusion loop, thus protecting the virus from premature fusion with other cellular membranes [40]. After furin cleaves E3, acidification of the virus during entry causes E2 domain B to move away from its neutral pH position and exposes the fusion loop [4]. We have demonstrated that the mCherry tag did not adversely affect any of these functions of E3 and E2.

We determined the 3D cryoEM structure of the mCherry-E2 virus to assess the effects, if any, of the 236-residue insertion on the structural integrity of the virus and its potential to alter the virus lifecycle. The cryo-reconstruction of the mCherry-E2 virus at 11 Å resolution revealed that the overall size of the virion and the icosahedral arrangement of the E1, E2, and CP proteins remained essentially unaffected despite the presence of the large FP insertion. The 240 copies of the mCherry tag wedge tightly between neighboring spikes, and this arrangement causes a slight rearrangement of the spikes as well as the nucleocapsid protein (NCP) pentamers and hexamers and small conformational changes in the membrane bilayer. This confirms our previous observation that minor conformational adjustments of the viral glycoprotein spikes get transmitted radially to the NC via the strong interactions that occur between the inner NC and the outer glycoprotein layers [6]. These small alterations, coupled with the delay in surface expression of the viral glycoproteins as demonstrated by flow cytometry analysis, contribute to the 10-fold reduction in mCherry-E2 virus growth. However, the presence of the mCherry tag did not affect the receptor binding or fusion functions of E2 and E1, respectively. The cryoEM structure of the FP-tagged SINV also confirmed that there is a 1:1 ratio of mCherry to E2 in every virion, which leads to a strong fluorescence signal and thus greatly facilitates single-particle tracking experiments.

We examined mCherry-E2 SINV in live cell imaging primarily to demonstrate virus assembly and budding in real-time using a FP-tagged virus that has been characterized as structurally stable. At times as early as 3 h p.i. we detected budding of FP-tagged virus particles from infected BHK cells, and we gleaned additional information about virus budding and dissemination by examining virus budding and entry mutants. By moving budding viral particles away from the PM of infected cells, the filopodia may act to suppress superinfection, possibly by reducing the re-attachment into the infected cells. As the first step to probe SINV entry and fusion in late endosomes and to study the mechanism of virus fusion, we generated a G91D fusion loop mutation in E1, which abrogates low-pH-triggered fusion and infection [43]. The mCherry-E2 with the G91D fusion loop mutation in E1 was released from the transfected cells, but was unable to fuse and became trapped presumably in the endosome after entry. Using cdE2-NC interaction-deficient, non-budding mCherry-E2 mutants $_{400}$YAL$_{402}$/A3 and $_{416}$CC$_{417}$/A2 we show that non-budding, FP-tagged cdE2 mutations are sufficient to stop fluorescent particle budding from transfected cells. Additionally, using live imaging, we describe that virus budding occurs at the PM for both wild-type and mCherry-E2 virus by the interaction of surface glycoproteins that are transported to the PM via cytopathic vacuoles. We characterized these cytopathic vacuoles using TEM, and live imaging has shown that they contain E2 glycoproteins on their membranes. NCs were also found on the outer membrane of these vacuoles by TEM analysis. As the virus assembly sites are established on the PM, the budded virions utilize filopodial extensions for spreading away from the infected cells.

Similar to the WT SINV, in the mCherry-E2 virus construct, the furin cleavage occurs after the E3 coding sequence, but before mCherry-E2. Data from our virus characterization and imaging experiments of the mCherry-E2 virus suggest that the presence of mCherry after the furin cleavage site on pE2 does not cause significant virus assembly and entry defects. While the E3 protein is cleaved in the Golgi from pE2 to yield the mature E2 protein, E3 stays associated with Venezuelan equine encephalitic virus (VEEV) even after furin cleavage, as evidenced from the cryoEM structure of mature VEEV [44]. Although, in this structure, densities could be attributed to the two alpha-helices of E3, due to disconnected densities, a high resolution E3 density map was not obtained for the cryoEM map of mature VEEV containing cleaved E3. Nevertheless, the observed E3 density decorating the outermost portion of E2 above subdomains A and B was similar to the position of E3 in the pE2 cleavage-impaired, immature SINV mutant virus [45]. These observations have suggested that E3 functions to maintain the relative orientation between E2 subdomains A and B, so as to protect the E1 fusion loop from premature exposure to the host membranes [4,40]. However, E3 does not stay associated with mature SINV after cleavage [6,40]. In our FP-tagged virus, the mCherry density is buried between neighboring glycoprotein spikes and does not occupy the position of E3 over the E2 acid-sensitive region. We hypothesize that this property of the FP-tagged virus is possibly because of the flexible linker region between E3 and E2 (between E3 and mCherry in the FP-tagged virus) that allows sufficient movement of E3 to still maintain its position on E2 to protect the acid-sensitive region of E2 during glycoprotein maturation of the mCherry-E2 virus. Thus, our cryoEM structure explains the unusual stability of the FP-tagged virus.

The density map of the mCherry-E2 virus reveals strong density extending from the N-terminus of E2 which is absent in the wild-type virus (Figure 4C). This density of mCherry can be seen near the five-fold axis between adjacent spikes in a close-up surface view of the virus. The shape and volume of the extra density closely fit the red fluorescent protein (RFP) crystal structure of a dimer but come from two different adjacent E2 molecules (Figure 4D) from two different spikes. Additionally, mCherry appears to make several contacts with the glycoprotein spikes, possibly adding to the stability of the tag. Similar observations were reported for a cryoEM density map of HSV-1 with a GFP-labeled UL17 capsid protein where the freedom of movement of the GFP tag was restricted due to the contact between the GFP tag and capsid density that was sufficient to prevent delocalization of the tag density but without abrogating formation of the capsid vertex-specific component heterodimer [26]. The mCherry tag thus gives additional stability for the FP-tagged virus and explains the accommodation of 240 copies of the mCherry molecule without increasing the diameter of the particles.

Correlative light and electron microscopy (CLEM) studies using fluorescently tagged SINV have indicated the importance of filopodial extensions as preferred sites for alphavirus production, and they appear to mediate cell-cell virus particle transfer [18]. By live imaging, Martinez *et al.* have shown that long cellular extensions are involved in alphavirus cell-to-cell particle transfer [18]. Importantly, using fluorescent SINV virions, we demonstrate single-particle budding that spread from infected cells via filopodial extensions. Such virus budding was absent from cells transfected with RNA from an FP-tagged, cdE2 budding mutant. Using live imaging experiments that utilize FP-tagged viruses, we demonstrate that, in infected BHK cells, fluorescent vesicles containing glycoproteins are transported to the PM. These vesicles presumably originate from late Golgi and we provide evidence for the association of glycoprotein E2 with Golgi using a live imaging-compatible Golgi stain.

In mammalian cells, the viral glycoproteins reside on the membranes of ER, Golgi, CPV-II, and PM, and the virus eventually buds from the PM. We showed that the released FP-tagged virus could be immobilized onto the PM of the infected cells by low pH-mediated fusion at pH 5, confirming that the virus particles are outside the cell. The budded virions that fuse to the PM lost their fluorescence when the cells were treated at pH 4, which is below the pK_a of mCherry. Along with the fused virions, fluorescent glycoprotein spikes present on the PM also lost their fluorescence whereas the mCherry-E2 molecules inside the infected cells were protected from the low pH 4 treatment of the cell. Consistent with our previous findings [5,15], we show that the interaction of NC with the cell-surface glycoproteins generates virions that get propelled by the filopodial extensions, and we hypothesize that this process facilitates viral dissemination while preventing superinfection. SINV attachment factors such as heparan sulfate [46] and entry receptors such as NRAMP (divalent metal, ion transporter natural resistance-associated macrophage protein) [47] have been shown to enhance viral infection. These host factors and the mechanism of receptor-mediated endocytosis can be further investigated with this FP-tagged virus. Exploiting the FP-tagged mutant viruses generated in this study in conjunction with live imaging-compatible stains and labeled host proteins, high-resolution live imaging studies are ongoing with an aim to understand the various molecular interactions between viral glycoproteins and host proteins that are required for productive alphavirus receptor binding, entry, and fusion. Such high-resolution live imaging studies will provide new spatial and temporal information regarding various steps in the alphavirus lifecycle.

Acknowledgments: We thank Laurie Mueller for TEM experiments and Anita Robinson for administrative support. We thank Roger Y. Tsien for supplying mCherry plasmid and James H. Strauss for polyclonal anti-E2 and anti-E1 antibodies. We acknowledge the use of the Bioscience Imaging Facility, and the Flow Cytometry and Cell Separation Facility of the Bindley Bioscience Center and Life Science Microscopy Facility, Purdue University. This work was supported in part by NIH grants GM56279 to R.J.K. and R37-GM33050 to T.S.B., and support from the University of California San Diego and the Agouron Foundation (to T.S.B.) was used to establish the cryoTEM facilities used in this study.

Author Contributions: Experiments were conceived and designed by J.J., J.T., T.S.B and R.J.K.; Experiments were performed by J.J., and J.T.; J.J., J.T., A.B.T., T.S.B and R.J.K. analyzed the data; J.J., J.T., T.S.B and R.J.K. wrote the paper.

Conflicts of Interest: The authors declare no conflict of interest.

References

1. Shirako, Y.; Strauss, J.H. Regulation of Sindbis virus RNA replication: Uncleaved P123 and nsP4 function in minus-strand RNA synthesis, whereas cleaved products from P123 are required for efficient plus-strand RNA synthesis. *J. Virol.* **1994**, *68*, 1874–1885.
2. Soonsawad, P.; Xing, L.; Milla, E.; Espinoza, J.M.; Kawano, M.; Marko, M.; Hsieh, C.; Furukawa, H.; Kawasaki, M.; Weerachatyanukul, W.; *et al.* Structural evidence of glycoprotein assembly in cellular membrane compartments prior to Alphavirus budding. *J. Virol.* **2010**, *84*, 11145–11151.
3. Strauss, J.H.; Strauss, E.G. The alphaviruses: Gene expression, replication, and evolution. *Microbiol. Rev.* **1994**, *58*, 491–562.
4. Li, L.; Jose, J.; Xiang, Y.; Kuhn, R.J.; Rossmann, M.G. Structural changes of envelope proteins during alphavirus fusion. *Nature* **2010**, *468*, 705–708.
5. Jose, J.; Przybyla, L.; Edwards, T.J.; Perera, R.; Burgner, J.W., 2nd; Kuhn, R.J. Interactions of the cytoplasmic domain of Sindbis virus E2 with nucleocapsid cores promote alphavirus budding. *J. Virol.* **2012**, *86*, 2585–2599.
6. Tang, J.; Jose, J.; Chipman, P.; Zhang, W.; Kuhn, R.J.; Baker, T.S. Molecular links between the E2 envelope glycoprotein and nucleocapsid core in Sindbis virus. *J. Mol. Biol.* **2011**, *414*, 442–459.

7. Snyder, J.E.; Kulcsar, K.A.; Schultz, K.L.; Riley, C.P.; Neary, J.T.; Marr, S.; Jose, J.; Griffin, D.E.; Kuhn, R.J. Functional characterization of the alphavirus TF protein. *J. Virol.* **2013**, *87*, 8511–8523.

8. Grimley, P.M.; Berezesky, I.K.; Friedman, R.M. Cytoplasmic structures associated with an arbovirus infection: Loci of viral ribonucleic acid synthesis. *J. Virol.* **1968**, *2*, 1326–1338.

9. Friedman, R.M.; Levin, J.G.; Grimley, P.M.; Berezesky, I.K. Membrane-associated replication complex in arbovirus infection. *J. Virol.* **1972**, *10*, 504–515.

10. Froshauer, S.; Kartenbeck, J.; Helenius, A. Alphavirus RNA replicase is located on the cytoplasmic surface of endosomes and lysosomes. *J. Cell Biol.* **1988**, *107 Pt 1*, 2075–2086.

11. Zhao, H.; Lindqvist, B.; Garoff, H.; von Bonsdorff, C.H.; Liljeström, P. A tyrosine-based motif in the cytoplasmic domain of the alphavirus envelope protein is essential for budding. *EMBO J.* **1994**, *13*, 4204–4211.

12. Griffiths, G.; Fuller, S.D.; Back, R.; Hollinshead, M.; Pfeiffer, S.; Simons, K. The dynamic nature of the Golgi complex. *J. Cell Biol.* **1989**, *108*, 277–297.

13. Garoff, H.; Wilschut, J.; Liljeström, P.; Wahlberg, J.M.; Bron, R.; Suomalainen, M.; Smyth, J.; Salminen, A.; Barth, B.U.; Zhao, H.; *et al.* Assembly and entry mechanisms of Semliki Forest virus. *Arch. Virol. Suppl.* **1994**, *9*, 329–338.

14. Acheson, N.H.; Tamm, I. Replication of Semliki Forest virus: An electron microscopic study. *Virology* **1967**, *32*, 128–143.

15. Owen, K.E.; Kuhn, R.J. Alphavirus budding is dependent on the interaction between the nucleocapsid and hydrophobic amino acids on the cytoplasmic domain of the E2 envelope glycoprotein. *Virology* **1997**, *230*, 187–196.

16. Orvedahl, A.; MacPherson, S.; Sumpter, R., Jr.; Tallóczy, Z.; Zou, Z.; Levine, B. Autophagy protects against Sindbis virus infection of the central nervous system. *Cell Host Microbe* **2010**, *7*, 115–127.

17. Tsvetkova, I.B.; Cheng, F.; Ma, X.; Moore, A.W.; Howard, B.; Mukhopadhyay, S.; Dragnea, B. Fusion of mApple and Venus fluorescent proteins to the Sindbis virus E2 protein leads to different cell-binding properties. *Virus Res.* **2013**, *177*, 138–146.

18. Martinez, M.G.; Snapp, E.L.; Perumal, G.S.; Macaluso, F.P.; Kielian, M. Imaging the Alphavirus Exit Pathway. *J. Virol.* **2014**, *88*, 6922–6933.

19. Snyder, J.E.; Azizgolshani, O.; Wu, B.; He, Y.; Lee, A.C.; Jose, J.; Suter, D.M.; Knobler, C.M.; Gelbart, W.M.; Kuhn, R.J. Rescue of infectious particles from preassembled alphavirus nucleocapsid cores. *J. Virol.* **2011**, *85*, 5773–5781.

20. Dai, H.S.; Liu, Z.; Jiang, W.; Kuhn, R.J. Directed evolution of a virus exclusively utilizing human epidermal growth factor receptor as the entry receptor. *J. Virol.* **2013**, *87*, 11231–11243.

21. Zheng, Y.; Kielian, M. Imaging of the alphavirus capsid protein during virus replication. *J. Virol.* **2013**, *87*, 9579–9589.

22. Spuul, P.; Balistreri, G.; Kääriäinen, L.; Ahola, T. Phosphatidylinositol 3-kinase-, actin-, and microtubule-dependent transport of Semliki Forest Virus replication complexes from the plasma membrane to modified lysosomes. *J. Virol.* **2010**, *84*, 7543–7557.

23. Jouvenet, N.; Bieniasz, P.D.; Simon, S.M. Imaging the biogenesis of individual HIV-1 virions in live cells. *Nature* **2008**, *454*, 236–240.

24. Lambert, C.; Thomé, N.; Kluck, C.J.; Prange, R. Functional incorporation of green fluorescent protein into hepatitis B virus envelope particles. *Virology* **2004**, *330*, 158–167.

25. Kratz, P.A.; Bottcher, B.; Nassal, M. Native display of complete foreign protein domains on the surface of hepatitis B virus capsids. *Proc. Natl. Acad. Sci. USA* **1999**, *96*, 1915–1920.

26. Toropova, K.; Huffman, J.B.; Homa, F.L.; Conway, J.F. The herpes simplex virus 1 UL17 protein is the second constituent of the capsid vertex-specific component required for DNA packaging and retention. *J. Virol.* **2011**, *85*, 7513–7522.

27. Murooka, T.T.; Deruaz, M.; Marangoni, F.; Vrbanac, V.D.; Seung, E.; von Andrian, U.H.; Tager, A.M.; Luster, A.D.; Mempel, T.R. HIV-infected T cells are migratory vehicles for viral dissemination. *Nature* **2012**, *490*, 283–287.

28. Owen, K.E.; Kuhn, R.J. Identification of a region in the Sindbis virus nucleocapsid protein that is involved in specificity of RNA encapsidation. *J. Virol.* **1996**, *70*, 2757–2763.

29. Kielian, M.; Klimjack, M.R.; Ghosh, S.; Duffus, W.A. Mechanisms of mutations inhibiting fusion and infection by Semliki Forest virus. *J. Cell Biol.* **1996**, *134*, 863–872.

30. Warrier, R.; Linger, B.R.; Golden, B.L.; Kuhn, R.J. Role of sindbis virus capsid protein region II in nucleocapsid core assembly and encapsidation of genomic RNA. *J. Virol.* **2008**, *82*, 4461–4470.

31. Baker, T.S.; Olson, N.H.; Fuller, S.D. Adding the third dimension to virus life cycles: Three-dimensional reconstruction of icosahedral viruses from cryo-electron micrographs. *Microbiol. Mol. Biol. Rev.* **1999**, *63*, 862–922.

32. RobEM. Available online: http://cryoem.ucsd.edu/programDocs/runRobem.txt (accessed on 11 March 2008).

33. Yan, X.; Sinkovits, R.S.; Baker, T.S. AUTO3DEM—An automated and high throughput program for image reconstruction of icosahedral particles. *J. Struct. Biol.* **2007**, *157*, 73–82.

34. Pettersen, E.F.; Goddard, T.D.; Huang, C.C.; Couch, G.S.; Greenblatt, D.M.; Meng, E.C.; Ferrin, T.E. UCSF Chimera—A visualization system for exploratory research and analysis. *J. Comput. Chem.* **2004**, *25*, 1605–1612.

35. Shu, X.; Shaner, N.C.; Yarbrough, C.A.; Tsien, R.Y.; Remington, S.J. Novel chromophores and buried charges control color in mFruits. *Biochemistry* **2006**, *45*, 9639–9647.

36. Pierce, M.W.; Coombs, K.; Young, M.; Avruch, J. Control by insulin and insulin-related growth factor 1 of protein synthesis in a cell-free translational system from chick-embryo fibroblasts. *Biochem. J.* **1987**, *244*, 239–242.

37. Reynolds, E.S. The use of lead citrate at high pH as an electron-opaque stain in electron microscopy. *J. Cell Biol.* **1963**, *17*, 208–212.

38. Atasheva, S.; Gorchakov, R.; English, R.; Frolov, I.; Frolova, E. Development of Sindbis viruses encoding nsP2/GFP chimeric proteins and their application for studying nsP2 functioning. *J. Virol.* **2007**, *81*, 5046–5057.

39. Klimstra, W.B.; Williams, J.C.; Ryman, K.D.; Heidner, H.W. Targeting Sindbis virus-based vectors to Fc receptor-positive cell types. *Virology* **2005**, *338*, 9–21.

40. Voss, J.E.; Vaney, M.C.; Duquerroy, S.; Vonrhein, C.; Girard-Blanc, C.; Crublet, E.; Thompson, A.; Bricogne, G.; Rey, F.A. Glycoprotein organization of Chikungunya virus particles revealed by X-ray crystallography. *Nature* **2010**, *468*, 709–712.

41. Shaner, N.C.; Campbell, R.E.; Steinbach, P.A.; Giepmans, B.N.; Palmer, A.E.; Tsien, R.Y. Improved monomeric red, orange and yellow fluorescent proteins derived from Discosoma sp. red fluorescent protein. *Nat. Biotechnol.* **2004**, *22*, 1567–1572.

42. Shaner, N.C.; Steinbach, P.A.; Tsien, R.Y. A guide to choosing fluorescent proteins. *Nat. Methods* **2005**, *2*, 905–909.

43. Duffus, W.A.; Levy-Mintz, P.; Klimjack, M.R.; Kielian, M. Mutations in the putative fusion peptide of Semliki Forest virus affect spike protein oligomerization and virus assembly. *J. Virol.* **1995**, *69*, 2471–2479.

44. Zhang, R.; Hryc, C.F.; Cong, Y.; Liu, X.; Jakana, J.; Gorchakov, R.; Baker, M.L.; Weaver, S.C.; Chiu, W. 4.4 A cryo-EM structure of an enveloped alphavirus Venezuelan equine encephalitis virus. *EMBO J.* **2011**, *30*, 3854–3863.

45. Paredes, A.M.; Heidner, H.; Thuman-Commike, P.; Prasad, B.V.; Johnston, R.E.; Chiu, W. Structural localization of the E3 glycoprotein in attenuated Sindbis virus mutants. *J. Virol.* **1998**, *72*, 1534–1541.

46. Klimstra, W.B.; Ryman, K.D.; Johnston, R.E. Adaptation of Sindbis virus to BHK cells selects for use of heparan sulfate as an attachment receptor. *J. Virol.* **1998**, *72*, 7357–7366.

47. Rose, P.P.; Hanna, S.L.; Spiridigliozzi, A.; Wannissorn, N.; Beiting, D.P.; Ross, S.R.; Hardy, R.W.; Bambina, S.A.; Heise, M.T.; Cherry, S. Natural resistance-associated macrophage protein is a cellular receptor for sindbis virus in both insect and mammalian hosts. *Cell Host Microbe* **2011**, *10*, 97–104.

Efficient Co-Replication of Defective Novirhabdovirus

Ronan N. Rouxel, Emilie Mérour, Stéphane Biacchesi and Michel Brémont

Abstract: We have generated defective Viral Hemorrhagic Septicemia Viruses (VHSV) which express either the green fluorescent protein (GFP) or a far-red fluorescent protein (mKate) by replacing the genes encoding the nucleoprotein N or the polymerase-associated P protein. To recover viable defective viruses, rVHSV-ΔN-Red and rVHSV-ΔP-Green, fish cells were co-transfected with both deleted cDNA VHSV genomes, together with plasmids expressing N, P and L of the RNA-dependent RNA polymerase. After one passage of the transfected cell supernatant, red and green cell foci were observed. Viral titer reached 10^7 PFU/mL after three passages. Infected cells were always red and green with the very rare event of single red or green cell foci appearing. To clarify our understanding of how such defective viruses could be so efficiently propagated, we investigated whether (i) a recombination event between both defective genomes had occurred, (ii) whether both genomes were co-encapsidated in a single viral particle, and (iii) whether both defective viruses were always replicated together through a complementation phenomenon or even as conglomerate. To address these hypotheses, genome and viral particles have been fully characterized and, thus, allowing us to conclude that rVHSV-ΔN-Red and rVHSV-ΔP-Green are independent viral particles which could propagate only by simultaneously infecting the same cells.

Reprinted from *Viruses*. Cite as: Rouxel, R.N.; Mérour, E.; Biacchesi, S.; Brémont, M. Efficient Co-Replication of Defective Novirhabdovirus. *Viruses* **2016**, *8*, 69.

1. Introduction

The presence of defective interfering (DI) viral particles during replication in infected cells is a well-known phenomenon in the virology field [1]. Genomes of DI particles are deleted during the replication and can further replicate into the infected cells only in the presence of co-infection with complete wild-type helper virus. DI genomes are encapsidated into neo-particles and propagate by interfering with wild-type virus replication [2]. Generally, the appearance of DI is the result of cell infection at high multiplicity of infection [3]. During passages in the cell culture of a mixture containing wild-type virus and DI particles, the viral titers progressively decrease. One plausible explanation for this observation is that the kinetics of replication of DI, due to the smaller size of the genome, is faster than for wild-type virus. DI and helper virus exist for most of the virus families including Novirhabdovirus, although they are understudied [4,5]. Apart from DI, viral RNA genome may in some cases be rearranged. For the positive strand RNA virus, that rearrangement is very frequent, mainly because, during viral replication, RNA genome is naked in the cytoplasm and the RNA polymerase may jump from one replicative genome to another one [6,7]. In contrast, for negative strand RNA virus, recombination event has never been described, with a single exception for Respiratory Syncytial Virus for which a co-infection with two replication-competent viruses, knock-out for NS1/NS2 and G genes, resulted in the generation of a virus with a rearranged genome [8]. A number of papers have described recombination events for another negative-strand RNA virus, Newcastle Disease Virus [9–16], however, these descriptions are controversial since in several examples these recombination events are only artificial and due to errors in the deposited Genbank sequences [17]. Finally, as it has been described mainly for Measles Virus, in some cases, a single virus particle may encapsidate more than one genome and stably propagate the different genomes [18,19]. This might be due to the pleomorphic structure of Paramyxoviruses. In contrast, viruses belonging to the Rhabdovirus family like Vesicular Stomatitis Virus or rabies virus and Novirhabdovirus present a rigid bullet shape containing a single RNA molecule [20]. Novirhabdovirus like the Viral Hemorrhagic Septicemia Virus (VHSV) are fish rhabdovirus infecting a large spectrum of fish species, mostly trout, thus replicating at low temperatures. The genome of Novirhabdovirus is about 11 Kbases consisting of a negative sense single-stranded RNA molecule which encodes five structural proteins, the nucleoprotein N, a polymerase-associated P protein, the matrix M protein, a unique G glycoprotein and the large L RNA-dependent RNA polymerase. In addition, located between the G and L genes, Novirhabdovirus genomes present an additional short gene encoding a non-structural NV protein which has been shown to be involved in the viral pathogenicity [21,22]. As demonstrated by reverse-genetics, only N, P and L proteins are needed for formation of transcriptionally active rhabdovirus nucleocapsids [23]. In the current study, we have generated by reverse genetics two recombinant replication-defective Novirhabdoviruses, deleted each for a gene essential for the replication encoding either the N or the P protein. Thus, VHSV-derived cDNA genomes have been engineered such as the N or P genes have been exchanged with reporter genes encoding the green fluorescent protein (GFP) or the red monomeric mKate protein, respectively [24,25]. The aim of this study was to investigate if a complementation phenomenon allowing efficient replication of both defective VHSV might exist.

2. Materials and Methods

2.1. Cells and Virus

Recombinant wild-type like VHSV designated rVHSV [26] and rVHSV-GFP (see below) were propagated in monolayer cultures of *Epithelioma papulosum cyprinid* (EPC) cells at 14 °C as previously described [27]. Virus titers were determined by plaque assay on EPC cells under an agarose overlay (0.35% in Glasgow's modified Eagle's medium-25 mM HEPES medium supplemented with 2% fetal bovine serum and 2 mM L-glutamine). At 5–7 days post-infection, cell monolayers were fixed with 10% formol and stained with crystal violet. Recombinant vaccinia virus expressing the T7 RNA polymerase, vTF7-3 [28], was kindly provided by B. Moss (National Institutes of Health, Bethesda, MD, USA).

2.2. Recovery of rVHSV-GFP

A plasmid construct pVHSV-dtTomato containing VHSV-derived genomic cDNA with an additional dtTomato gene (Clontech) between N and P VHSV genes [26] was digested with *SpeI/SnaBI* restriction enzymes to remove and to replace the dtTomato gene with the GFP gene derived from the pMAX-GFP expression plasmid (Amaxa). The rVHSV-GFP was rescued following transfection of pVHSV-GFP together with pT7-N, pT7-P and pT7-L into vTF7-3-infected EPC cells as previously described [26]. All the restriction enzymes are from Thermo Fisher Scientific (Villebon-sur-Yvette, France).

2.3. Plasmid Constructs Encoding Defective VHSV-Derived cDNA Genomes

A pVHSV plasmid construct containing the full length VHSV-derived genomic cDNA [26] was used as a DNA matrix to amplify by PCR two DNA fragments *SacII/PsiI* (containing the N gene) and *PsiI/MfeI* (containing the P gene); using pairs of primers SACPSIF/SACPSIR and SPEMFEF/SPEMFER, respectively (Table 1).

Both DNA fragments were cloned into a pJet1.2 (Thermo Fisher Scientific). *SpeI* and *SnaBI* restriction enzyme sites were introduced at the start and stop codons, respectively, of the N and P genes by site-directed mutagenesis using QuikChange® Site-Directed Mutagenesis Kit (Agilent, Les Ullis, France) and specific primers SPEMUTN and SNABIMUTN (for N) and SPEMUTP and SNABIMUTP (for P) (Table 1). Each of the mutagenized DNA fragments were digested with *SpeI/SnaBI* restriction enzymes to remove N and P genes and exchanged them by mKate and GFP genes, respectively. *SacII/PsiI* and *PsiI/MfeI* DNA fragments containing either mKate or GFP genes were reintroduced back into two separate pVHSV constructs leading to pVHSV-ΔN-Red and pVHSV-ΔP-Green plasmids, respectively.

Table 1. Primer sequences used in the study.

Primer Name	Primer Sequence (5'to 3')
VHSCDNA	GTATCATAAAAGATGATGAGTTATGTTACAAGGG
VHSPCRF	GTTGAACACAGAGTCATATCTCATAATCG
VHSPCRR	GGTGGAGACACGGTCCTCATCATTGGACGTGAGG
PJETFOR	CGACTCACTATAGGGAGAGCGGC
PJETREV	AAGAACATCGATTTTCCATGGCAG
VHSNF	GATGACGACTACCCCGAGGACTCTGAC
SACPSIF	CTCCACCGCGGTAATACGACTCACTATAGG
SACPSIR	GCTTTGATCAAAGAGAAATTCTTATAATCGTGCCG
PSIMFEF	CGGCACGATTATAAGAATTTCTCTTTG
PSIMFER	GGCCTGCCACAATTGCCTTGACCACC
SPEMUTN	CGTTGAACAAAAGAACTCAGTACTAGTATGGAAGGAGGAATTCGTGCAGCG
SNABIMUTN	CGACTACCCCGAGGACTCTGACTAATACGTACTCCCGTCTCATAACCAACATAG
SPEMUTP	GACAAACACTGAGATACTAGTATGGCTGATATTGAGATGAGCGAGTCC
SNABIMUTP	CGATCAAGGCGGAGCTGGACAAGCTAGAGTAGTACGTACACAACGCATCACAC

2.4. Recovery of Defective rVHSV Expressing mKate and GFP Genes

pVHSV-ΔN-Red and pVHSV-ΔP-Green plasmid constructs (1 µg each) were transfected into vTF7-infected EPC cells together with pT7-N, pT7-P and pT7-L as previously described [26]. One week later, cell supernatant was used to infect fresh EPC cells. Three days post-infection cells were observed under UV-light microscope (Leica, Nanterre, France).

2.5. Plaque Purification

Recombinant VHSV-ΔN-Red and VHSV-ΔP-Green were titered under an agarose overlay as described above. At 6 days post-infection, plaques were observed under UV-light microscope. Plaques appearing exclusively red or green were recovered by aspiration with micropipette and resuspended in 200 µL 2% complete medium before EPC cells infection. Complete medium (2%) was added after infection and cells were incubated at 15 °C until apparition of cytopathic effect.

2.6. RT-PCR on Plaque-Purified Viruses

Genomic viral RNA was extracted from infected-cell supernatant after total cytopathic effect using the QIAamp Viral RNA Purification Kit (QIAGEN, Courtaboeuf, France) according to the manufacturer's instructions. Viral RNA was then reverse-transcribed using Reverse Transcriptase IV (Thermo Fisher Scientific) with VHSCDNA primer and then amplified by PCR using primers VHSPCRF and VHSPCRR (Table 1). PCR products of roughly 1800 and 2300 nucleotides (nt) for rVHSV-ΔN-Red and for rVHSV-ΔP-Green RNA genomes, respectively, were purified with PCR Purification Kit (QIAGEN) following the manufacturer's instructions and cloned into pJet1.2 (Thermo Fisher scientific). Positive clones were selected and subjected to sequencing with PJETFOR, PJETREV and VHSNF primers (Table 1).

2.7. Purification of rVHSV and rVHSV-ΔN-Red + rVHSV-ΔP-Green Particles on Sucrose Gradient

All the viruses were mass produced in EPC cells and supernatants of the infected cells were clarified by low speed centrifugation (4000 rpm 15 min). Supernatants (35 mL) were first concentrated by ultracentrifugation in a SW28 Beckman rotor (Beckman Coulter, Villepinte, France) at 25,000 rpm for 90 min at 4 °C, resuspended in cell culture medium without fetal serum and then loaded on a 25% sucrose cushion in TEN buffer 1× (10 mM Tris-HCl pH 7.5, 150 mM NaCl, 1 mM EDTA pH 8). After ultracentrifugation at 36,000 rpm in a SW41 Beckman rotor (Beckman Coulter, Villepinte, France) for 4 h at 4 °C, viral pellets were resuspended in 100 µL of TEN, loaded onto a 15%–45% discontinuous sucrose gradient, and ultracentrifuged overnight at 25,000 rpm in a Beckman SW55 rotor (Beckman Coulter) at 4 °C. Unique bands of purified viral particles were collected, diluted in TEN and pelleted by ultracentrifugation in a SW41 Beckman rotor at 25,000 rpm for 90 min at 4 °C. Final pellets were resuspended in TEN (60 µL) and stored at −80 °C until further use.

2.8. Virus Preparation and Electron Microscopy Observations

As above, rVHSV, rVHSV-ΔN-Red + rVHSV-ΔP-Green and rVHSV-GFP were mass produced in EPC cells, clarified by low centrifugation at 4000 RPM for 15 min and were ultracentrifugated at 25,000 rpm in a SW28 Beckman rotor (Beckman Coulter) for 90 min at 4 °C. Pellets were carefully resuspended in TEN buffer in appropriated volume to obtain 100 fold concentration rate. Four microliters of purified viruses were added to a Formvar-coated EM grid (300 meshs), incubated 5 min and then contrasted with 1% aluminium molybdate pH: 8 for 20–30 s. All steps were performed at room temperature. The grids were observed using a transmission electron microscope HITACHI HT7700 (Elexience-France, Verrières-le-buisson, France) operated at 80 kV. Microphotographies were acquired with a charge-coupled device CCD camera 8 million pixels and analyses were done with Hitachi HT7700-associated program (Advanced Microscopy Techniques Corp, Woburn, MA, USA). Viral particles were measured with for length and width (nm) and data were represented by scatter plots (Graphpad Prism 5) with statistical Tukey's Multiple Comparison Test analysis.

3. Results

3.1. Recovery of Defective Recombinant Viral Hemorrhagic Septicemia Virus

Following EPC cell infection with vTF7 and transfection of the two plasmid-constructs pVHSV-ΔN-Red (mKate) and pVHSV-ΔP-Green (GFP) together with the helper expression plasmids pT7-N, pT7-P and pT7-L, the cell supernatant (P0) was used to infect fresh EPC cells. Three days later, when cells were observed under UV-light microscope, a large number of red and green cell foci could be seen, the vast majority of them being yellow after overlapping of Red and Green fluorescence, although the intensity of each signal was variable from one infected cell to the other (Figure 1). Comparison of the panels A and B clearly emphasizes that infected cells are always both green and red.

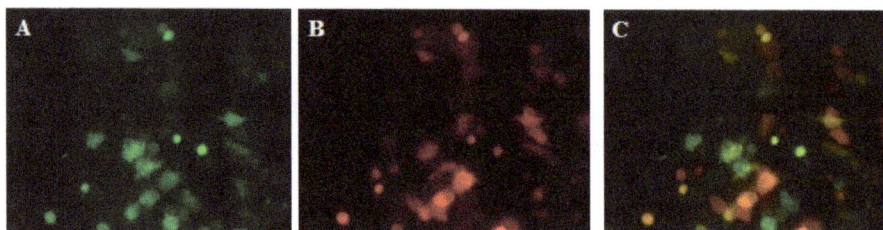

Figure 1. Observation under UV-light microscope of defective rVHSV-infected cells. EPC cells were infected with the supernatant P0 from vTF7-3-infected cells transfected with pVHSV-ΔN-Red and pVHSV-ΔP-Green plasmid together with pT7-N, pT7-P and pT7-L. Infected cells were observed under UV-light microscope at wavelength of 509 nm for green fluorescence (**A**) and 633 nm for far-red fluorescence (**B**). Overlapping picture of green and red infected-cell foci (**C**).

When supernatant of infected cells (P1) was further passaged several times, a similar observation was made, except for the appearance of some cell foci exclusively Red or Green (see below). The rescue of replication-defective viruses rVHSV-ΔN-Red + rVHSV-ΔP-Green was a very efficient process since for three out of four independent transfection assays the recovery of recombinant viruses was successful.

3.2. Genome Characterization of Virus from Red and Green Infected-Cells

Supernatant of rVHSV-ΔN-Red- and rVHSV-ΔP-Green-infected cells was collected and viral genomic RNA was extracted and served to amplify using specific primers by RT-PCR part of the genomes containing either mKate or GFP genes (Figure 2A).

As a positive control, the initial pVHSV-ΔN-Red and pVHSV-ΔP-Green plasmid constructs were used as DNA template and amplified by PCR with the same primers as above. Analysis on agarose gel of the PCR and RT-PCR products indicated that the expected size for the PCR products of 1761 nt and 2280 nt for rVHSV-ΔN-Red and rVHSV-ΔP-Green, respectively, have been amplified (Figure 2B). That evidenced that a mix of the initial defective RNA genomes were still present in the viral RNA extracted from the infected-cell supernatant and also that no apparent genome rearrangement has occurred.

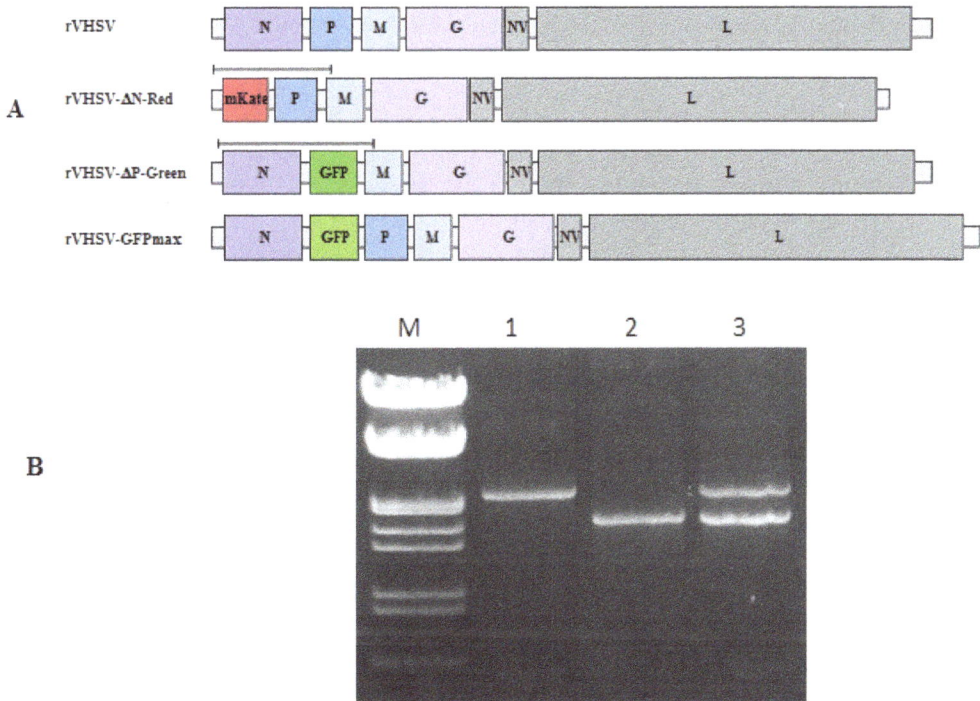

Figure 2. Schematic representation of the various recombinant VHSV genomes and RT-PCR products analysis. A schematic representation of rVHSV, rVHSV-ΔN-Red (mKate), rVHSV-ΔP-Green (GFP) or rVHSV-GFP genomes is shown. Part of the genomes containing mKate or GFP genes was amplified through RT-PCR with specific primers (Table 1). The black line above genomes indicates the parts of the genomes amplified (**A**). Agarose gel analysis of the PCR or RT-PCR products amplified from either the plasmid constructs pVHSV-ΔP-Green (1) and pVHSV-ΔN-Red (2) or from supernatant of yellow infected-cell foci (3), respectively (**B**). M: DNA molecular weight marker (ThermoFisher scientific).

3.3. Characterization of Red and Green Cell Foci

As indicated above, after three to five passages, some infected cell foci were exclusively Red or Green (Figure 3).

These cell foci were plaque purified under agarose, visualized under UV-light microscope and isolated by picking them and propagated in 24-well plaques. That confirmed that indeed some infected cells after observations under UV-light microscope were exclusively Red or Green. Supernatant of these infected cell cultures were collected and viral genomic RNA was extracted and used to amplify by RT-PCR, as above, part of the genomes containing the mKate or GFP genes. Analysis of agarose gel of the PCR products showed as above that two products of 1761 nt and

31

2280 nt have been amplified, indicating that the two distinct viral genomes (ΔN and ΔP) were present. The nucleotide sequencing of these purified RT-PCR products indicated that several mutations leading to non-conservative amino acid changes appeared in the mKate and GFP genes (Table 2).

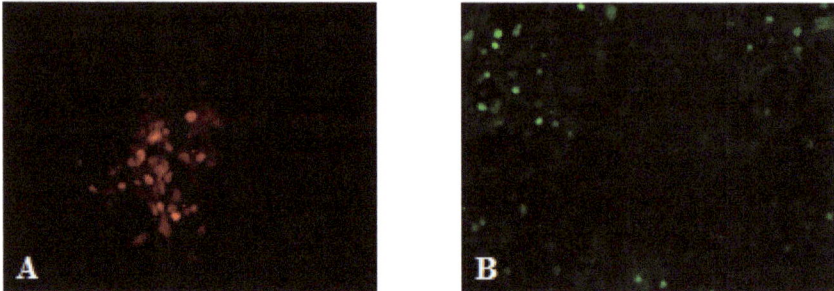

Figure 3. Fluorescence of Red or Green infected-cell foci on same field. At passage five, EPC cells were infected with rVHSV-ΔN-Red + rVHSV-ΔP-Green (m.o.i = 0.0001). Three days later, the same field of infected cells was examined for Red and Green fluorescence. Some of the infected cell foci appeared exclusively Red (**A**) or Green (**B**). A counting, mean of two independent experiments, of the monochromatic infected cell foci indicated an estimate of 9.6% of the total foci.

Table 2. Amino acid changes observed in fluorescent proteins in Red and Green plaque purified viruses.

Red Plaque-Purified Virus	rVHSV-ΔP-Green Clone	F60L/Y211H (GFP)
	rVHSV-ΔN-Red Clone	-
Green Plaque-Purified Virus	rVHSV-ΔP-Green Clone	-
	rVHSV-ΔN-Red Clone	L116P/V120A/F127L/L175S/Y179H/Y195H/Y231H (mKate)

These mutations may explain that although being co-infected with the two rVHSV-ΔN-Red + rVHSV-ΔP-Green, some cells appeared exclusively Green or Red. To ascertain that these mutations leaded up to deleterious effect on the reporter protein fluorescence, both mutated mKate and GFP genes were cloned into a CMV-driven eukaryotic expression vector (ThermoFisher Scientific) and transfected into fish cells. Two days later when transfected-cells where examined under UV-light microscope, no fluorescence could be observed in contrast to the cells transfected with non-mutated reporter genes. That demonstrated that the mutations observed in both reporter genes completely abolished the fluorescence of the expressed proteins.

3.4. Viral Particles Content and Morphology

To further characterize the defective rVHSV, supernatant of infected cells was loaded on sucrose cushion as described in Materials and Methods. Protein contents of the resuspended virus pellets were analyzed through separation on a 4%–12% SDS-PAGE. Following Coomassie-blue staining, a similar viral protein pattern could be observed between rVHSV and the mix of rVHSV-ΔN-Red + rVHSV-ΔP-Green (Figure 4A).

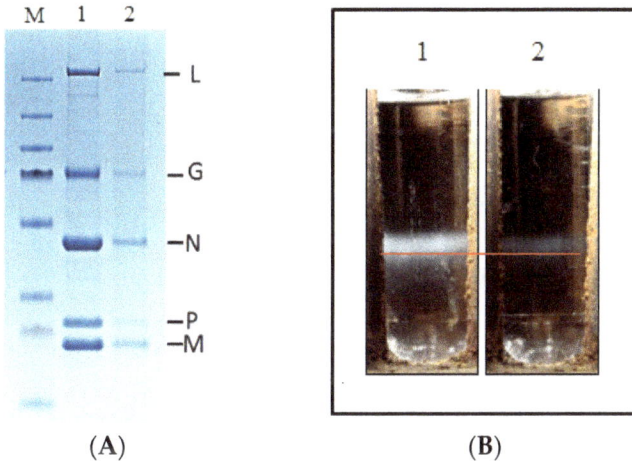

(A)   (B)

Figure 4. SDS-PAGE analysis and purification on sucrose gradient of recombinant VHSV. Semi-purified rVHSV and rVHSV-ΔN-Red + rVHSV-ΔP-Green on 25% sucrose cushion were analyzed through protein separation onto a 4%–12% SDS-PAGE (**A**). After Coomassie blue staining, a similar viral protein pattern for rVHSV (panel 1) and VHSV-ΔN-Red + rVHSV-ΔP-Green (panel 2) could be observed. On the right part, name of VHSV proteins (L, G, N, P and M) are indicated; M: molecular weight marker (ThermoFisher scientific). Discontinuous 15%–45% sucrose gradients were loaded with semi-purified viral stocks (**B**). After ultracentrifugation, unique bands, sedimenting at the same position in the gradient (red line), were visible for rVHSV (1) and rVHSV-ΔN-Red + rVHSV-ΔP-Green (2).

To compare the density of the viral particles rVHSV-ΔN-Red + rVHSV-ΔP-Green to rVHSV, aliquots of semi-purified virus on sucrose cushion were loaded on a discontinuous sucrose gradient. Figure 4B shows that although less material is present for rVHSV-ΔN-Red + rVHSV-ΔP-Green compared to rVHSV, those particles sedimented at the same position in the gradient, indicating that density of rVHSV-ΔN-Red + rVHSV-ΔP-Green and rVHSV was equal. As above, analysis on SDS-PAGE of the viruses contained in the sucrose gradient bands revealed a similar pattern of viral proteins between rVHSV and mixture of rVHSV-ΔN-Red + rVHSV-ΔP-Green. In addition, we showed that sucrose gradient-purified rVHSV-ΔN-Red + rVHSV-ΔP-Green were still infectious when used to infect EPC cells (Figure 5).

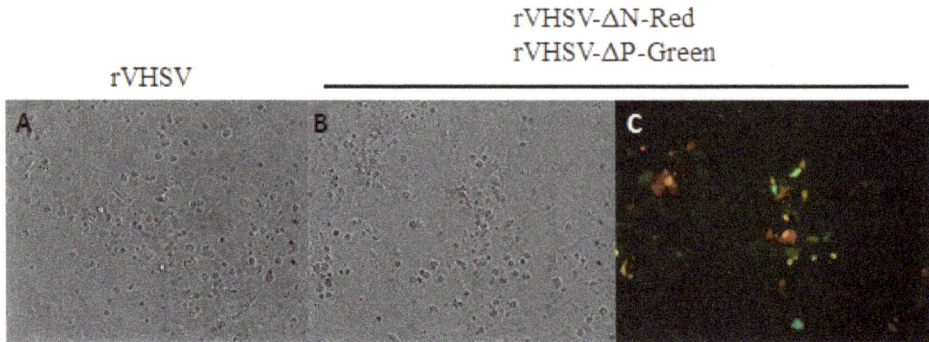

rVHSV

rVHSV-ΔN-Red
rVHSV-ΔP-Green

Figure 5. Infection of EPC cells with sucrose gradient-purified recombinant viruses. EPC cells in 24-well plates were infected (m.o.i. = 1) with rVHSV or rVHSV-ΔN-Red + rVHSV-ΔP-Green purified on sucrose gradient (see Figure 4B). Infected cells were observed two days post-infection either under phase-contrast (rVHSV and rVHSV-ΔN-Red + rVHSV-ΔP-Green, **A** and **B**, respectively) or UV-light microscope (rVHSV-ΔN-Red + rVHSV-ΔP-Green, **C**). Overlapping of Red and Green fluorescence is presented showing that majority of the infected cells were yellow (Red + Green).

Semi-purified viruses were analyzed through electron microscopy observation. rVHSV-ΔN-Red + rVHSV-ΔP-Green were compared to the rVHSV but also to a non-defective rVHSV expressing the GFP (rVHSV-GFP) from an additional expression cassette in the viral genome (see Materials and Methods). The size of the genome of the rVHSV-GFP is increased of 750 nt compared to rVHSV, from 11,165 to 11,915 nucleotides. For rhabdovirus, the size in length of the particle increases proportionally to the size of the genome [29,30]. Figure 6A shows electron microscopy pictures of rVHSV, rVHSV-ΔN-Red + rVHSV-ΔP-Green and rVHSV-GFP. While rVHSV and rVHSV-ΔN-Red + rVHSV-ΔP-Green appear similar in size, the rVHSV-GFP seems to be larger, reflecting the larger size of the genome.

To indeed confirm this observation, individual viral particles were measured in length and in width. As shown in Figure 6B, rVHSV and rVHSV-ΔN-Red + rVHSV-ΔP-Green are strictly similar in size while rVHSV-GFP is significantly larger. That observation reinforced the idea that individual rVHSV-ΔN-Red and rVHSV-ΔP-Green particles contain a single RNA genome having the expected size in length.

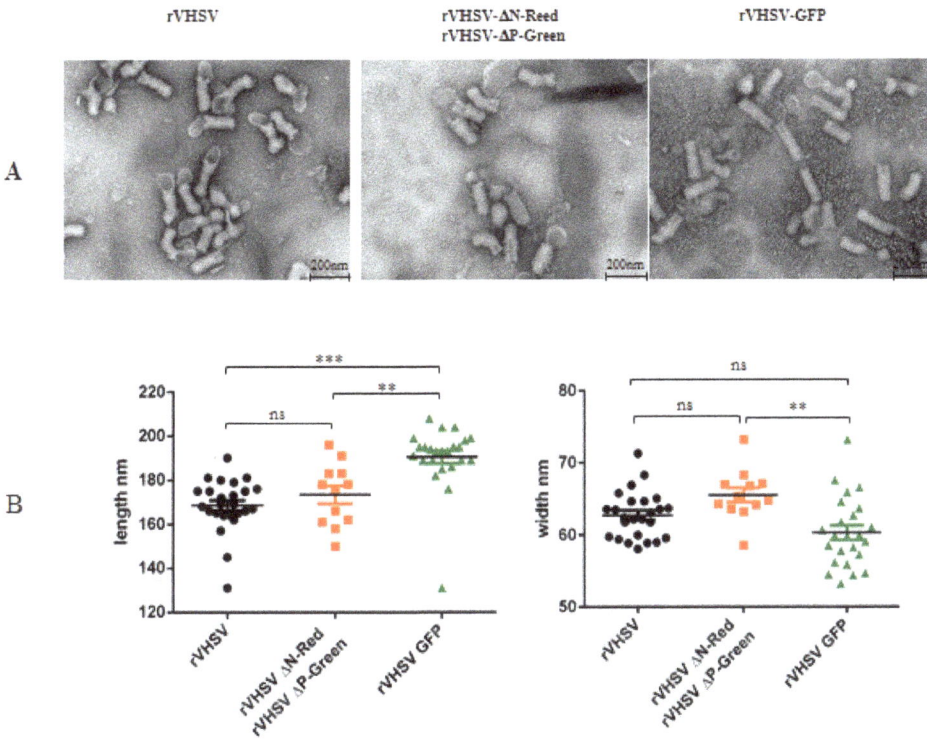

Figure 6. Electron microscopy observations of various recombinant VHSV. Recombinant rVHSV, rVHSV-GFP, rVHSV-ΔN-Red + rVHSV-ΔP-Green were semi-purified onto a 25% sucrose-cushion as described in Materials and Methods. Aliquots (5 µL) of the various recombinant viruses were deposited on Formvar-coated EM grid (300 meshs). Following staining with 1% aluminium molybdate, grids were observed with HITACHI HT7700 electron microscope at 25,000× (Bar = 200 nm) (**A**). Recombinant rVHSV and rVHSV-ΔN-Red + rVHSV-ΔP-Green are morphologically indistinguishable while rVHSV-GFP appeared larger in size. Individual recombinant viral particles were measured in length and in width (**B**). Data were represented by scatter plots (Graphpad Prism 5) with statistical Tukey's Multiple Comparison Test analysis. Groups that are not significantly different from each other are noted ns ($p > 0.05$), whereas those that are significantly different are noted ** ($p < 0.01$) or *** ($p < 0.001$).

4. Discussion

In the current study, we have shown that under particular conditions, defective Novirhabdovirus could efficiently propagate together in cell culture. For successful replication, those defective Novirhabdovirus need to be complemented for the lacking proteins, like in the current study the nucleoprotein N and the polymerase-associated P protein. To investigate if that was indeed feasible, we have generated defective recombinant VHSV expressing reporter genes, mKate a far red-fluorescent protein and GFP a green fluorescent protein easy to observe under UV-light microscope. Previous attempts to rescue defective rVHSV using stable clone cell lines expressing viral N or P proteins always failed. As a defective virus cannot by itself replicate, helper virus is needed. Thus, to generate helper virus, we reasoned that if we could provide in a single cell, defective viral genomes deleted from the N or the P genes, the defective genomes will complement each other. To achieve that, two VHSV-derived cDNA genomes were constructed—pVHSV-ΔN-Red and pVHSV-ΔP-Green—in which the N and P genes are replaced by mKate and GFP genes, respectively. Thanks to the reverse genetics system previously established in our laboratory [26] when the two plasmid constructs were transfected into fish EPC cells, yellow overlapping red and green cell foci could be visualized under UV-light microscope. This observation, potentially indicated, that recombinant defective rVHSV-ΔN-Red + rVHSV-ΔP-Green have been generated and were able to replicate and to express the reporter genes. Titer of these dual recombinant viruses was around 10^7 PFU/mL (two log lower than the rVHSV). More interestingly, these recombinant defective viruses could be propagated and passaged in cell culture up to seven times, without a decrease in viral titers and infected-cells were still mainly yellow (Red + Green) with the exceptions of some infected cell foci exclusively Red or Green. To explain how this very efficient complementation phenomenon leading to the propagation of two defective recombinant virus expressing a reporter gene could be possible, we speculated three main hypotheses: (i) a recombination event leading to a rearranged viral genome in a single particle has occurred; (ii) two defective genomes encapsidated in a single viral particle; (iii) two distinct recombinant viruses (ΔN and ΔP) always co-infect and replicate together into the same cells. To address these questions, a series of assays aiming to genetically and morphologically characterize the defective recombinant VHSV expressing the Red and Green fluorescent proteins. While the recombination is a well-known and frequent phenomenon during replication of positive-sense RNA virus, it is extremely rare and even non-existent for *Mononegavirales*. A single study on Respiratory Syncytial Virus (RSV) described the appearance of rearranged viral RNA genomes following experimental cell infection with two-defective but replication-competent recombinant RSV [8]. It was a very rare event since only in one out of six co-infection assays a rearranged genome was generated. Thus, to investigate whether a recombination event might explain the successful recovery of rVHSVs able to co-express Red and Green fluorescent proteins simultaneously in the same cell, viral RNA genomes were extracted from infected-cell supernatant and characterized. The successful amplification through RT-PCR with primers localized at the beginning of the viral genome and in the M gene leading to two PCR products of the expected sizes strongly suggested that no recombination event during viral replication has occurred. That has also been confirmed by the nucleotide sequencing of each of the PCR product reflecting that the initial defective genomes' organization was maintained during passages of rVHSV-ΔN-Red + rVHSV-ΔP-Green in cell culture. At that step, it became clear that the permanent co-expression of Red and Green fluorescent proteins during the virus passages in cell culture was not the result of a recombination phenomenon between both defective viral genomes. Another possible explanation for the effective co-existence of both rVHSV-ΔN-Red and rVHSV-ΔP-Green could be that instead of having two distinct viruses each containing a defective genome, particles containing two genomes are produced. That has been observed, for example, for Measles virus in which mutated and wild type genomes are present in the same viral particle [18,19]. To solve this question, two studies have been conducted aiming to characterize the viral particles. If two genomes (Red and Green) are co-encapsidated in a single particle, those particles should sediment at a different position in a sucrose gradient compared to particles containing a single genome. The observation that rVHSV and rVHSV-ΔN-Red + rVHSV-ΔP-Green sedimented similarly at the same position in the gradient, reinforced the hypothesis that each of the rVHSV-ΔN-Red and rVHSV-ΔP-Green particles contains a single genome. Also, this observation excluded the possible explanation that these viruses replicated while being associated and forming conglomerates. Anyway, the hypothesis of conglomerates as infectious entities has been invalidated by infecting cells with viral supernatants before and after sonication. No significant decrease of the viral titer was observed after sonication. Viruses extracted from the purified bands from the sucrose gradient were shown to be as infectious as non-purified viral supernatant, confirming that extracted band corresponded to rVHSV-ΔN-Red + rVHSV-ΔP-Green. To try to get more evidence for that "single genome" hypothesis, rVHSV-ΔN-Red + rVHSV-ΔP-Green particles were visualized by electron microscopy, allowing also to measure those viral particles. Morphologic appearance was strictly similar between rVHSV, VHSV-ΔN-Red + rVHSV-ΔP-Green and the size of the various particles identical with the exception of rVHSV-GFP. The size of rVHSV-GFP is larger in terms of length as its genome was increased in size by the addition of the GFP expression cassette. Together, these observations lead to the conclusion that the two defective recombinant rVHSV-ΔN-Red + rVHSV-ΔP-Green are co-replicating together through co-infection of the same cell. During the course of this study, we observed the appearance of infected cell foci which were either exclusively Red or Green. It was demonstrated that both rVHSV-ΔN-Red and rVHSV-ΔP-Green still co-replicated but expression of either Red or Green fluorescent proteins was abolished due to the introduction of deleterious mutations. In conclusion, in the current study, we showed for the first time, to our knowledge, that two defective *Novirhabdovirus* can very efficiently co-infect and co-replicate in the same cell and produce progenies at high titer. It will be of interest in future studies to use those recombinant viruses to infect their natural host, rainbow trout, and to follow whether these viruses are still pathogenic and whether they continue to co-replicate in the animal.

5. Conclusions

In the current work, we clearly demonstrated for the first time, that two defective negative strand RNA viruses can co-replicate and produce progenies at high titer. This phenomenon is independent from recombination or co-encapsidation events. If ever these recombinant defective viruses are shown to be attenuated in fish, this approach could represent an attractive alternative for the development of live-attenuated vaccine.

Acknowledgments: We thank the MIMA2 Platform UR1196, INRA-CRJ (www6.jouy.inra.fr/mima2). A special thanks to Christine Longin (INRA, Jouy-en-Josas, France).

Author Contributions: Ronan N. Rouxel, Emilie Mérour and Michel Brémont conceived and designed the experiments; Ronan N. Rouxel, Emilie Mérour, and Stéphane Biacchesi performed the experiments; Ronan N. Rouxel, Emilie Mérour, Stéphane Biacchesi and Michel Brémont analyzed the data; Stéphane Biacchesi and Michel Brémont wrote the paper.

Conflicts of Interest: The authors declare no conflict of interest.

References

1. Dimmock, N.J.; Easton, A.J. Cloned defective interfering influenza RNA and a possible pan-specific treatment of respiratory virus diseases. *Viruses* **2015**, *7*, 3768–3788.
2. Frensing, T. Defective interfering viruses and their impact on vaccines and viral vectors. *Biotechnol. J.* **2015**, *10*, 681–689.
3. Cole, C.N.; Smoler, D.; Wimmer, E.; Baltimore, D. Defective interfering particles of poliovirus. I. Isolation and physical properties. *J. Virol.* **1971**, *7*, 478–485.
4. Drolet, B.S.; Chiou, P.P.; Heidel, J.; Leong, J.C. Detection of truncated virus particles in a persistent RNA virus infection *in vivo*. *J. Virol.* **1995**, *69*, 2140–2147.
5. Kim, C.H.; Dummer, D.M.; Chiou, P.P.; Leong, J.A. Truncated particles produced in fish surviving infectious hematopoietic necrosis virus infection: Mediators of persistence? *J. Virol.* **1999**, *73*, 843–849.
6. González-Candelas, F.; López-Labrador, F.X.; Bracho, M.A. Recombination in hepatitis C virus. *Viruses* **2011**, *3*, 2006–2024.
7. Casal, P.E.; Chouhy, D.; Bolatti, E.M.; Perez, G.R.; Stella, E.J.; Giri, A.A. Evidence for homologous recombination in Chikungunya Virus. *Mol. Phylogenet. Evol.* **2015**, *85*, 68–75.
8. Spann, K.M.; Collins, P.L.; Teng, M.N. Genetic recombination during coinfection of two mutants of human respiratory syncytial virus. *J. Virol.* **2003**, *77*, 11201–11211.
9. Chare, E.R.; Gould, E.A.; Holmes, E.C. Phylogenetic analysis reveals a low rate of homologous recombination in negative-sense RNA viruses. *J. Gen. Virol.* **2003**, *84*, 2691–2703.

10. Chong, Y.L.; Padhi, A.; Hudson, P.J.; Poss, M. The effect of vaccination on the evolution and population dynamics of avian paramyxovirus-1. *PLoS Pathog.* **2010**, *6*, e1000872.

11. Han, G.Z.; He, C.Q.; Ding, N.Z.; Ma, L.Y. Identification of a natural multirecombinant of Newcastle disease virus. *Virology* **2008**, *371*, 54–60.

12. Han, G.Z.; Worobey, M. Homologous recombination in negative sense RNA viruses. *Viruses* **2011**, *3*, 1358–1373.

13. Miller, P.J.; Kim, L.M.; Ip, H.S.; Afonso, C.L. Evolutionary dynamics of Newcastle disease virus. *Virology* **2009**, *391*, 64–72.

14. Qin, Z.; Sun, L.; Ma, B.; Cui, Z.; Zhu, Y.; Kitamura, Y.; Liu, W. F gene recombination between genotype II and VII Newcastle disease virus. *Virus Res.* **2008**, *131*, 299–303.

15. Yin, Y.; Cortey, M.; Zhang, Y.; Cui, S.; Dolz, R.; Wang, J.; Gong, Z. Molecular characterization of Newcastle disease viruses in Ostriches (*Struthio camelus* L.): Further evidences of recombination within avian paramyxovirus type 1. *Vet. Microbiol.* **2011**, *149*, 324–329.

16. Zhang, R.; Wang, X.; Su, J.; Zhao, J.; Zhang, G. Isolation and analysis of two naturally-occurring multi-recombination Newcastle disease viruses in China. *Virus Res.* **2010**, *151*, 45–53.

17. Song, Q.; Cao, Y.; Li, Q.; Gu, M.; Zhong, L.; Hu, S.; Wan, H.; Liu, X. Artificial recombination may influence the evolutionary analysis of Newcastle disease virus. *J. Virol.* **2011**, *85*, 10409–10414.

18. Rager, M.; Vongpunsawad, S.; Duprex, W.P.; Cattaneo, R. Polyploid measles virus with hexameric genome length. *EMBO J.* **2002**, *21*, 2364–2372.

19. Shirogane, Y.; Watanabe, S.; Yanagi, Y. Cooperation between different RNA virus genomes produces a new phenotype. *Nat. Commun.* **2012**, *3*.

20. Ge, P.; Tsao, J.; Schein, S.; Green, T.J.; Luo, M.; Zhou, Z.H. Cryo-EM model of the bullet-shaped vesicular stomatitis virus. *Science* **2010**, *327*, 689–693.

21. Thoulouze, M.I.; Bouguyon, E.; Carpentier, C.; Brémont, M. Essential role of the NV protein of Novirhabdovirus for pathogenicity in rainbow trout. *J. Virol.* **2004**, *78*, 4098–5107.

22. Ammayappan, A.; Vakharia, V.N. Nonvirion protein of Novirhabdovirus suppresses apoptosis at the early stage of virus infection. *J. Virol.* **2011**, *85*, 8393–8402.

23. Schnell, M.J.; Mebatsion, T.; Conzelmann, K.K. Infectious rabies viruses from cloned cDNA. *EMBO J.* **1994**, *13*, 4195–4203.

24. Chalfie, M.; Tu, Y.; Euskirchen, G.; Ward, W.W.; Prasher, D.C. Green fluorescent protein as a marker for gene expression. *Science* **1994**, *263*, 802–805.

25. Shcherbo, D.; Merzlyak, E.M.; Chepurnykh, T.V.; Fradkov, A.F.; Ermakova, G.V.; Solovieva, E.A.; Lukyanov, K.A.; Bogdanova, E.A.; Zaraisky, A.G.; Lukyanov, S.; *et al.* Bright far-red fluorescent protein for whole-body imaging. *Nat. Methods* **2007**, *4*, 741–746.

26. Biacchesi, S.; Lamoureux, A.; Mérour, E.; Bernard, J.; Brémont, M. Limited interference at the early stage of infection between two recombinant Novirhabdoviruses: Viral hemorrhagic septicemia virus and infectious hematopoietic necrosis virus. *J. Virol.* **2010**, *84*, 10038–10050.

27. Fijan, N.; Sulimanovic, M.; Bearzotti, M.; Muzinic, D.; Zwillenberg, L.O.; Chilmonczyk, S.; Vautherot, J.; de Kinkelin, P. Some properties of the *Epithelioma papulosum cyprini* (EPC) cell line from carp *cyprinus carpio. Ann. Inst. Pasteur Virol.* **1983**, *134*, 207–220.

28. Fuerst, T.R.; Niles, E.G.; Studier, F.W.; Moss, B. Eukaryotic transient-expression system based on recombinant vaccinia virus that synthesizes bacteriophage T7 RNA polymerase. *Proc. Natl. Acad. Sci. USA* **1986**, *83*, 8122–8126.

29. Schnell, M.J.; Buonocore, L.; Kretzschmar, E.; Johnson, E.; Rose, J.K. Foreign glycoproteins expressed from recombinant vesicular stomatitis viruses are incorporated efficiently into virus particles. *Proc. Natl. Acad. Sci. USA* **1996**, *15*, 11359–11365.

30. Soh, T.K.; Whelan, S.P. Tracking the fate of genetically distinct vesicular stomatitis virus matrix proteins highlights the role for late domains in assembly. *J. Virol.* **2015**, *89*, 11750–11760.

Recombinant Pseudorabies Virus (PRV) Expressing Firefly Luciferase Effectively Screened for CRISPR/Cas9 Single Guide RNAs and Antiviral Compounds

Yan-Dong Tang, Ji-Ting Liu, Qiong-Qiong Fang, Tong-Yun Wang, Ming-Xia Sun, Tong-Qing An, Zhi-Jun Tian and Xue-Hui Cai

Abstract: A Pseudorabies virus (PRV) variant has emerged in China since 2011 that is not protected by commercial vaccines, and has not been well studied. The PRV genome is large and difficult to manipulate, but it is feasible to use clustered, regularly interspaced short palindromic repeats (CRISPR)/Cas9 technology. However, identification of single guide RNA (sgRNA) through screening is critical to the CRISPR/Cas9 system, and is traditionally time and labor intensive, and not suitable for rapid and high throughput screening of effective PRV sgRNAs. In this study, we developed a recombinant PRV strain expressing firefly luciferase and enhanced green fluorescent protein (EGFP) as a reporter virus for PRV-specific sgRNA screens and rapid evaluation of antiviral compounds. Luciferase activity was apparent as soon as 4 h after infection and was stably expressed through 10 passages. In a proof of the principle screen, we were able to identify several PRV specific sgRNAs and confirmed that they inhibited PRV replication using traditional methods. Using the reporter virus, we also identified PRV variants lacking US3, US2, and US9 gene function, and showed anti-PRV activity for chloroquine. Our results suggest that the reporter PRV strain will be a useful tool for basic virology studies, and for developing PRV control and prevention measures.

Reprinted from *Viruses*. Cite as: Tang, Y.-D.; Liu, J.-T.; Fang, Q.-Q.; Wang, T.-Y.; Sun, M.-X.; An, T.-Q.; Tian, Z.-J.; Cai, X.-H. Recombinant Pseudorabies Virus (PRV) Expressing Firefly Luciferase Effectively Screened for CRISPR/Cas9 Single Guide RNAs and Antiviral Compounds. *Viruses* **2016**, *8*, 90.

1. Introduction

Pseudorabies virus (PRV) belongs to the Herpesviridae family [1,2] and is the etiological agent of pseudorabies (PR), also known as Aujeszky's disease. PR causes substantial economic losses to the global swine industry [1], but has been largely controlled for at least 30 years using the Bartha-K61 vaccine. However, a novel PRV variant has emerged in China, and the Bartha-K61 vaccine has failed to provide complete protection [3–6]. Full-length genomic sequencing demonstrated that the PRV variant causing the outbreak belonged to a novel genotype [7]. Given the urgency of the outbreak and the need for a new vaccine, additional studies of PRV variants are critical.

Due to its large genome, PRV has been manipulated for basic virology studies using homologous recombination (HR) [8] or bacterial artificial chromosome (BAC) techniques [9]. Both traditional methods have considerable drawbacks. For example, the frequency and efficacy of the expected recombination by HR is quite low; and BAC mutagenesis is only available for virus isolates for which a useful BAC has been produced. Therefore, CRISPR/Cas9 can be applied to new isolates as they emerge in nature, with the only delay being the need to have sequence data for design of guide RNAs. In addition, inserting drug selection markers or parts of BAC plasmids into the viral genome may affect viral function [10]. However, genetic manipulation is essential for identifying gene function and for vaccine development.

With the development of alternative technologies such as zinc-finger nucleases (ZFNs), transcription activator-like effector nucleases (TALENs) and CRISPR/Cas9 [11,12], genome editing has become significantly less complicated. These approaches use a nuclease to specifically target a gene, cleaving the DNA to induce double-stranded breaks at the target site. The DNA break triggers cellular DNA repair mechanisms, including error-prone non-homologous end joining (NHEJ) and homology-directed repair (HDR) [13]. Customizing gene disruption using either ZFNs or TALENs requires the design of specific proteins to target each dsDNA site [14,15], which requires several weeks and is labor and time intensive. However, gene knock out or gene knock in recombinants can be obtained within a short period by simply transfecting the CRISPR/Cas system, which is both efficient and convenient [16–18].

The CRISPR/Cas9 system—derived from the bacterial adaptive immune system—has been used to successfully edit many viruses [18–21]. One key factor affecting DNA editing using the CRISPR/Cas9 system is effective screening for single guide RNAs (sgRNAs), which has to be validated by amplified fragment length polymorphism (AFLP), T7 endonuclease I assay (T7E1), surveyor mismatch cleavage assays, or DNA sequencing. These methods are time consuming, laborious, and minimally sensitive [22]. The PRV genome contains a high GC content, which is challenging for PCR amplification. Therefore, all of the traditional approaches are suboptimal for effectively screening for PRV sgRNAs.

Here, to address the shortcomings inherent in screening for PRV sgRNAs, we have taken a novel approach and use a PRV containing firefly luciferase to validate PRV-specific sgRNA screening. We then applied the reporter virus to develop novel inactivated PRV strains and to demonstrate its utility in antiviral screening assays.

2. Materials and Methods

2.1. Cell Lines and Viruses

Vero cells and MARC 145 cells were cultured in Dulbecco's modified Eagle's medium (DMEM; GIBCO, Grand Island, NY, USA). All culture media was supplemented with 10% heat-inactivated fetal bovine serum (FBS, GIBCO, Life technologies) and antibiotics (0.1 mg/mL streptomycin and 100 IU/mL penicillin). The PRV HeN1 strain (GenBank accession number: KP098534.1) was described previously [5,7]. PRRSV HuN4 strain (GenBank accession number: EF635006.1).

2.2. Generation of the HR Plasmid and CRISPR/Cas9 sgRNA Plasmids

The HR plasmid was constructed in the pcDNA3.1 (+) expression vector (Clontech, PaloAlto, CA, USA). First, pcDNA3.1 (+) was digested using *PmeI* to remove the multiple cloning site (MCS) between it (from *Afl II* to *Apa I*). Then, site mutagenesis was utilized to insert the 5′ *BamHI* and 3′ *NotI* sites that flank the neomycin resistance gene. In addition, sites for *EcoRI*, *SalI*, and *XhoI* were created followed by an SV40 poly(A) signal sequence. Next, *BamHI* and *NotI* were used to remove the *EGFP* gene from the pEGFP-N1 plasmid (Clontech, PaloAlto, CA, USA) and used to replace the neomycin gene (Tang-EGFP vector). The firefly luciferase gene was then amplified from the pGL3-Basic vector (Promega, Madison, WI, USA) and cloned into the Tang-EGFP vector between the *NheI* and *PmeI* sites. The right and left homologous arms were cloned from the PRV HeN1 strain by PCR and inserted between the CMV promoter and SV40 poly (A) signal. The final plasmid was termed the Tang-Luc-EGFP-HR vector. sgRNAs involved in this study were designed using the online CRISPR Design Tool [23], and target the *gE*, *US2*, *US3* and *US9* genes open reading frames. A human codon-optimized SpCas9 and the chimeric guide RNA expression plasmid PX330 were gifts from Feng Zhang [17,24]. The PX330 plasmid was digested using BbsI (Thermo scientific fermentas, Waltham, MA, USA) and the CRISPR/Cas9 constructs were constructed. All of the constructs in this study were verified by sequencing. The primers used are provided in Table 1.

Table 1. Sequences of the Primers and sgRNAs Utilized in this Study.

Primers and sgRNAs	Sequences
CreBamHI	5′-CTTTTGCAAAAAGCTCCCGGGATCCTGTATATCCATTTTCG-3′ 5′-CGAAAATGGATATACAGGATCCCGGGAGCTTTTTGCAAAAG-3′
CreNotI	5′-GGATGATCCTCCAGCGGCCGCATCTCATGCTGGAG-3′ 5′-CTCCAGCATGAGATGCGGCCGCTGGAGGATCATCC-3′
CreEcoRI XhoI	5′-CTTATCATGTCTGAATTCCGTCGACCTCTAGCTCGAGCTTGG-3′ 5′-CAAGCTCGAGCTAGAGGTCGACGGAATTCAGACATGATAAG-3′
Luciferase	5′-CACGCTAGCCACCATGGAAGATGCCAAAAAC-3′ 5′-AGCTTTGTTTAAAC TTACACGGCGATCTTGCCGC-3′
HR L arm	5′-ACAAGATCTCCGGTCCGTAGCCTCCGCAGTA-3′ 5′-ACAACGCGTCGAAGCTCGGCCAACGTCATC-3′
HR R arm	5′-CCGGAATTCGGGCCGTGTTCTTTGTGGC-3′ 5′-CGGCTCGAGACTCGCTGGGCGTCTCGTTG-3′
sgRNA-gE1	5′-CACCGGGGCAGGAACGTCCAGATCC-3′ 5′-AAACGGATCTGGACGTTCCTGCCCC-3′
sgRNA-US3-1	5′-CACCGCCCCGACGAGATCCTGTACT-3′ 5′-AAACAGTACAGGATCTCGTCGGGGC-3′
sgRNA-US3-2	5′-CACCGGAGATCATCATCGACGGCGA-3′ 5′-AAACTCGCCGTCGATGATGATCTCC-3′
sgRNA-US3-3	5′-CACCGGAGATCATCATCGACGGCGA-3′ 5′-AAACTCGCCGTCGATGATGATCTCC-3′
sgRNA-US2-1	5′-CACCGACCGTGGTCACGCTGATGGA-3′ 5′-AAACTCCATCAGCGTGACCACGGTC-3′
sgRNA-US2-2	5′-CACCGGGGCGCATCCCCGCCTTCGT-3′ 5′-AAACACGAAGGCGGGGATGCGCCCC-3′
sgRNA-US2-3	5′-CACCGGGCGCACCCGGACCTGTGGA-3′ 5′-AAACTCCACAGGTCCGGGTGCGCCC-3′
sgRNA-US9-1	5′-CACCGCGACGTCCTGCTGGCCCCCA-3′ 5′-AAACTGGGGGCCAGCAGGACGTCGC-3′
sgRNA-US9-2	5′-CACCGGCCAGCAGGACGTCGGCGGC-3′ 5′-AAACGCCGCCGACGTCCTGCTGGCC-3′
sgRNA-US9-3	5′-CACCGGGGGGTCCCTTGGGGGCCAGC-3′ 5′-AAACGCTGGCCCCCAAGGGACCCCC-3′

2.3. Recombination and Purification of the Luciferase Tagged PRV

The PRV HeN1 genome was extracted as previously described [7,25]. Vero cells were co-transfected with 1μg of the PRV genome, 1 μg of cas9 plasmid gRNA-gE1, and 3 μg of the Tang-Luc-EGFP-HR plasmid using the X-tremeGENE HP DNA transfection reagent (Roche, Basel, Switzerland) according to the manufacturer's instructions. Forty-eight hours post transfection, the cells were collected and then subjected to three freeze-thaw cycles. Recombinant PRV was purified from the cell lysates by plaque purification (EGFP + selected) in Vero cells overlaid with 1% low-melting point agarose and 2% FBS in DMEM. After 10 rounds of purification, all of the plaques were EGFP+ upon examination by fluorescent microscopy.

2.4. In Vitro Growth Properties

Viral titers were determined by plaque forming unit (PFU) or the 50% tissue culture infection dose (TCID$_{50}$) in Vero cells. To compare the growth kinetics of the parental and EGFP expressing viruses, Vero cells were infected with wild type PRV HeN1 or recombinant PRV-Luc-EGFP at a dose of 200 TCID$_{50}$. The cells were collected at 24, 48, 72, and 96 h post infection (hpi), and the viral titer was determined by PFU or TCID$_{50}$ at the time points indicated.

2.5. Transfection and Western Blot

Cells were transiently transfected with the indicated plasmids using the X tremeGENE HP DNA transfection reagent (Roche, Basel, Switzerland) according to the manufacturer's instructions. Forty-eight (48) hpi, the cells were collected and washed once with PBS, and then lysed in RIPA Lysis Buffer containing a protease inhibitor cocktail (Roche, Basel, Switzerland). The proteins in the cell lysates were separated by SDS-PAGE, transferred to PVDF membranes (Millipore, Milford, MA, USA), and probed with the indicated antibodies for detection.

2.6. CRISPR/Cas9 sgRNA Screening and Antiviral Assay

The CRISPR/Cas9 plasmids were transfected into Vero cells, and 12 h later the cells were infected with 0.01 multiple of infection(MOI) PRV HeN1 or PRV-Luc-EGFP. The expression levels of the indicated proteins were assessed at the indicated time points post infection by Western blot, or luciferase activity was measured using the luciferase assay system (Promega, Madison, WI, USA). For the antiviral assay, monolayers of 90% confluent Vero cells in 12-well plates were treated with Cyclosporine A (CsA) (Beyotime, Jiangsu, China) or chloroquine (Sigma, St. Louis, MO, USA) for 4 h and then infected with PRV HeN1 or PRV Luc-EGFP. The effects of the antiviral compounds were determined based on viral protein expression and luciferase activity as described above. For CsA inhibit porcine reproductive and respiratory syndrome virus (PRRSV) assays, MARC 145 cells were infected with 0.1 MOI PRRSV HuN4 strain and incubated 12 h at 37 °C. Cells were fixed with 80% ethanol for 30 min and stained with SR-30 FITC conjugated mAb (Rural Technologies, Brookings, SD, USA) for 3 h. and images were acquired with a fluorescent microscope. The effect of CsA and chloroquine on cell viability was determined by Trypan Blue staining (Sigma, St. Louis, MO, USA).

3. Results

3.1. PRV-Luc-EGFP was Successfully Constructed

As glycoprotein E (gE) is unnecessary for PRV replication *in vitro* [26,27], we chose this region to insert the firefly luciferase gene. First, plasmids were constructed containing both firefly luciferase and EGFP flanked by homologous arms designated Tang-Luc-EGFP-HR vector (Figure 1A). To generate the recombinant viruses, Vero cells were transfected with the purified PRV HeN1 genome, Tang-Luc-EGFP-HR, and gRNA-gE1 (a gE specific CRISPR/Cas9 construct that was used to improve the efficiency of the HR, our unpublished data). Forty-eight (48) hpi, the viruses were collected and plaque purified. After 10 rounds of purification, the expression of EGFP was confirmed by fluorescence microscopy (Figure 1B) and firefly luciferase expression was confirmed by luciferase activity (Figure 1C). The luciferase activity was detected as early as 4 hpi and reached a higher level at 36 h (Figure 1C). To determine whether the expression of firefly luciferase and EGFP influenced the replication of the PRV-Luc-EGFP virus, one-step growth curves were performed for PRV-HeN1 and PRV-Luc-EGFP. We found that PRV-Luc-EGFP replicated significantly slower than PRV-HeN1 (Figure 1D). We also evaluated the stability of the recombinant virus, and found that the luciferase activity from PRV-Luc-EGFP was stable through at least 10 passages (Figure 1E).

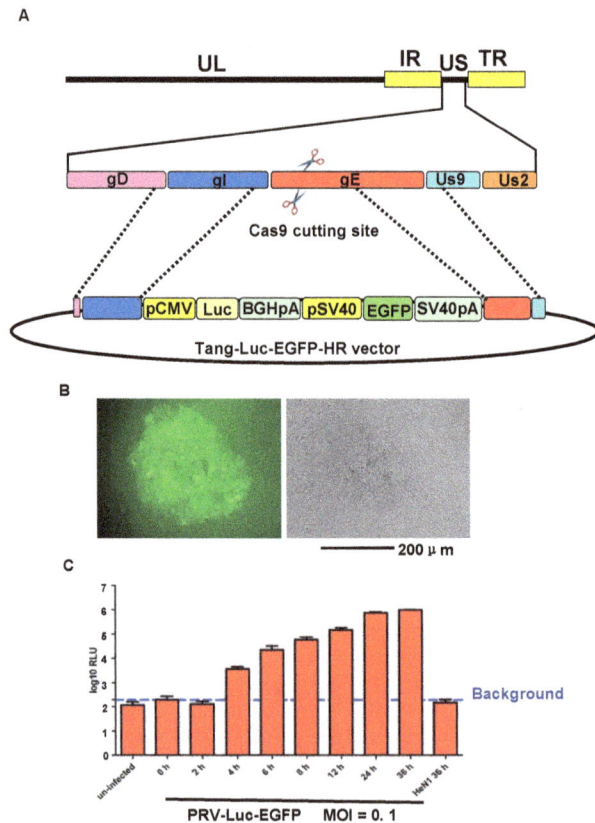

Figure 1. *Cont.*

46

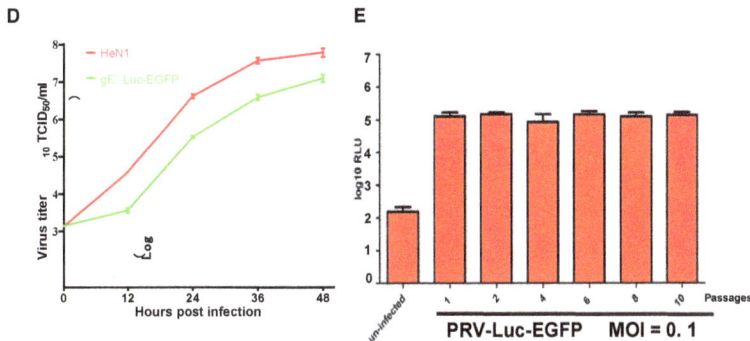

Figure 1. Construction and identification of PRV-Luc-EGFP. (**A**) Schematic showing the recombination sites used to modify the PRV genome and insert the Tang Luc-EGFP-HR donor vector using the CRISPR/Cas9 system; (**B**) Representative image showing the plaque assay for PRV-Luc-EGFP purification. The left panel shows fluorescent green plaque formation after seven rounds of plaque purification. The right panel shows the same area in bright field; (**C**) Luciferase expression in Vero cells at different time points post infection with PRV-Luc-EGFP (0.1 MOI); (**D**) One-step growth curve of the recombinant virus (PRV-Luc-EGFP; green) compared to the parental virus (HeN1; red); (**E**) Stability of PRV-Luc-EGFP. Luciferase activity was measured from passage 1 through passage 10.

3.2. PRV-Luc-EGFP Was an Effective Tool for CRISPR/Cas9 sgRNA Screening

To test whether PRV-Luc-EGFP could be used to screen for sgRNA, we designed and constructed several CRISPR/Cas9 constructs that were specific for *US3*, *US2*, and *US9*, respectively (Table 1). In short, effective CRISPR/Cas9 constructs would disrupt the PRV genome and inhibit viral replication, thereby decreasing luciferase activity. To test the designed sgRNAs, Vero cells were transfected with the CRISPR/Cas9 constructs, infected with PRV-Luc-EGFP 12 h later, and then luciferase activity was measured 24 hpi. As shown in Figure 2A, the sgRNAs US3-L3, US2-3, and US9-2 significantly reduced the amount of luciferase activity. To confirm the reduction in viral replication, we also assessed the effects of the sgRNAs on viral protein expression by Western blot. Consistent with the luciferase assay results, the cells transfected with the sgRNAs US3-L3, US2-3, and US9-2 had significantly decreased levels of US3 (Figure 2B). DNA breaks trigger the NHEJ repair pathway, which sometimes result in the introduction of crippling mutations and then inactivates gene expression. We next tested whether the sgRNAs would inactivate specific *PRV* genes. PRV-HeN1 treated with US3-L1, US2-3, and US9-2 was collected for plaque purification. Five (5) purified plaques were randomly selected to determine whether they had been inactivated based on US3 expression, and of these 60% (3/5) had reduced or no US3 expression (Figure 2C). No disruption was observed in the gI proteins, indicating the target specificity of sgRNA (Figure 2C). One of the plaques lacking US3 expression (plaque 4) was sequenced and found to be lacking two base pairs in the *US3* region (Figure 2D). PRV strains lacking expression of US2 and US9 were generated using similar methods (Figure 2E,F). These results indicated that PRV-Luc-EGFP was a powerful tool for effectively screening for CRISPR/Cas9 sgRNA.

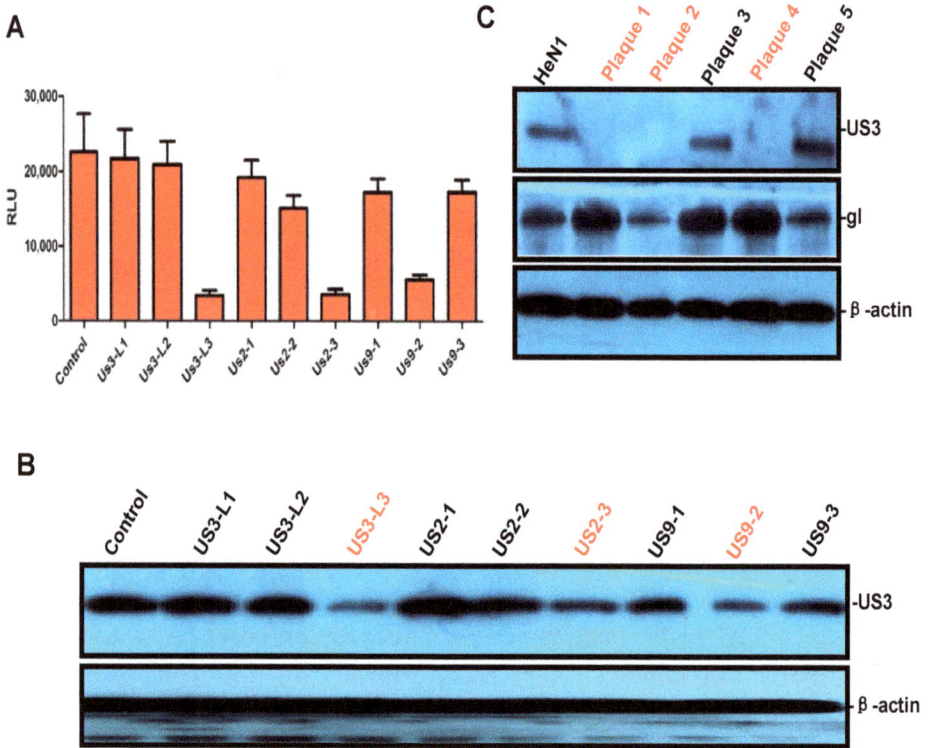

Figure 2. *Cont.*

D

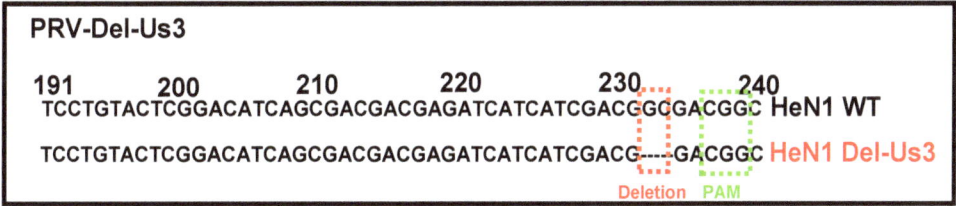

E

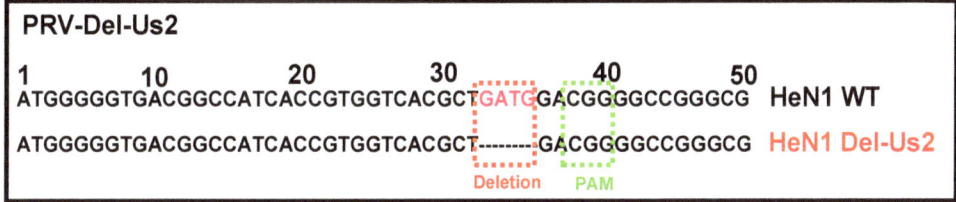

F

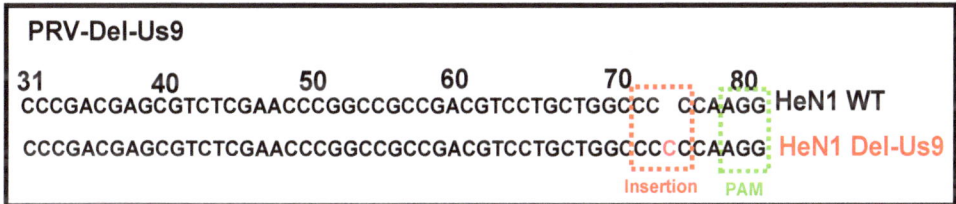

Figure 2. Effective sgRNA screening using PRV-Luc-EGFP and identification of PRV strains with the *US3*, *US2*, or *US9* genes inactivated. (**A**) sgRNA screening using PRV-Luc-EGFP. Cas9 constructs and a vector control were transfected individually into Vero cells. Twelve (12) h later, the cells were infected with PRV Luc-EGFP and 24 hpi luciferase activity was detected; (**B**) Representative image showing US3 expression by Western blot in PRV-HeN1 infected cells screening the same panel of sgRNAs as in (**A**); (**C**) Representative image showing US3 and gI expression in five randomly selected plaques that were treated with sgRNA US3-L3 by Western blot. Inactivated viruses were identified based on lack of US3 expression (e.g., plaque 4). Sequencing was used to confirm that the viruses considered inactivated had non-function genes for *US3* (**D**); *US2* (**E**); and *US9* (**F**). The images in A, B and C show representative results from three independent experiments.

3.3. Evaluating the Antiviral Activity of Chloroquine and CsA

There are very few antiviral compounds that are effective against PRV *in vitro* and none *in vivo*. Thus, an effective *in vitro* assay for screening antiviral compounds against PRV would likely contribute to controlling PR. Therefore, as proof of principle, we evaluated the antiviral effects of chloroquine and CsA using PRV-Luc-EGFP and confirmed the results with PRV HeN1. Vero cells were pretreated for 4 h

49

with chloroquine or CsA and then infected with PRV-Luc-EGFP for 24 h. The luciferase activity decreased in a dose dependent fashion with increasing chloroquine (Figure 3A–C).

In contrast, CsA did not have any inhibitory activity against PRV-Luc-EGFP (Figure 4A). Consistent with this, PRV HeN1 infection was not inhibited by CsA (Figure 4B,C). We simultaneously used PRRSV as a positive control for the antiviral effects of CsA. Using a FITC conjugated antibody directed against the N protein in PRRSV, CsA powerfully inhibited PRRSV replication (Figure 4D). Taken together, these results indicated that PRV-Luc-EGFP was a useful tool for screening antiviral compounds.

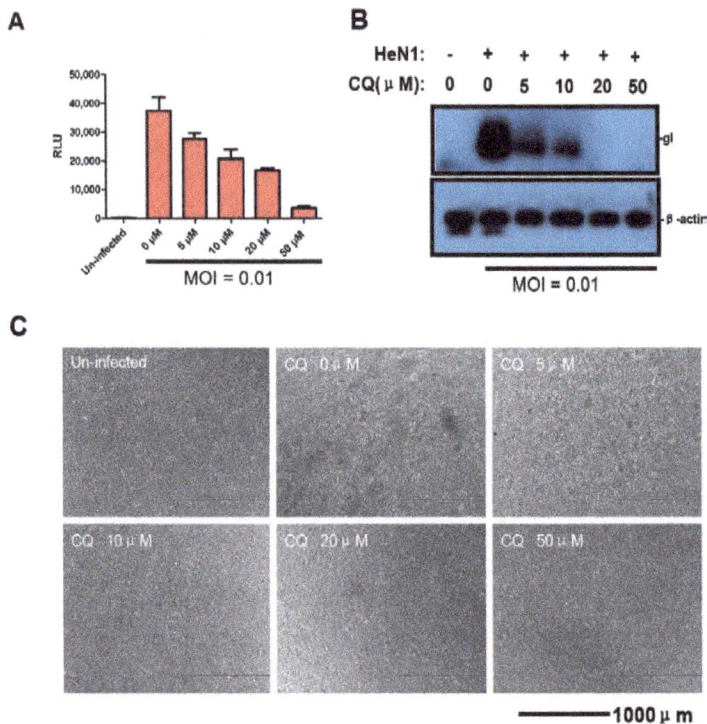

Figure 3. Chloroquine inhibits PRV replication. (**A**) Chloroquine inhibits PRV Luc-EGFP replication. Vero cells were pretreated with the indicated amounts of chloroquine and 4 h later infected with PRV-Luc-EGFP (0.01 MOI). After 24 h luciferase activity was evaluated; Chloroquine inhibits PRV HeN1, which is detected by gI expression (**B**); and CPE (**C**). The images show representative results from three independent experiments.

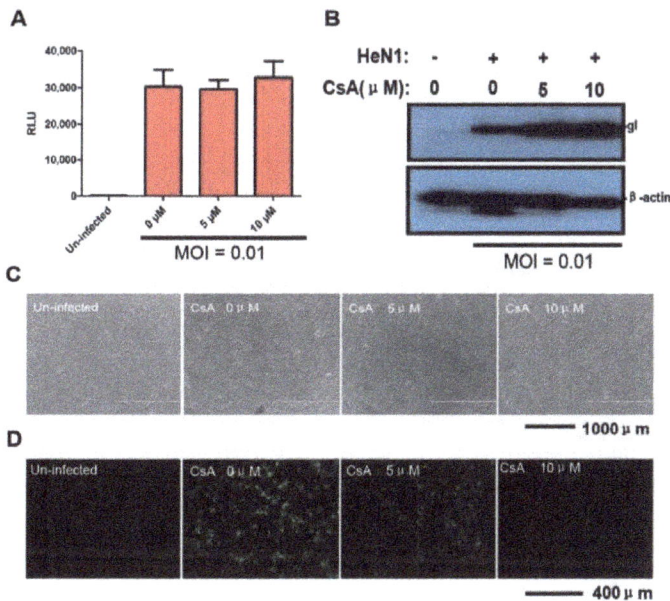

Figure 4. Cyclosporin A (CsA) has no effect on PRV replication. (**A**) Vero cells were pretreated with the indicated amounts of CsA and 4 h later were infected with PRV-Luc-EGFP (0.01MOI). After 24 h, luciferase activity was evaluated; (**B**) Vero cells were pretreated with the indicated amounts of CsA and 4 h later were infected with PRV HeN1 (0.01MOI). After 24 h, gI expression was evaluated by Western blot (representative image); (**C**) Representative images showing the cytopathic effect of HeN1 on Vero cells pretreated with different amount of CsA; (**D**) Marc145 cells were pretreated with the indicated amounts of CsA and then infected with 0.1 MOI PRRSV. Twelve (12) hpi, the N protein was detected by immunofluorescence. The images show representative results from three independent experiments.

4. Discussion

Genetic manipulation of large, complex DNA viruses such as PRV is critical for understanding their biology and developing novel control strategies. Therefore, in this study, we developed a novel PRV strain expressing firefly luciferase for use as a tool in future studies. Luciferase expression was detectable as early as 4 hpi, suggesting that the reporter virus will be able to shorten assay time for antiviral screens and improve the assay sensitivity. In addition, when PRV genomes were cut by the CRISPR/Cas9 system, the luciferase expression was inhibited, indicating that the reporter virus is a powerful tool for identifying PRV specific sgRNAs, which will help us edit PRV genome more feasible. Finally, using the reporter virus, we successfully developed US2, US3, and US9 inactivated PRV strains, and demonstrated that chloroquine inhibits PRV replication, whereas CsA does not.

The CRISPR/Cas9 system has many distinct advantages as a genome editing tool including being both efficient and convenient compared to other methods [16–18]. We have previously utilized the DNA breaks caused by the CRISPR/Cas9 system that inhibit PRV replication to develop the gE⁻/gI⁻/TK⁻ PRV strain to screen for PRV sgRNA using plaque formation assays or Western blot; however, these methods require 3–4 days (our unpublished data revised in Virology). The current luciferase based reporter system yields results in a much shorter timeframe (4 h), allowing a more rapid pace of vaccine development and basic research. For large genome viruses, the CRISPR/Cas9 system was a usable tool to insert foreign reporter genes in the genome as well [18,21].

Chloroquine is a 9-aminoquinoline that is well-known for antimalarial activity and antiviral activity against some viral infections [28]. Exposure to a low intracellular pH is required for successful entry of many viruses. Chloroquine exerts direct antiviral effects by inhibiting pH-dependent steps of the virus lifecycle for some members of the flaviviruses, retroviruses, and coronaviruses [28]. Chloroquine has been showed to inhibit herpes simplex virus type 1 (HSV-1) replication [29], and may inhibit PRV may by same mechanisms, but this remains to be confirmed. CsA is a fungal metabolite that exerts profound effects on the immune system and is potentially a selective immuno-suppressive agent [30]. The antiviral effects of CsA originate from inhibiting virus replication by blocking cyclophilins, which are beneficial for some viruses [31,32]. CsA has been reported to inhibit replication of Human Immunodeficiency Virus-1 [33], Hepatitis C Virus [34,35], PRRSV, and Equine arteritis virus [31]. With regard to herpes viruses in particular, CsA has been shown to inhibit HSV-1 replication [36,37]. Interestingly, in our study, CsA did not show any inhibitory activity against PRV in Vero cells. The lack of antiviral CsA-mediated antiviral activity may be because PRV replication is not associated with cyclophilins in Vero cells.

In conclusion, we developed a luciferase tagged PRV, which was a powerful tool for screening for sgRNA and antiviral compounds. The reporter virus likely has utility for basic research into emerging PRV variants, and developing virus control and prevention measures.

Acknowledgments: This study was supported by grants from the National Natural Science Foundation of China (Grant No. 31302065), and the National High-Tech Research and Development Program of China (No. 2011AA10A212).

Author Contributions: Yan-Dong Tang, Xue-Hui Cai, Tong-Qing An and Zhi-Jun Tian designed the experiment and wrote the paper. Yan-Dong Tang, Ji-Ting Liu, Qiong-Qiong Fang, Tong-Yun Wang and Ming-Xia Sun performed the experiments. All authors read and approved the final manuscript.

Conflicts of Interest: The authors declare no conflict of interest.

References

1. Muller, T.; Hahn, E.C.; Tottewitz, F.; Kramer, M.; Klupp, B.G.; Mettenleiter, T.C.; Freuling, C. Pseudorabies virus in wild swine: A global perspective. *Arch. Virol.* **2011**, *156*, 1691–1705.

2. Mettenleiter, T.C. Aujeszky's disease (pseudorabies) virus: The virus and molecular pathogenesis—State of the art, June 1999. *Vet. Res.* **2000**, *31*, 99–115.

3. Tong, W.; Liu, F.; Zheng, H.; Liang, C.; Zhou, Y.J.; Jiang, Y.F.; Shan, T.L.; Gao, F.; Li, G.X.; Tong, G.Z. Emergence of a Pseudorabies virus variant with increased virulence to piglets. *Vet. Microbiol.* **2015**, *181*, 236–240.

4. Luo, Y.; Li, N.; Cong, X.; Wang, C.H.; Du, M.; Li, L.; Zhao, B.; Yuan, J.; Liu, D.D.; Li, S.; *et al.* Pathogenicity and genomic characterization of a pseudorabies virus variant isolated from Bartha-K61-vaccinated swine population in China. *Vet. Microbiol.* **2014**, *174*, 107–115.

5. An, T.Q.; Peng, J.M.; Tian, Z.J.; Zhao, H.Y.; Li, N.; Liu, Y.M.; Chen, J.Z.; Leng, C.L.; Sun, Y.; Chang, D.; *et al.* Pseudorabies virus variant in Bartha-K61-vaccinated pigs, China, 2012. *Emerg. Infect. Dis.* **2013**, *19*, 1749–1755.

6. Hu, D.; Zhang, Z.; Lv, L.; Xiao, Y.; Qu, Y.; Ma, H.; Niu, Y.; Wang, G.; Liu, S. Outbreak of variant pseudorabies virus in Bartha-K61-vaccinated piglets in central Shandong Province, China. *J. Vet. Diagn. Investig.* **2015**, *27*, 600–605.

7. Ye, C.; Zhang, Q.Z.; Tian, Z.J.; Zheng, H.; Zhao, K.; Liu, F.; Guo, J.C.; Tong, W.; Jiang, C.G.; Wang, S.J.; *et al.* Genomic characterization of emergent pseudorabies virus in China reveals marked sequence divergence: Evidence for the existence of two major genotypes. *Virology* **2015**, *483*, 32–43.

8. Wang, C.H.; Yuan, J.; Qin, H.Y.; Luo, Y.; Cong, X.; Li, Y.; Chen, J.; Li, S.; Sun, Y.; Qiu, H.J. A novel gE-deleted pseudorabies virus (PRV) provides rapid and complete protection from lethal challenge with the PRV variant emerging in Bartha-K61-vaccinated swine population in China. *Vaccine* **2014**, *32*, 3379–3385.

9. Zhang, C.; Guo, L.; Jia, X.; Wang, T.; Wang, J.; Sun, Z.; Wang, L.; Li, X.; Tan, F.; Tian, K. Construction of a triple gene-deleted Chinese Pseudorabies virus variant and its efficacy study as a vaccine candidate on suckling piglets. *Vaccine* **2015**, *33*, 2432–2437.

10. Suenaga, T.; Kohyama, M.; Hirayasu, K.; Arase, H. Engineering large viral DNA genomes using the CRISPR-Cas9 system. *Microbiol. Immunol.* **2014**, *58*, 513–522.

11. Gaj, T.; Gersbach, C.A.; Barbas, C.F., III. ZFN, TALEN, and CRISPR/Cas-based methods for genome engineering. *Trends Biotechnol.* **2013**, *31*, 397–405.

12. Kim, H.; Kim, J.S. A guide to genome engineering with programmable nucleases. *Nat. Rev. Genet.* **2014**, *15*, 321–334.

13. Wyman, C.; Kanaar, R. DNA double-strand break repair: All's well that ends well. *Annu. Rev. Genet.* **2006**, *40*, 363–383.

14. Urnov, F.D.; Rebar, E.J.; Holmes, M.C.; Zhang, H.S.; Gregory, P.D. Genome editing with engineered zinc finger nucleases. *Nat. Rev. Genet.* **2010**, *11*, 636–646.

15. Bogdanove, A.J.; Voytas, D.F. TAL effectors: Customizable proteins for DNA targeting. *Science* **2011**, *333*, 1843–1846.

16. Mali, P.; Yang, L.; Esvelt, K.M.; Aach, J.; Guell, M.; DiCarlo, J.E.; Norville, J.E.; Church, G.M. RNA-guided human genome engineering via Cas9. *Science* **2013**, *339*, 823–826.

17. Cong, L.; Ran, F.A.; Cox, D.; Lin, S.; Barretto, R.; Habib, N.; Hsu, P.D.; Wu, X.; Jiang, W.; Marraffini, L.A.; Zhang, F. Multiplex genome engineering using CRISPR/Cas systems. *Science* **2013**, *339*, 819–823.

18. Bi, Y.; Sun, L.; Gao, D.; Ding, C.; Li, Z.; Li, Y.; Cun, W.; Li, Q. High-efficiency targeted editing of large viral genomes by RNA-guided nucleases. *PLoS Pathog.* **2014**, *10*, e1004090.

19. Ebina, H.; Misawa, N.; Kanemura, Y.; Koyanagi, Y. Harnessing the CRISPR/Cas9 system to disrupt latent HIV-1 provirus. *Sci. Rep.* **2013**, *3*.

20. Liao, H.K.; Gu, Y.; Diaz, A.; Marlett, J.; Takahashi, Y.; Li, M.; Suzuki, K.; Xu, R.; Hishida, T.; Chang, C.J.; *et al.* Use of the CRISPR/Cas9 system as an intracellular defense against HIV-1 infection in human cells. *Nat. Commun.* **2015**, *6*.

21. Xu, A.; Qin, C.; Lang, Y.; Wang, M.; Lin, M.; Li, C.; Zhang, R.; Tang, J. A simple and rapid approach to manipulate pseudorabies virus genome by CRISPR/Cas9 system. *Biotechnol. Lett.* **2015**, *37*, 1265–1272.

22. Wang, K.; Mei, D.Y.; Liu, Q.N.; Qiao, X.H.; Ruan, W.M.; Huang, T.; Cao, G.S. Research of methods to detect genomic mutations induced by CRISPR/Cas systems. *J. Biotechnol.* **2015**, *214*, 128–132.

23. Blue Heron Biotechnology. Available online: https://wwws.blueheronbio.com/external/tools/gRNASrc.jsp (accessed on 1 September 2015).

24. Ran, F.A.; Hsu, P.D.; Wright, J.; Agarwala, V.; Scott, D.A.; Zhang, F. Genome engineering using the CRISPR-Cas9 system. *Nat. Protoc.* **2013**, *8*, 2281–2308.

25. Smith, G.A.; Enquist, L.W. Construction and transposon mutagenesis in *Escherichia coli* of a full-length infectious clone of pseudorabies virus, an alphaherpesvirus. *J. Virol.* **1999**, *73*, 6405–6014.

26. Quint, W.; Gielkens, A.; Van Oirschot, J.; Berns, A.; Cuypers, H.T. Construction and characterization of deletion mutants of pseudorabies virus: A new generation of "live" vaccines. *J. Gen. Virol.* **1987**, *68*, 523–534.

27. Pomeranz, L.E.; Reynolds, A.E.; Hengartner, C.J. Molecular biology of pseudorabies virus: Impact on neurovirology and veterinary medicine. *Microbiol. Mol. Biol. Rev.* **2005**, *69*, 462–500.

28. Savarino, A.; Boelaert, J.R.; Cassone, A.; Majori, G.; Cauda, R. Effects of chloroquine on viral infections: An old drug against today's diseases? *Lancet Infect. Dis.* **2003**, *3*, 722–727.

29. Koyama, A.H.; Uchida, T. Inhibition of multiplication of herpes simplex virus type 1 by ammonium chloride and chloroquine. *Virology* **1984**, *138*, 332–335.

30. Borel, J.F.; Feurer, C.; Magnee, C.; Stahelin, H. Effects of the new anti-lymphocytic peptide cyclosporin A in animals. *Immunology* **1977**, *32*, 1017–1025.

31. De Wilde, A.H.; Li, Y.; van der Meer, Y.; Vuagniaux, G.; Lysek, R.; Fang, Y.; Snijder, E.J.; van Hemert, M.J. Cyclophilin inhibitors block arterivirus replication by interfering with viral RNA synthesis. *J. Virol.* **2013**, *87*, 1454–1464.

32. Nakagawa, M.; Sakamoto, N.; Tanabe, Y.; Koyama, T.; Itsui, Y.; Takeda, Y.; Chen, C.H.; Kakinuma, S.; Oooka, S.; Maekawa, S.; *et al.* Suppression of hepatitis C virus replication by cyclosporin a is mediated by blockade of cyclophilins. *Gastroenterology* **2005**, *129*, 1031–1041.

33. Niu, M.T.; Stein, D.S.; Schnittman, S.M. Primary human immunodeficiency virus type 1 infection: Review of pathogenesis and early treatment intervention in humans and animal retrovirus infections. *J. Infect. Dis.* **1993**, *168*, 1490–1501.

34. Watashi, K.; Hijikata, M.; Hosaka, M.; Yamaji, M.; Shimotohno, K. Cyclosporin A suppresses replication of hepatitis C virus genome in cultured hepatocytes. *Hepatology* **2003**, *38*, 1282–1288.

35. Nakagawa, M.; Sakamoto, N.; Enomoto, N.; Tanabe, Y.; Kanazawa, N.; Koyama, T.; Kurosaki, M.; Maekawa, S.; Yamashiro, T.; Chen, C.H.; *et al.* Specific inhibition of hepatitis C virus replication by cyclosporin A. *Biochem. Biophys. Res. Commun.* **2004**, *313*, 42–47.

36. McKenzie, R.C.; Epand, R.M.; Johnson, D.C. Cyclosporine A inhibits herpes simplex virus-induced cell fusion but not virus penetration into cells. *Virology* **1987**, *159*, 1–9.

37. Vahlne, A.; Larsson, P.A.; Horal, P.; Ahlmen, J.; Svennerholm, B.; Gronowitz, J.S.; Olofsson, S. Inhibition of herpes simplex virus production *in vitro* by cyclosporin A. *Arch. Virol.* **1992**, *122*, 61–75.

Development of a Triple-Color Pseudovirion-Based Assay to Detect Neutralizing Antibodies against Human Papillomavirus

Jianhui Nie, Yangyang Liu, Weijin Huang and Youchun Wang

Abstract: Pseudovirion-based neutralization assay is considered the gold standard method for evaluating the immune response to human papillomavirus (HPV) vaccines. In this study, we developed a multicolor neutralization assay to simultaneously detect the neutralizing antibodies against different HPV types. FluoroSpot was used to interpret the fluorescent protein expression instead of flow cytometry. The results of FluoroSpot and flow cytometry showed good consistency, with $R^2 > 0.98$ for the log-transformed IC_{50} values. Regardless of the reporter color, the single-, dual-, and triple-color neutralization assays reported identical results for the same samples. In low-titer samples from naturally HPV-infected individuals, there was strong agreement between the single- and triple-color assays, with kappa scores of 0.92, 0.89, and 0.96 for HPV16, HPV18, and HPV58, respectively. Good reproducibility was observed for the triple-color assay, with coefficients of variation of 2.0%–41.5% within the assays and 8.3%–36.2% between the assays. Three triple-color systems, HPV16-18-58, HPV6-33-45, and HPV11-31-52, were developed that could evaluate the immunogenicity of a nonavalent vaccine in three rounds of the assay. With the advantages of an easy-to-use procedure and less sample consumption, the multiple-color assay is more suitable than classical assays for large sero-epidemiological studies and clinical trials and is more amenable to automation.

Reprinted from *Viruses*. Cite as: Nie, J.; Liu, Y.; Huang, W.; Wang, Y. Development of a Triple-Color Pseudovirion-Based Assay to Detect Neutralizing Antibodies against Human Papillomavirus. *Viruses* **2016**, *8*, 107.

1. Introduction

Cervical cancer is the second most frequently diagnosed cancer and the third leading cause of cancer death among women in developing countries [1], and is strongly associated with high-risk human papillomavirus (HPV) infection. To date, more than 200 HPV genotypes have been identified [2], among which 15 are classified as high-risk types (HPV16, 18, 31, 33, 35, 39, 45, 51, 52, 56, 58, 59, 68, 73, and 82) [3,4]. HPV vaccines are an effective prophylactic strategy implemented in many countries, and a bivalent HPV type 16/18 vaccine (Cervarix®) and a quadrivalent HPV type 6/11/16/18 vaccine (Gardasil®) have been licensed for use [3,5]. In clinical trials, the vaccines have shown almost 100% efficacy against precancerous lesions associated with the vaccine HPV types, which account for 70% of cervical cancer cases worldwide [3]. In 2014, a nonavalent HPV vaccine, Gardasil 9, directed against HPV6, 11, 16, 18, 31, 33, 45, 52, and 58, was approved by the U.S. Food and Drug Administration (FDA), increasing the potential prevention of cervical cancer from 70% to 90% [6].

In HPV vaccinology and epidemiology, serological testing methods are used to analyze the antibody levels in naturally infected or vaccinated individuals. A variety of methods have been developed for testing antibody levels, including the competitive Luminex® immunoassay (cLIA) [7], the virus-like-particle (VLP)-based multiplex immunoassay (VLP-MIA) [8], and the *in situ* purified glutathione S-transferase L1-based MIA (GST-L1-MIA) [9]. Depending on the preselected monoclonal antibody, the cLIA results are interpreted to the overlapping epitopes, which might not predict the *in vivo* protection well when the dominant antibodies in some serum samples are not the selected epitopes. VLP-MIA and GST-L1-MIA detect the total VLP-specific antibodies, which include some non-neutralizing antibodies, and might therefore overestimate the efficacy of the vaccine.

Although the level of protection afforded by neutralizing antibodies has not been determined, neutralizing antibodies are accepted as the primary mediator of a vaccine's potency. The HPV life cycle is strictly dependent on the differentiation stage of the host cell [10]. Native HPV virions cannot be produced in conventional culture, and it is almost impossible to detect neutralizing antibodies using authentic virions, especially in large-scale analyses of naturally infected or vaccinated cohorts. The cotransfection of mammalian cells with two HPV capsid genes, L1 and L2, together with a reporter plasmid produced high infectious titers of pseudovirions, which presented surface conformational epitopes similar to those of the native virions [11]. The pseudovirion-based neutralization assay (PBNA) is recognized as the gold standard method for the analysis of the functional humoral immune response to HPV. Several PBNAs have been developed using different reporter genes, including fluorescent reporter genes (such as enhanced green fluorescent protein (EGFP) [12,13]), or chemiluminescent reporter genes (such as secreted alkaline phosphatase (SEAP) [12] and *Gaussia* luciferase (Gluc) [14,15]). The infected status of the target cells in the PBNA is detected using fluorescent microscopy and/or flow cytometry (FCM) for EGFP and with a chemiluminescence reader for SEAP and GLuc. Because the interpretation with microscopy is subjective and the procedure for FCM is laborious, the EGFP method has not been widely used. Although the SEAP- and GLuc-based method can enhance the throughput to a certain extent, the throughput is still lower than that of MIAs [7,16], which can simultaneously quantitate antibodies directed against different HPV types. Recent advances in fluorescent reporter genes and the ELISPOT reader allow the simultaneous detection of several differently colored fluorescent proteins [17]. Using these innovations, we have established a new multicolor PBNA to simultaneously quantify the neutralizing antibodies directed against different types of HPV.

2. Materials and Methods

2.1. Cells, Plasmids, Serum Samples, HPV Antibody Standards, and HPV Vaccines

The 293FT cell line (Invitrogen, Carlsbad, CA, USA) was maintained in growth medium (high-glucose Dulbecco's modified Eagle's medium with 10% fetal bovine serum, 1% penicillin- streptomycin solution, 1% nonessential amino acids, 2% HEPES). HPV L1/L2 expressing plasmids (p16LLw [18], p18LLw [12], p6LLw2 [19], p11Lw [20], p11∫w [21], p31sheLL [22], p45sheLL [23], p52LLw [24] and p58LLw [22]) and the red fluorescent protein (RFP) reporter plasmid pRwB [25] were kindly provided by John Schiller (National Cancer Institute, Bethesda, MD, USA). E2-CFP (Clontech, Mountain View, CA, USA) expresses the E2-crimson fluorescent protein (CFP). The EGFP reporter plasmid was constructed by inserting the *EGFP* gene into the pCDNA3.1 vector (Invitrogen, Carlsbad, CA, USA), as described previously [26].

2.2. Serum Samples

Naturally HPV-infected serum samples were selected from a large number of samples from a blood bank (Shanghai RAAS Blood Products Co. Ltd, Shanghai, China). Fifty post-vaccinated human serum samples that had been collected in a phase I clinical trial of a bivalent HPV16/18 vaccine, produced in *Pichia pastoris* (ClinicalTrials.gov ID: 2011L01085) were kindly provided by Shanghai Zerun Biotechnology (Shanghai, China). Written informed consent was obtained from all donors. Mouse sera were collected from mice intraperitoneally administered a candidate nonavalent HPV vaccine (Innovax, Xiamen, China). Rabbit serum samples were collected from rabbits immunized with a monovalent vaccine kindly provided by Qiming Li (National Vaccine and Serum Institute, Beijing, China), as described previously [15]. This study was approved by the Institutional Animal Care and Use Committee of the National Institutes for Food and Drug Control, China. All animals were housed and maintained in accordance with the relevant guidelines and regulations.

2.3. Preparation and Titration of HPV Pseudovirions

HPV pseudovirions were produced in 293FT cells using a previously described method, with modification [11,27]. Briefly, 293FT cells were cotransfected with HPV-L1/L2-expressing plasmids together with a reporter plasmid, using Lipofectamine 2000 (Invitrogen), according to the manufacturer's instructions.

The cells were harvested 48 h post-transfection and suspended in Dulbecco's phosphate-buffered saline (Invitrogen) supplemented with 0.5% Triton X-100, 0.1% Benzonase® (Sigma-Aldrich, St. Louis, MO, USA), 0.1% Plasmid-Safe ATP-Dependent DNase (Epicentre, Madison, WI, USA), and 1/40 volume of 1 M ammonium sulfate. The pseudovirions were matured by incubating the cell lysates overnight at 37 °C. The lysates were then transferred to an ice bath for 5 min and the NaCl concentration was adjusted to 0.85 M. The lysates were clarified by centrifugation at $5000 \times g$ for 10 min. The three types of pseudovirions (expressing reporter protein EGFP, RFP, or E2-CFP) were stored as aliquots at $-80\,^\circ$C until titration.

The pseudovirion stocks were titrated as follows: 293FT cells were trypsinized and placed in the inner wells of 96-well tissue culture plates at 1.5×10^4 cells/100 μL/well for 3–6 h before the pseudovirions were added. Each type of pseudovirion was prediluted 100-fold, and nine serial five-fold dilutions were prepared. After incubation for 72 h at 37 °C under 5% CO_2, the percentages of fluorescence-positive cells and fluorospots were determined with FCM (BD, Franklin Lakes, NJ, USA) and an ImmunoSpot reader (CTL, Shaker Heights, OH, USA). The pseudoviron titers were defined as 50% tissue culture infective dose (TCID50), calculated with the Reed-Muench method [28].

2.4. Pseudovirion-Based Neutralization Assay

Once the pseudovirion titer had been determined, the neutralizing antibody titers of the sera were determined with a PBNA, as in previous studies [12,13, 15]. Briefly, 293FT cells were placed in 96-well flat-bottom cell culture plates at 15,000 cells/well with 100 μL of growth medium and incubated for 3–6 h before the pseudovirions were added. The pseudovirion stocks were diluted to 4000–6000 TCID50/mL [13]. The diluted pseudovirions (60 μL) and serially diluted sera (60 μL) were mixed in 96-well round-bottom plates and placed on ice for 1 h. The pseudovirion-serum mixtures (100 μL) were transferred into cell culture plates preseeded with 293FT cells, and incubated for 68–72 h. After incubation, the numbers of fluorospots were counted with the ImmunoSpot reader. The serum neutralization titers were defined as the 50% maximal inhibitory concentration (IC_{50}), calculated with the Reed-Muench method [28].

3. Results

3.1. Selection of Fluorescent Proteins

To select the appropriate fluorescent proteins for the multiplex PBNA, 293FT cells were transfected with different fluorescence-expressing plasmids: AmCyan1 (blue), EGFP (green), enhanced yellow fluorescent protein (EYFP; yellow), RFP (red), and E2-CFP (far red). The cells were scanned with different excitation and emission wavelengths using ImmunoSpot (CTL, Shaker Heights, OH, USA). Only EGFP, RFP, and E2-CFP were detected with unique excitation-emission wavelength pairs. Therefore, the three reporter plasmids were cotransfected with the HPV structural genes to generate pseudovirions producing different colors (Figure 1). No mutual interference was observed with the specific excitation-emission wavelength pairs used to detect the pseudovirions.

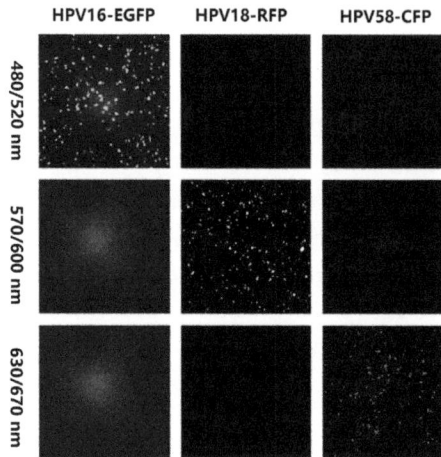

Figure 1. Fluorescence of three fluorescent proteins under different excitation and emission filters. The excitation and emission filter wavelengths for row 1 were 480 nm and 520 nm, respectively. For row 2, the excitation and emission filter wavelengths were 570 nm and 600 nm, respectively. For row 3, the excitation and emission filter wavelengths were 630 nm and 670 nm, respectively.

3.2. Correlation of the FluoroSpot and FCM Results

The classic GFP-pseudovirion assay was detected using the FCM [12]. To compare the results of FluoroSpot and FCM, 293FT cells were transfected with an HPV16 or HPV18 structural genes, together with the reporter gene *EGFP* or *RFP*, to yield four pseudovirions: HPV16-EGFP, HPV18-EGFP, HPV16-RFP, and HPV18-RFP. The 293FT cells were then infected with the serially diluted pseudovirions and detected with FluoroSpot and FCM 72 h post-infection. The FluoroSpot counts and the percentages of positive cells correlated well, with $R^2 > 0.97$ (Figure 2A–D). The neutralizing antibodies against HPV16-EGFP and HPV18-RFP in 50 post-vaccination human serum samples were detected with both FluoroSpot and FCM. The log-transformed IC_{50} values showed good consistency, with $R^2 > 0.98$ (Figure 2E,F).

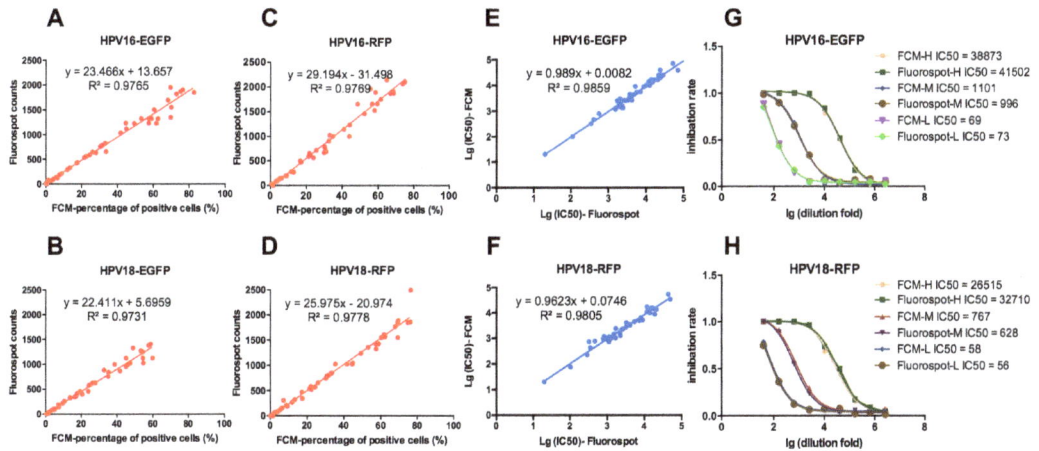

Figure 2. Correlation between FluoroSpot and FCM analyses of EGFP- and RFP-PBNA. (**A–D**) Correlation of the Fluorospot counts and FCM-percentage of positive cells when tested for HPV16-EGFP, HPV18-EGFP, HPV16-RFP and HPV18-RFP, respectively; (**E,F**) correlation of the lg(IC50) values for Fluorospot counts and FCM-percentage of positive cells when tested for HPV16-EGFP and HPV18-RFP, respectively; (**G,H**) the titration of three serum samples (H, M, and L with high, medium, and low neutralizing antibody titers collected from the dual-valent vaccine clinical tiral) against HPV16-EGFP and HPV18-RFP, respectively. The sensitivity parameters used in these assay was 215.

3.3. Impact of the Sensitivity Parameter on the Interpretation of the FluoroSpot Results

The sensitivity of the FluoroSpot counts can be adjusted by changing the sensitivity parameter. Too high a sensitivity could yield high background and low specificity, whereas too low a sensitivity might miss specific spots. To identify the optimal sensitivity parameter, 293FT cells were infected with HPV16-EGFP, HPV18-EGFP, HPV16-RFP, or HPV18-RFP and detected with different sensitivity parameters (150–250), and the results were compared with the FCM results. The linear correlation R^2 increased initially and then decreased as the sensitivity parameter increased (Figure 3). The correlation was most easily disturbed by altering the sensitivity parameter for pseudovirions containing EGFP. Only when the sensitivity parameter was 200–230, the R^2 reached more than 0.99. For the pseudovirions expressing RFP, $R^2 > 0.99$ when the sensitivity parameter was 180–250. Therefore, 215 was used as the optimal sensitivity parameter for this assay. At this sensitivity, 300 and 400 fluorescent spots were detected for EGFP and RFP, respectively, if the rate of positive cells reached 15%, which was the most suitable pseudoviral dose for FCM [13].

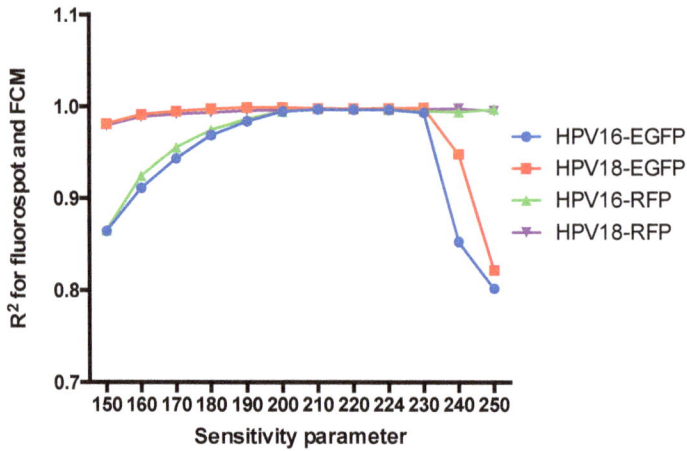

Figure 3. Effect of the sensitivity parameter on the correlation between the FluoroSpot and FCM results.

3.4. Comparison of Single- and Dual-Color PBNA

To determine whether the reporter protein affects the PBNA results, green (EGFP) and red (RFP) HPV16 pseudovirions were tested against rabbit sera immunized with HPV16. The almost totally overlapping inhibition curves yielded similar IC_{50} values for HPV16 with both reporter proteins (Figure 4A). A similar phenomenon was observed for HPV18 (Figure 4B). To investigate whether the mixed pseudovirions affected the PBNA results, the same serum samples were tested with two dual-color pseudovirion combinations (HPV16-EGFP + HPV18-RFP and HPV16-RFP + HPV18-EGFP) (Figure 4C,D). The similar inhibition curves and IC_{50} values indicated that there was no distinct difference between these two combinations. The results for dual-color PBNA were also consistent with those for single-color PBNA (Figure 4). In addition, no cross-talk was observed between HPV16 and HPV18 in all the PBNAs.

3.5. Comparison of Single- and Triple-Color PBNA

To further improve the throughput of the PBNA, a third color (CFP: far red) was introduced to the assay. Nine kinds of pseudovirions were prepared: HPV16-EGFP, HPV16-RFP, HPV16-CFP, HPV18-EGFP, HPV18-RFP, HPV18-CFP, HPV58-EGFP, HPV58-RFP, and HPV58-CFP. The differently colored pseudovirions and different combinations of pseudovirion were tested against pooled serum samples from mice immunized with a nonavalent (HPV6/11/16/18/31/33/45/52/58) HPV candidate vaccine. The similar inhibition curves suggest that there were no significant differences between the single-color PBNA and triple-color PBNA in testing HPV16, HPV18, and HPV58 (Figure 5).

62

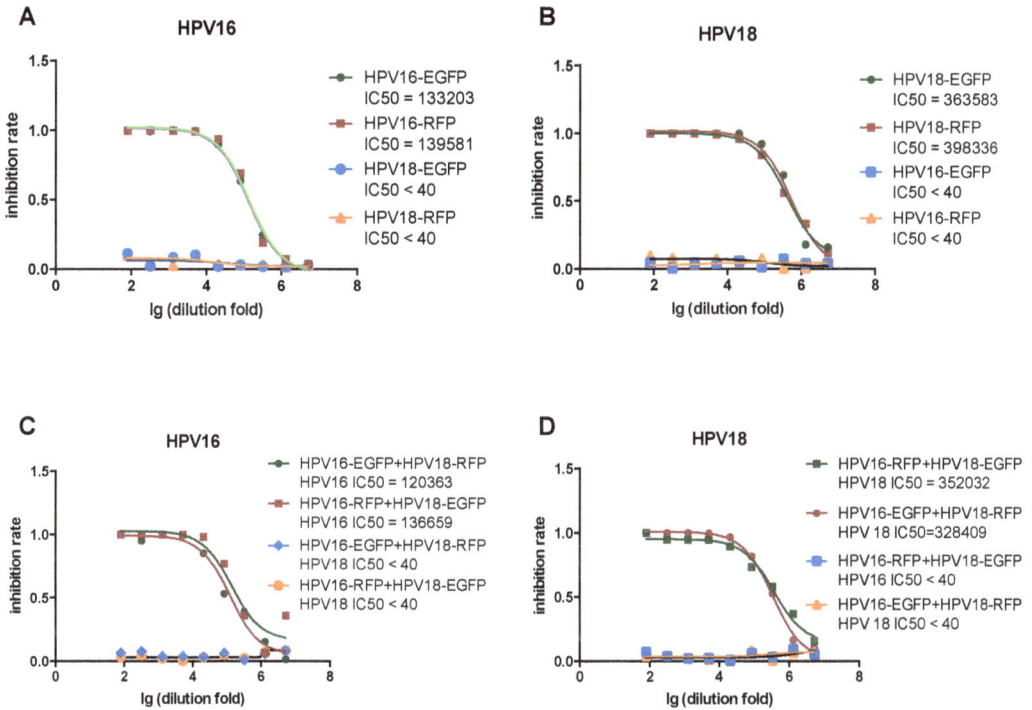

Figure 4. Comparison of single- and dual-color PBNA in detecting inhibition rates. (**A**) Single-color HPV16 and HPV18 pseudovirions were tested against mono-valent HPV16 vaccine immunized rabbit serum; (**B**) Single-color HPV16 and HPV18 pseudovirions were tested against monovalent HPV18 vaccine immunized rabbit serum; (**C**) Dual-color HPV16 and HPV18 pseudovirions mixtures were tested against monovalent HPV16 vaccine immunized rabbit serum; (**D**) Dual-color HPV16 and HPV18 pseudovirion mixtures were tested against monovalent HPV18 vaccine immunized rabbit serum.

The single-color and triple-color PBNA were also compared using different kinds of samples, including sera from naturally infected humans, sera from immunized mice, and WHO International Standards for antibodies. For HPV16, HPV18, and HPV58, the coefficients of variation (CVs) were 18–34%, 16–51%, and 28–43%, respectively (Table 1). The HPV16 standard specifically neutralized HPV16 in the triple-color PBNA, with IC_{50} values of 118, 97, and 55, respectively, which are similar to those reported previously [29]. The HPV18 standard specifically neutralized HPV18 in the triple-color PBNA, with IC_{50} values of 801, 1027, and 628, which are similar to those reported previously [14]. Most importantly, no cross-reaction was observed in the triple-color PBNA, confirming the consistency between the triple- and single-color PBNAs.

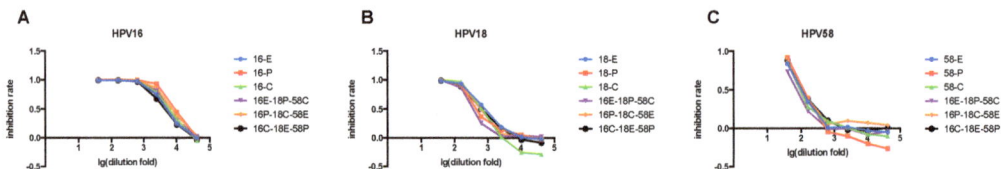

Figure 5. Comparison of single- and three-color PBNAs against HPV16, HPV18, and HPV58. (**A**) Comparison of single- and three-color PBNAs for HPV16; (**B**) Comparison of single- and three-color PBNAs for HPV18; (**C**) Comparison of single- and three-color PBNAs for HPV58. 16-E: HPV16 pseudovirion packaged with the *EGFP* reporter gene; 16-R: HPV16 pseudovirion packaged with the *RFP* reporter gene; 16-C: HPV16 pseudovirion packaged with the *CFP* reporter gene; 16E-18R-58C: HPV16 EGFP + HPV18 RFP + HPV CFP; 16R-18C-58E: HPV16 RFP + HPV18 CFP + HPV58 EGFP; 16C-18E-58R: HPV16 CFP + HPV18 RFP + HPV58 EGFP. The sample tested contained mixed sera from mice immunized with a candidate nonavalent HPV vaccine.

Table 1. Comparison of the single- and three-color PBNAs in detecting antibody titers.

		Sample				
	Pseudovirion	**A**	**B**	**C**	**D**	**E**
HPV16	16E	113	967	5025	99	<40
	16P	95	1945	8608	109	<40
	16C	75	1225	5786	75	<40
	16E-18P-58C	93	1016	5866	118	<40
	16P-18C-58E	74	1328	7039	97	<40
	16C-18E-58P	<40	761	4381	55	<40
	CV	18%	34%	25%	25%	0%
HPV18	18E	120	16,957	15,630	<40	826
	18P	80	5641	14,820	<40	824
	18C	78	6875	13,912	<40	674
	16E-18P-58C	102	6500	15,366	<40	801
	16P-18C-58E	98	11,296	5106	<40	1067
	16C-18E-58P	98	18,427	9519	<40	628
	CV	16%	51%	34%	0%	19%
HPV58	58E	103	1778	8799	<40	<40
	58P	121	2029	7422	<40	<40
	58C	93	1098	3077	<40	<40
	16E-18P-58C	76	1395	3686	<40	<40
	16P-18C-58E	189	952	10,764	<40	<40
	16C-18E-58P	115	1608	7599	<40	<40
	CV	34%	28%	43%	0%	0%

A, B: Human sera obtained from naturally infected individuals; C: Serum sample from mice immunized with the candidate nonavalent HPV vaccine; D: WHO Anti-HPV16 serum standard (National Institute for Biological Standards and Control (NIBSC) code: 05/134); E: WHO Anti-HPV18 serum standard (NIBSC code: 10/140).

Because the antibody titers are low in individuals naturally infected with HPV, the agreement between the single- and triple-color PBNAs can be assessed most stringently with this kind of sample. Samples from 48 individuals naturally infected with HPV were used in both assays for HPV16, HPV18, and HPV58, using an IC_{50} of 40 as the cut-off value. The sero-status assignment of the samples with the two assays is cross-tabulated in Table 2. For each of the three HPV genotypes, the agreement rates in qualitative analyses were greater than 95.8% (HPV16, 95.8%; HPV18, 95.8%; and HPV58, 97.9%). The kappa scores between the two assays were 0.92, 0.89, and 0.96 for the three HPV types, respectively, suggesting good agreement between the two assays.

Table 2. Agreement between the single- and triple-color PBNAs in a naturally infected cohort.

Single-Color PBNA	Triple-Color PBNA								
	HPV16			HPV18			HPV58		
	−	+	Total	−	+	Total	−	+	Total
−	26	0	26	35	0	35	29	1	30
+	2	20	22	2	11	13	0	18	18
Total	28	16	48	37	11	48	29	19	48
Agreement (%)		95.8			95.8			97.9	
Negative Predictive Value (%)		92.9			94.6			100	
Positive Predictive Value (%)		100			100			94.7	
Kappa score		0.92			0.89			0.96	

3.6. Reproducibility of the Triple-Color PBNA

To determine the reproducibility of the three-color PBNA, three samples for each type were tested: two from naturally infected individuals (H16.1, H16.2 for HPV16; H18.1, H18.2 for HPV18; H58.1, H58.2 for HPV58) and one from a rabbit immunized three times with the corresponding monovalent HPV vaccines (R16 for HPV16, R18 for HPV18, R58 for HPV58). Each serum was tested nine times in three independent runs (Figure 6). When the serum titers were calculated, the intra-assay and inter-assay coefficients of variation were 2.0%–41.5% and 8.3%–36.2%, respectively, indicating good reproducibility.

3.7. Triple-Color PBNA for HPV6-33-45 and HPV11-31-52

A nonavalent HPV vaccine targeting HPV6/11/16/18/31/33/45/52/58 (Gardasil 9; Merck) was approved by the U.S. FDA in 2014 [6]. To test all nine HPV types, another two triple-color pseudovirion mixtures were examined, HPV6-33-45 and HPV11-31-52. Three serum samples from mice immunized with different amounts of the nonavalent vaccines were used to test the consistency between the

65

single-color and triple-color PBNAs. Good concordance was observed between the single- and triple-color PBNAs for both the HPV6-33-45 assay and the HPV11-31-52 assay (Figure 7). The neutralizing antibody profiles were detected for the samples immunized with the nonavalent vaccine with three runs of the assay using the triple-color PBNA.

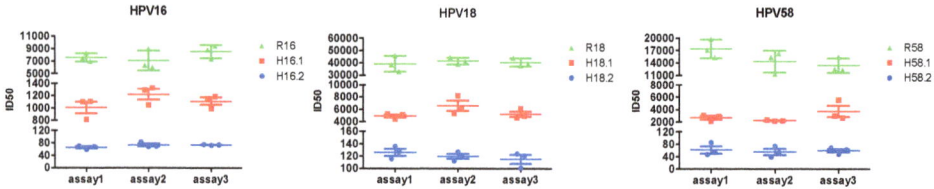

Figure 6. Reproducibility of the triple-color PBNA. R16, R18, and R58 were rabbit serum samples immunized with corresponding mono-valent HPV vaccines. H16.1, H16.2, H18.1, H18.2, H58.1, and H58.2 were serum samples from naturally infected individuals.

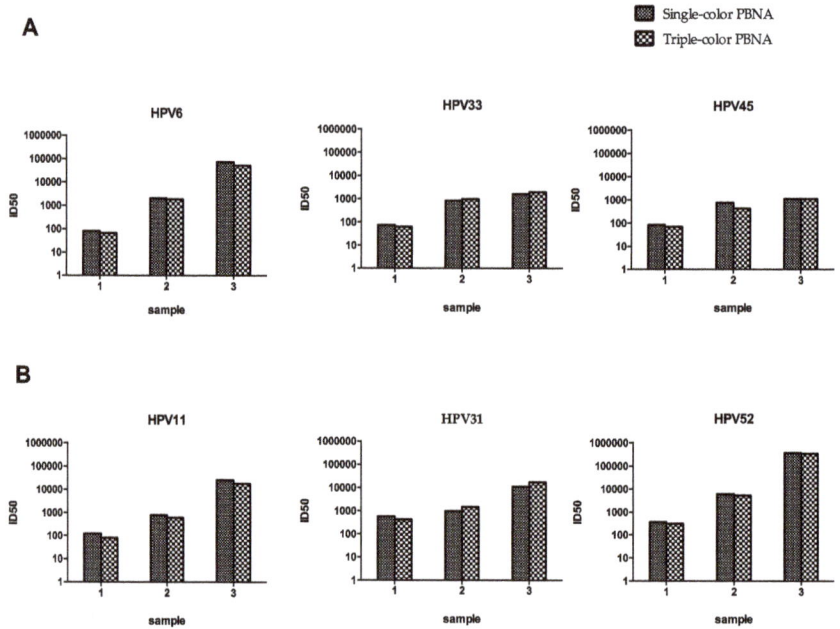

Figure 7. Comparison of the single- and triple-color PBNA for HPV6-33-45 and HPV11-31-52 PBNAs. (**A**) Comparison of the single- and triple-color PBNA for HPV6-33-45; (**B**) Comparison of the single- and triple-color PBNA for HPV11-31-52. Three serum samples from mice immunized with different amounts of the nonavalent vaccines were tested using single-color and triple-color PBNAs, respectively.

4. Discussion

Neutralizing antibodies are considered the primary protective mechanisms for most prophylactic vaccines, and PBNA is considered the gold standard method for evaluating the immune response raised against HPV vaccines [12,14,30]. However, PBNA is not widely used in clinical trials or sero-epidemiological analysis because it is laborious and complex. To date, a series of optimized PBNAs has been developed and the throughput of PBNAs has been improved to some extent [14]. However, compared with antibody-binding detection assays, especially the multiplex cLIA based on Luminex, the PBNA procedures require considerable simplification. In this study, a new assay based on a fluorescent reporter gene was developed as an economical and easy-to-use method for simultaneously comparing the neutralizing antibodies raised against different HPV types.

FluoroSpot, which is used in the ELISPOT assay, allows the analysis of single cells secreting several cytokines [31]. Here, FluoroSpot directly detected target cells infected with the HPV pseudovirion, without cell digestion or the addition of further substrates. The detection time was less than 5 min for a 96-well plate. The traditional detection strategy for fluorescent proteins is fluorescent microscopy and/or FCM. The results of the EGFP-based pseudovirion infection assay showed good consistency between the FCM and FluoroSpot detection methods. When the single-color method was extended to include multiple colors, any possible cross-interference of the different reporters had to be considered. For example, YFP can be detected by the EGFP wavelength pair. Furthermore, the specificity of the assay decreases with decreasing wavelengths. For example, cell debris and unexpected fibers will present nonspecific blue light at the wavelengths used for the blue fluorescent protein reporter, which is therefore unsuitable for this assay.

Although the mechanisms have not been fully revealed, cross-protection has been observed with bivalent and quadrivalent vaccines [32–34]. Therefore, to avoid the detection of false positives, the mixed pseudovirions should be distantly phylogenetically related. HPV16 and HPV18, which belong to the alpha 9 and alpha 7 genotypes, respectively, were selected for the dual-color combination. Based on this principle, closely related HPV types were assigned to different combinations. Therefore, for the immunological analysis of the nonavalent vaccine, mixtures HPV16-18-58, HPV6-33-45, and HPV11-31-52 were used to avoid cross-reaction.

To date, neutralizing methods have been used in few clinical HPV vaccine trials. Instead, antibody binding is used as a surrogate index for the immune response. The coating antigens and principles of detection differ completely between these two types of assays. No comparison data for different clinical trials are available. Placebo controls cannot be used in future clinical studies, especially for the development of second-generation vaccines. Therefore, the development of high-throughput easy-to-use protection-related PBNAs is urgently required for future clinical trials.

5. Conclusions

Three triple-color PBNAs, HPV16-18-58, HPV6-33-45, and HPV11-31-52, were developed to evaluate a nonavalent vaccine. This multicolor PBNA provided rapid and precise profiles of the protection-related immune response in vaccine-immunized or naturally infected cohorts. The new assay has advantage over the FCM assay for the easy-to-use procedure, which shortens the detection time from more than 3 h to 5 min for 96-well plate. Even compared to the recently developed high-throughput Gluc-based assay [15], triple-color assay only needs one third of the detection time and sample quantity for the same sample evaluation. With the development of fluorescent reporter proteins and the FluoroSpot technology, more colors for PBNA might become available to improve the analysis of immune responses in the future.

Acknowledgments: We would like to thank John Schiller for providing the plasmids encoding the HPV structural proteins (HPV L1 and L2). We are grateful to Shanghai RAAS for supplying the normal human sera and Shanghai Zerun Biotechnology for supplying the pre- and post-vaccinated human serum samples. This work was supported by the National High Technology Research and Development Program of China (2012AA02A402).

Author Contributions: Jianhui Nie, Yangyang Liu, and Youchun Wang wrote the manuscript; Youchun Wang, Jianhui Nie, Yangyang Liu, and Weijin Huang conceived and designed the experiments; Jianhui Nie and Yangyang Liu conducted the experiments. All the authors have read and approved the final manuscript.

Conflicts of Interest: The authors declare no conflict of interest. The funding bodies had no role in the study design, data collection and analysis, decision to publish, or preparation of the manuscript.

Abbreviations

PBNA	Pseudovirion-based neutralization assay
HPV	Human papillomavirus
US	United States
FCM	Flow cytometry
FDA	Food and Drug Administration
cLIA	Competitive Luminex® immunoassay
VLP	Virus-like particle
VLP-MIA	Virus-like particle-based multiplex immunoassay
GST-L1-MIA	Glutathione S-transferase L1-based MIA
EGFP	Enhanced green fluorescent protein
GLuc	*Gaussia* luciferase
SEAP	Secreted alkaline phosphatase
HEPES	4-(2-Hydroxyethyl)-1-piperazineethanesulfonic acid
RFP	Red fluorescent protein

E2-CFP	E2-crimson fluorescent protein
WHO	World Health Organization
NIBSC	National institute for Biological Standards and Control
YFP	Yellow fluorescent protein

References

1. Torre, L.A.; Bray, F.; Siegel, R.L.; Ferlay, J.; Lortet-Tieulent, J.; Jemal, A. Global cancer statistics, 2012. *CA Cancer J. Clin.* **2015**, *65*, 87–108.
2. Bravo, I.G.; Felez-Sanchez, M. Papillomaviruses: Viral evolution, cancer and evolutionary medicine. *Evol. Med. Public Health* **2015**, *2015*, 32–51.
3. Armstrong, E.P. Prophylaxis of cervical cancer and related cervical disease: A review of the cost-effectiveness of vaccination against oncogenic HPV types. *J. Manag. Care Pharm.* **2010**, *16*, 217–230.
4. Zhao, H.; Lin, Z.-J.; Huang, S.-J.; Li, J.; Liu, X.-H.; Guo, M.; Zhang, J.; Xia, N.-S.; Pan, H.-R.; Wu, T.; Li, C.G. Correlation between ELISA and pseudovirion-based neutralisation assay for detecting antibodies against human papillomavirus acquired by natural infection or by vaccination. *Hum. Vaccines Immunother.* **2014**, *10*, 740–746.
5. Cassidy, B.; Schlenk, E.A. Uptake of the human papillomavirus vaccine: A review of the literature and report of a quality assurance project. *J. Pediatr. Health Care* **2012**, *26*, 92–101.
6. Joura, E.A.; Giuliano, A.R.; Iversen, O.E.; Bouchard, C.; Mao, C.; Mehlsen, J.; Moreira, E.D., Jr.; Ngan, Y.; Petersen, L.K.; Lazcano-Ponce, E.; *et al.* A 9-valent HPV vaccine against infection and intraepithelial neoplasia in women. *N. Engl. J. Med.* **2015**, *372*, 711–723.
7. Opalka, D.; Lachman, C.E.; MacMullen, S.A.; Jansen, K.U.; Smith, J.F.; Chirmule, N.; Esser, M.T. Simultaneous quantitation of antibodies to neutralizing epitopes on virus-like particles for human papillomavirus types 6, 11, 16, and 18 by a multiplexed luminex assay. *Clin. Diagn. Lab. Immunol.* **2003**, *10*, 108–115.
8. Scherpenisse, M.; Mollers, M.; Schepp, R.M.; Boot, H.J.; de Melker, H.E.; Meijer, C.J.; Berbers, G.A.; van der Klis, F.R. Seroprevalence of seven high-risk HPV types in the Netherlands. *Vaccine* **2012**, *30*, 6686–6693.
9. Waterboer, T.; Sehr, P.; Michael, K.M.; Franceschi, S.; Nieland, J.D.; Joos, T.O.; Templin, M.F.; Pawlita, M. Multiplex human papillomavirus serology based on *in situ*-purified glutathione s-transferase fusion proteins. *Clin. Chem.* **2005**, *51*, 1845–1853.
10. Doorbar, J.; Quint, W.; Banks, L.; Bravo, I.G.; Stoler, M.; Broker, T.R.; Stanley, M.A. The biology and life-cycle of human papillomaviruses. *Vaccine* **2012**, *30*, F55–F70.
11. Buck, C.B.; Pastrana, D.V.; Lowy, D.R.; Schiller, J.T. Efficient intracellular assembly of papillomaviral vectors. *J. Virol.* **2004**, *78*, 751–757.

12. Pastrana, D.V.; Buck, C.B.; Pang, Y.Y.; Thompson, C.D.; Castle, P.E.; FitzGerald, P.C.; Kruger Kjaer, S.; Lowy, D.R.; Schiller, J.T. Reactivity of human sera in a sensitive, high-throughput pseudovirus-based papillomavirus neutralization assay for HPV16 and HPV18. *Virology* **2004**, *321*, 205–216.

13. Wu, X.; Zhang, C.; Feng, S.; Liu, C.; Li, Y.; Yang, Y.; Gao, J.; Li, H.; Meng, S.; Li, L.; *et al.* Detection of HPV types and neutralizing antibodies in Gansu province, China. *J. Med. Virol.* **2009**, *81*, 693–702.

14. Sehr, P.; Rubio, I.; Seitz, H.; Putzker, K.; Ribeiro-Muller, L.; Pawlita, M.; Muller, M. High-throughput pseudovirion-based neutralization assay for analysis of natural and vaccine-induced antibodies against human papillomaviruses. *PLoS ONE* **2013**, *8*, e75677.

15. Nie, J.; Huang, W.; Wu, X.; Wang, Y. Optimization and validation of a high throughput method for detecting neutralizing antibodies against human papillomavirus (HPV) based on pseudovirons. *J. Med. Virol.* **2014**, *86*, 1542–1555.

16. Panicker, G.; Rajbhandari, I.; Gurbaxani, B.M.; Querec, T.D.; Unger, E.R. Development and evaluation of multiplexed immunoassay for detection of antibodies to HPV vaccine types. *J. Immunol. Methods* **2015**, *417*, 107–114.

17. Hadjilaou, A.; Green, A.M.; Coloma, J.; Harris, E. Single-cell analysis of B cell/antibody cross-reactivity using a novel multicolor FluoroSpot assay. *J. Immunol.* **2015**, *195*, 3490–3496.

18. Leder, C.; Kleinschmidt, J.A.; Wiethe, C.; Muller, M. Enhancement of capsid gene expression: Preparing the human papillomavirus type 16 major structural gene L1 for DNA vaccination purposes. *J. Virol.* **2001**, *75*, 9201–9209.

19. Pastrana, D.V.; Gambhira, R.; Buck, C.B.; Pang, Y.Y.; Thompson, C.D.; Culp, T.D.; Christensen, N.D.; Lowy, D.R.; Schiller, J.T.; Roden, R.B. Cross-neutralization of cutaneous and mucosal *Papillomavirus* types with anti-sera to the amino terminus of L2. *Virology* **2005**, *337*, 365–372.

20. Mossadegh, N.; Gissmann, L.; Muller, M.; Zentgraf, H.; Alonso, A.; Tomakidi, P. Codon optimization of the human papillomavirus 11 (HPV 11) L1 gene leads to increased gene expression and formation of virus-like particles in mammalian epithelial cells. *Virology* **2004**, *326*, 57–66.

21. Kieback, E.; Muller, M. Factors influencing subcellular localization of the human papillomavirus L2 minor structural protein. *Virology* **2006**, *345*, 199–208.

22. Kondo, K.; Ishii, Y.; Ochi, H.; Matsumoto, T.; Yoshikawa, H.; Kanda, T. Neutralization of HPV16, 18, 31, and 58 pseudovirions with antisera induced by immunizing rabbits with synthetic peptides representing segments of the HPV16 minor capsid protein L2 surface region. *Virology* **2007**, *358*, 266–272.

23. Roberts, J.N.; Buck, C.B.; Thompson, C.D.; Kines, R.; Bernardo, M.; Choyke, P.L.; Lowy, D.R.; Schiller, J.T. Genital transmission of HPV in a mouse model is potentiated by nonoxynol-9 and inhibited by carrageenan. *Nat. Med.* **2007**, *13*, 857–861.

24. Kondo, K.; Ochi, H.; Matsumoto, T.; Yoshikawa, H.; Kanda, T. Modification of human papillomavirus-like particle vaccine by insertion of the cross-reactive L2-epitopes. *J. Med. Virol.* **2008**, *80*, 841–846.

25. Johnson, K.M.; Kines, R.C.; Roberts, J.N.; Lowy, D.R.; Schiller, J.T.; Day, P.M. Role of heparan sulfate in attachment to and infection of the murine female genital tract by human papillomavirus. *J. Virol.* **2009**, *83*, 2067–2074.

26. Wu, X.L.; Zhang, C.T.; Zhu, X.K.; Wang, Y.C. Detection of HPV types and neutralizing antibodies in women with genital warts in Tianjin City, China. *Virol. Sin.* **2010**, *25*, 8–17.

27. Buck, C.B.; Thompson, C.D. Production of *papillomavirus*-based gene transfer vectors. *Curr. Protoc. Cell Biol.* **2007**.

28. Matumoto, M. A note on some points of calculation method of LD50 by Reed and Muench. *Jpn. J. Exp. Med.* **1949**, *20*, 175–179.

29. Ferguson, M.; Wilkinson, D.E.; Heath, A.; Matejtschuk, P. The first international standard for antibodies to HPV 16. *Vaccine* **2011**, *29*, 6520–6526.

30. Stanley, M.; Pinto, L.A.; Trimble, C. Human *papillomavirus* vaccines—Immune responses. *Vaccine* **2012**, *30*, F83–F87.

31. Ahlborg, N.; Axelsson, B. Dual- and triple-color fluorospot. *Methods Mol. Biol.* **2012**, *792*, 77–85.

32. Wheeler, C.M.; Kjaer, S.K.; Sigurdsson, K.; Iversen, O.E.; Hernandez-Avila, M.; Perez, G.; Brown, D.R.; Koutsky, L.A.; Tay, E.H.; Garcia, P.; *et al.* The impact of quadrivalent human papillomavirus (HPV; types 6, 11, 16, and 18) L1 virus-like particle vaccine on infection and disease due to oncogenic nonvaccine HPV types in sexually active women aged 16–26 years. *J. Infect. Dis.* **2009**, *199*, 936–944.

33. Brown, D.R.; Kjaer, S.K.; Sigurdsson, K.; Iversen, O.E.; Hernandez-Avila, M.; Wheeler, C.M.; Perez, G.; Koutsky, L.A.; Tay, E.H.; Garcia, P.; *et al.* The impact of quadrivalent human papillomavirus (HPV; types 6, 11, 16, and 18) L1 virus-like particle vaccine on infection and disease due to oncogenic nonvaccine HPV types in generally HPV-naive women aged 16–26 years. *J. Infect. Dis.* **2009**, *199*, 926–935.

34. Paavonen, J.; Jenkins, D.; Bosch, F.X.; Naud, P.; Salmeron, J.; Wheeler, C.M.; Chow, S.N.; Apter, D.L.; Kitchener, H.C.; Castellsague, X.; *et al.* Efficacy of a prophylactic adjuvanted bivalent L1 virus-like-particle vaccine against infection with human papillomavirus types 16 and 18 in young women: An interim analysis of a phase III double-blind, randomised controlled trial. *Lancet* **2007**, *369*, 2161–2170.

Dengue Virus Reporter Replicon is a Valuable Tool for Antiviral Drug Discovery and Analysis of Virus Replication Mechanisms

Fumihiro Kato and Takayuki Hishiki

Abstract: Dengue, the most prevalent arthropod-borne viral disease, is caused by the dengue virus (DENV), a member of the *Flaviviridae* family, and is a considerable public health threat in over 100 countries, with 2.5 billion people living in high-risk areas. However, no specific antiviral drug or licensed vaccine currently targets DENV infection. The replicon system has all the factors needed for viral replication in cells. Since the development of replicon systems, transient and stable reporter replicons, as well as reporter viruses, have been used in the study of various virological aspects of DENV and in the identification of DENV inhibitors. In this review, we summarize the DENV reporter replicon system and its applications in high-throughput screening (HTS) for identification of anti-DENV inhibitors. We also describe the use of this system in elucidation of the mechanisms of virus replication and viral dynamics *in vivo* and *in vitro*.

Reprinted from *Viruses*. Cite as: Kato, F.; Hishiki, T. Dengue Virus Reporter Replicon is a Valuable Tool for Antiviral Drug Discovery and Analysis of Virus Replication Mechanisms. *Viruses* **2016**, *8*, 122.

1. Introduction

Dengue virus (DENV), which includes four serotypes (DENV1–4), is transmitted to humans by *Aedes* mosquitos and is the etiological agent of dengue fever and dengue hemorrhagic fever [1]. DENV causes an estimated 50–100 million cases of dengue fever, 500,000 cases of severe dengue (dengue hemorrhagic fever/dengue shock syndrome (DHF/DSS)), and more than 20,000 deaths each year in tropical and subtropical regions, representing a considerable public health threat in over 100 countries worldwide [2]. However, there are still no specific antiviral drugs or licensed vaccines against DENV infection.

DENV is an enveloped, positive-strand RNA virus belonging to the genus *Flavivirus* in the family *Flaviviridae*. Several other flaviviruses, including Japanese encephalitis virus (JEV), yellow fever virus (YFV), West Nile virus (WNV), tick-borne encephalitis virus (TBEV), and Zika virus (ZIKV), also are significant human pathogens [3,4]. The DENV genome consists of approximately 11 kb, containing one large open-reading frame (ORF). This viral RNA encodes a polyprotein that is processed by cellular and viral proteases into three structural proteins (capsid (C), pre-membrane (prM), and envelope (E)), which form the virus particle, and seven nonstructural (NS) proteins (NS1: essential for RNA replication, NS2A: inhibition of interferon signal, NS2B: cofactor of NS3 protease, NS3: protease and helicase activity, NS4A: induction of membrane rearrangements, NS4B: inhibition of interferon signal, and NS5: methyltransferase and RNA polymerase activity, inhibition of interferon signal). These NS proteins are responsible for replication of the viral genome but are not detectable in viral particles [4]. The ORF is flanked by highly structured 5′- and 3′-untranslated regions (UTRs), which play regulatory roles in translation of the viral proteins and viral RNA genome replication. In DENV and other flaviviruses, the presence of complementary sequences at the ends of the genome mediates long-range RNA-RNA interactions [5]. DENV RNA displays two pairs of complementary sequences (CS) required for genome circularization and viral replication [6,7]. The downstream 5′ CS pseudoknot (DCS-PK) elements enhance viral RNA replication by regulating cyclization [8].

The replicon system contains gene elements necessary for autonomous replication of the genome in cells. Expression of viral genes by replicon systems has been established in a number of positive-strand RNA viruses, such as Sindbis virus, poliovirus, Semliki Forest virus, human rhinovirus 14, coronavirus, and hepatitis C virus, and in various flaviviruses, including Kunjin virus, DENV, WNV, YFV, and TBEV [9–19]. In this review, we describe a replication-competent DENV subgenomic and full-length replicon system composed of reporter genes. This technology has improved dramatically in recent years and can be used for the screening of antiviral compounds and analysis of virus replication mechanisms.

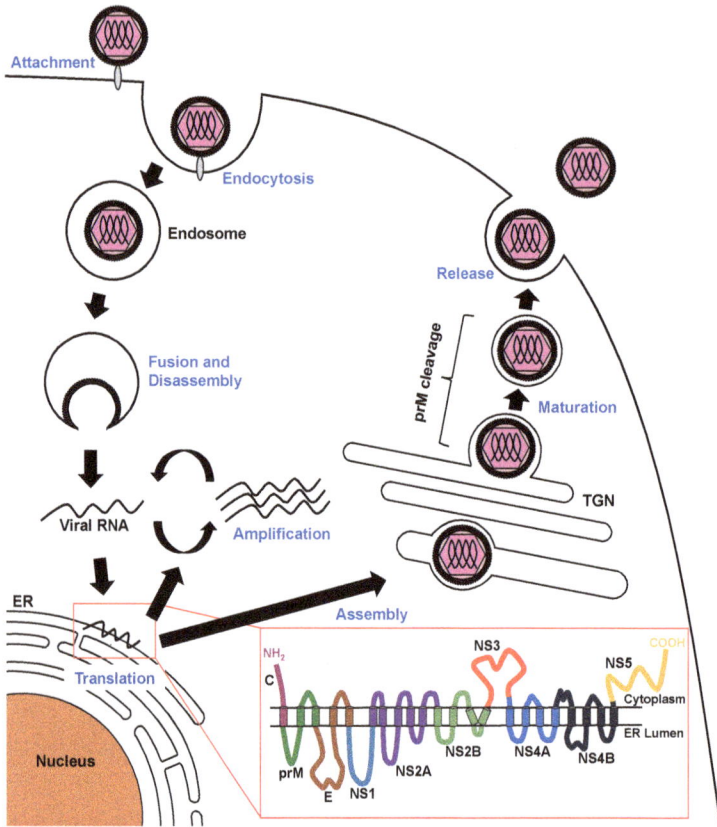

Figure 1. Schematic model of the dengue virus (DENV) life cycle. DENV particles bind to host cell factors and then enter the cell by clathrin-mediated endocytosis. After trafficking to endosomal compartments, envelope protein-mediated fusion of viral and cellular membranes occurs with changes in pH, allowing disassembly of the virus particles and release of single-stranded viral RNA into the cytoplasm, where translation occurs. The viral RNA is then translated to a polyprotein, which is processed by host cellular and viral proteases. Nonstructural (NS) proteins then replicate the viral RNA. Viral particle assembly occurs on the membrane of the endoplasmic reticulum (ER), and particles then bud into the ER as immature virus particles. During egress of the progeny virus particle through the secretory pathway, pre-membrane (prM) protein is cleaved by the cellular serine protease furin. Mature virus particles are released into the extracellular space. The red inset indicates the putative membrane topology of the viral proteins. TGN: trans-Golgi network.

2. DENV Life Cycle

DENV attaches to cells via interactions between the E proteins of viral particles and cellular factors, including heparan sulfate, mannose receptor, dendritic cell

(DC)-specific intercellular adhesion molecule 3-grabbing nonintegrin (DC-SIGN), and T-cell immunoglobulin and mucin domain (TIM) and Tyro3, Axl, and Mer (TAM) family proteins, on the target cell [20]. After binding, DENV is internalized into cells via clathrin-mediated endocytosis and traffic into the endosomal compartment, in which the low pH induces structural changes in the E protein, resulting in viral membrane fusion. The positive-stranded viral RNA is then released into the cytoplasm. The DENV genome is a single-stranded positive-sense RNA that functions as mRNA and is subsequently translated by the cell machinery, thus generating viral proteins in the endoplasmic reticulum (ER). DENV genome RNA replication is performed in a structure enclosed by a virus-induced intracellular membrane, called the replication complex (RC); the RC contains viral proteins, viral RNA, and host cell factors [21,22]. The assembly of DENV particles occurs in the ER, and the virions bud into the ER as immature virus particles that incorporate 60 trimeric spikes of the prM and E proteins. These immature virus particles are then transported through the trans-Golgi network (TGN). During egress, prM is cleaved by the cellular serine protease furin. Thereafter, infectious mature virus particles are released into the extracellular space (Figures 1 and 2).

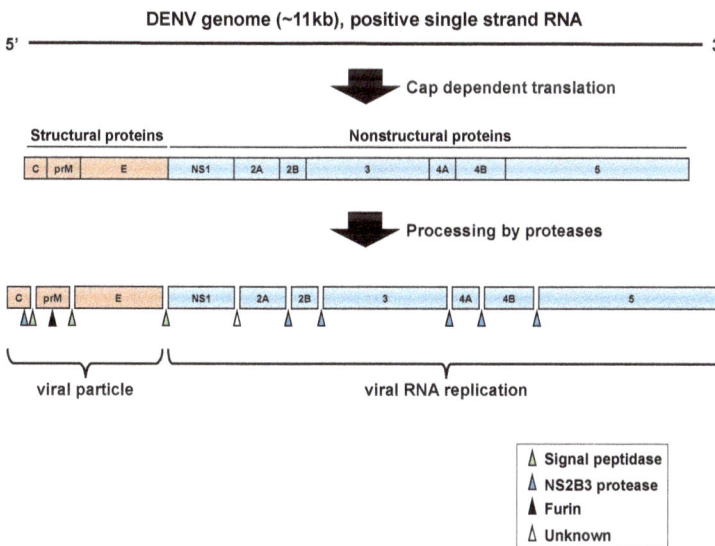

Figure 2. Schematic diagram of the DENV genome. The single-stranded viral RNA is translated by cap-dependent initiation scanning of the 5′- untranslated region (UTR). The translated polyprotein is processed by cellular and viral proteases into three structural proteins (capsid (C), pre-membrane (prM), and envelope (E) proteins) and seven NS proteins (NS1, NS2A, NS2B, NS3, NS4A, NS4B, and NS5). C, prM, and E proteins constitute the components of viral particles, whereas NS1–5 proteins function in the replication of RNA viral genome.

3. Subgenomic Reporter Replicon (Transient Expression of NS Proteins)

Subgenomic replicon systems including the coding region of NS proteins (NS1 to NS5) and the *cis*-acting element in the 5′- and 3′-UTR, which are needed for viral RNA translation and replication, are able to self-replicate in cultured cells. These replicon systems can be safely used to study many aspects of virus replication because of the lack of structural genes necessary for the production of virus particles. Consequently, subgenomic replicons are suitable for examination of viral genome replication independently of the process of viral particle assembly.

The DENV replicon system described by Pang and colleagues does not contain a reporter gene for analysis of the level of DENV RNA replication [16]. To improve this system, researchers have developed new replicon systems (Figure 3b) [23–32]. Among them, DVRep, which harbors a firefly luciferase (Fluc) gene to replace the structural proteins [28]. The C-terminal 24 amino acids of the E protein, corresponding to the transmembrane (TM) domain, were included in the system to maintain the topology of the viral protein NS1 inside of the ER compartment. The luciferase was fused to the N-terminal 34 amino acids of the C protein, which contained the *cis*-acting element of 11 nucleotides complementary to the 3′ CS [33,34]. Furthermore, to ensure appropriate cleavage of luciferase from the viral polyprotein, foot-and-mouth disease virus (FMDV) 2A protease cleavage sites were introduced between the C-terminus of luciferase and the beginning of the TM domain of the E protein (Figure 3b, upper panel) [35]. After transfection of replicon RNA into BHK-21 and C6/36 cells, DVRep RNA translation and amplification could be monitored by measurement of luciferase activity. Moreover, an RNA element was identified in the 3′- UTR that differentially modulates viral replication in mosquito and mammalian cells in this replicon system.

Puig-Basagoiti *et al.* developed two types of subgenomic replicons [29]. One replicon contained a *Renilla* luciferase (Rluc) gene, which was substituted with the viral structure genes (DEN-1 Rluc-Rep). The other replicon was a FMDV 2A cleavage sequence inserted into the C-terminal of the luciferase of DEN-1 Rluc-Rep (DEN-1 Rluc2A-Rep). After transfection of DEN-1 Rluc-Rep RNA into BHK-21 cells, only a single luciferase peak was observed during the initial 10 h, and no further luciferase activity was detected up to 96 h. In contrast, after transfection of DEN-1 Rluc2A-Rep RNA into BHK-21 cells, two luciferase peaks were observed; the first peak was observed during the initial 10 h, and the second peak was observed after 10 h. These results suggested that the first luciferase peak may represent the translation of input RNA, whereas the second peak may represent viral RNA replication. Similar observations regarding the importance of the FMDV 2A sequence for cleavage between luciferase and the C-terminal fragment of the E protein were reported by Alvarez and co-workers [28].

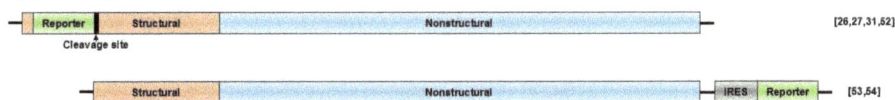

Figure 3. Schematic diagram of DENV reporter replicons. (**a**) Structure of the complete DENV genome; (**b**) subgenomic reporter replicons that are used for transient replication assays; (**c**) selectable subgenomic reporter replicons. The RNA supports stable expression of reporter and nonstructural (NS) proteins; (**d**) full-length reporter replicons that produce infectious viral particles. Reporter genes: *Renilla* (Rluc), firefly (Fluc), or *Gaussia* luciferase (Gluc) or green fluorescence protein (GFP); cleavage site: foot-and-mouth disease virus (FMDV) 2A or ubiquitin cleavage sequence; IRES: internal ribosome entry site; Drug^R: drug-resistance gene.

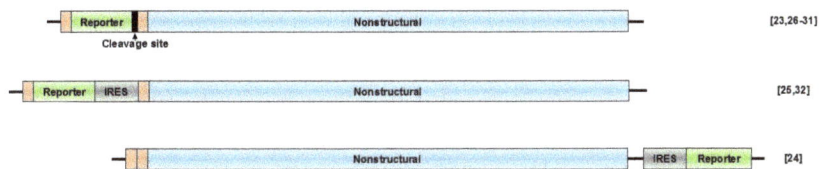

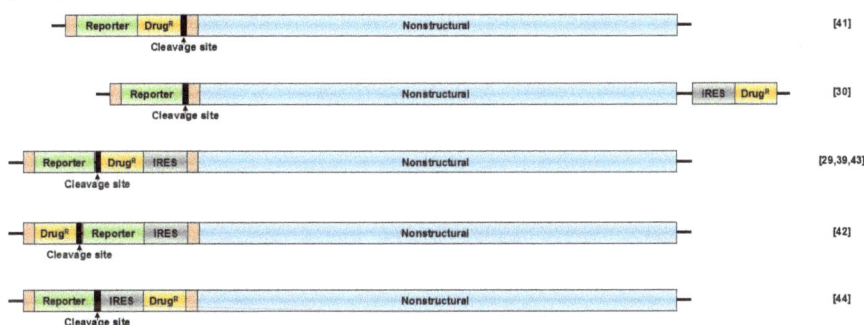

Generally, a replicon is established by transfection of *in vitro* transcribed RNA. However, some reports have described DNA-based replicons, in which transcription is controlled by a cytomegalovirus (CMV) promoter prepared from Venezuelan equine encephalitis virus (*Alphavirus*), porcine reproductive and respiratory syndrome virus (*Arterivirus*), WNV, and DENV [36–39]. Leardkamolkarn *et al.* also successfully developed four types of subgenomic replicons, in which the GFP reporter gene was inserted into the structural region [31]. The DNA-based replicon is stable compared with RNA-based constructs, and the DNA-based replicon can be directly transfected (without *in vitro* transcription). Therefore, DNA-based replicons are simple and convenient to use for examination of the mechanism of DENV replication and for high-throughput screening (HTS) of anti-DENV compounds. Recently, we developed a DNA-based transient DENV-1 replicon encoding the Gluc reporter gene [32]. *Gaussia* is smaller than firefly or *Renilla* luciferase and generates a stronger signal [40]. Thus, as a major advantage of our system, *Gaussia* luciferase activity can be analyzed in culture medium, without the requirement for cell lysis.

4. Subgenomic Reporter Replicon Cells (Stable Expression of NS Proteins)

Stable viral replicating systems, called replicon cells, enable continuous viral RNA replication in cell culture through introduction of a drug-resistance gene into the viral genome. DENV replicon cells are useful tools for analysis of the replication mechanisms of the DENV genome; however, such cells are not suitable for HTS of anti-DENV agents. Therefore, researchers modified this system to introduce a reporter gene and to develop an appropriate assay system that could be used to analyze the amount of DENV RNA in cells by measuring reporter activity, such as luciferase or GFP [29,30,39,41–44].

The reporter within the DENV replicon contained Rluc ubiquitin, a selectable neomycin-resistance (*neo*) gene, and an encephalomyocarditis virus (EMCV) internal ribosome entry site (IRES) fragment; this was substituted for the viral structural genes to construct a replicon fragment that retained the 37 N-terminal amino acids of the C protein and 31 C-terminal amino acids of the E protein [29]. After transfection of the DENV replicon RNA into Vero cells, the replicon polyprotein driven by the DENV 5'-UTR and EMCV IRES was processed through cellular and viral protease-mediated cleavage, resulting in individual *Renilla* luciferase, neomycin phosphotransferase II conferring resistance to various aminoglycoside antibiotics, and NS proteins. The transfected cells were selected by geneticin treatment, allowing continuously replicating replicon cells to survive. Examination of the established cells demonstrated that viral proteins were expressed in all cells. Additionally, high levels of *Renilla* luciferase activity were maintained more than four months.

Some researchers have conducted studies in which the luciferase gene is replaced with the enhanced green fluorescent protein (EGFP) gene for construction of DENV replicon cells [41,42,44]. In this replicon assay system, readouts are easy to monitor by measurement of the fluorescence intensity of living cells. Indeed, Leardkamolkarn *et al.* carried out HTS and identified new anti-DENV compounds using this DENV replicon system.

5. Virus-like Particles

In recent years, single-round infectious virus-like particles (VLPs) that express reporter genes have been used widely as tools for studying several flaviviruses, including YFV and WNV [45,46]. VLPs are composed of an RNA reporter replicon genome that is packaged into virus particles by the viral structural proteins (C, prM, and E proteins) when provided in *trans*. The VLPs exhibit a structure similar to that of infectious live virus particles and can be used to study the entry and replication steps of the viral life cycle.

Qing *et al.* confirmed that DENV-VLPs are susceptible to the neutralizing antibody 4G2, which recognizes the fusion loop of domain II of the E protein, and to the anti-DENV compound NITD008, which is a nucleoside inhibitor of DENV RNA-dependent RNA polymerase (RdRp) [47–49]. Validation and optimization of VLPs for HTS of DENV inhibitors in a 384-well format yielded consistent, strong signals. Moreover, consistent with previous studies, they also found that the infectivity of VLPs was influenced by temperature [50]. Furthermore, Mattia *et al.* demonstrated the usefulness of VLPs for identification and measurement of neutralizing antibodies in human serum samples against all four DENV serotypes in large-scale, long-term studies [51].

6. Full-Length Reporter Replicon

The use of subgenomic replicon systems has improved our knowledge of the mechanisms of DENV replication; however, the genomic sequences used in these systems exhibit large deletions in the structural region (C, prM, and E coding sequences). Therefore, the subgenomic replicon cannot be used to examine several virus life cycle steps, such as entry, assembly, and release, or to identify antiviral agents targeting structural proteins. To overcome this limitation, researchers have developed full-length replicon systems [26,27,31,52–54]. Mondotte and colleagues first reported the development of a full-length reporter infectious virus, DV-R, and described the roles of two glycans in the E protein during DENV infection [53]. In this construct, the Rluc gene was introduced into the 3'-UTR of the DENV genome, with translation dependent on the EMCV IRES. After transfection of DV-R RNA into BHK-21 cells, the genome becomes self-replicative, resulting in production of infectious virus, although the DV-R RNA is longer than that of the parent strain, whereas growth curve analysis indicated that DV-R replication decreased in comparison with that of the parent strain; indeed, the titers at 24 and 48 h were significantly lower than those of the parent strain. Furthermore, DV-R infection is suppressed by heparin, an entry inhibitor, and replication abolished by an adenosine analog, also known as viral polymerase inhibitor. Therefore, DV-R provides a useful tool for investigating all steps of the virus life cycle.

Leardkamolkarn *et al.* generated a full-length (FL) replicon, FL-DENV/GFP, in which the GFP gene was inserted into the DENV genome between C and prM coding sequences [31]. However, infectious virus was not detectable in the cell culture supernatant, although GFP expression was detected in cells transfected with replicon RNA. Additionally, the amount of intracellular viral RNA was significantly decreased compared with that in the parent strain. These results suggested that GFP insertion in the viral RNA genome in this position affected the amplification of the virus genome and the infectivity of virus particles.

Reporter viruses are unstable; indeed, reporter genes are often deleted after a few rounds of viral replication [55]. To overcome this limitation, a DENV strain stably expressing luciferase, Luc-DENV, was developed. In this construct, the N-terminal 38 amino acids of the C protein were fused to the *Rluc* gene, and an FMDV 2A cleavage sequence was introduced into the 5'-UTR of DENV. After transfection of the Luc-DENV RNA transcript into BHK-21 cells, the viral titer from culture supernatants was lower than the parent virus [27]. To investigate the stability of Luc-DENV, researchers passaged the virus in Vero cells five times. As a result, luciferase activity was increased from the third to fifth rounds of passage compared with that during the first two rounds of passage virus. Furthermore, researchers showed that adaptive mutation in the NS4B gene could enhance viral RNA replication in a cell type-specific manner.

Schoggins *et al.* reported two types of full-length reporter replicons, which included luminescent or fluorescent reporters within the viral genome [52]. The first 25 amino acids of DENV C protein were repeated and introduced upstream of the reporter gene (Fluc or GFP), which was fused to a sequence encoding the FMDV 2A cleavage site. Both DENV constructs produced infectious viruses in culture supernatants through the *in vitro* transcribed RNA of the DENV reporter construct introduced into Vero cells. Interestingly, the infectious viral productivity of DENV-GFP was lower than that of the parent DENV strain. Furthermore, serial passage of the virus in the cell culture supernatant results in reduced GFP fluorescence, suggesting that the *GFP* gene was unstable during the DENV life cycle. Similarly, Zou *et al.* also reported that the Rluc incorporated into the full-length DENV genome was stable, whereas the GFP gene incorporated into the full-length DENV genome was unstable [27]. These results suggested that certain RNA elements within the GFP gene may interfere with DENV replication, thereby resulting in deletion of GFP during replication of GFP-DENV. Despite this limitation, the GFP-expressing replicon system still provides benefits for the study of DENV. For example, the replication level can be quantified by live-cell fluorescence imaging, a technique that provides rapid, simple screening. Furthermore, using this GFP-DENV replicon system, it is possible to use fluorescence-activated cell sorting (FACS) to measure and sort cells. Indeed, several antiviral effectors have been identified from an interferon-stimulated gene library using DENV-GFP.

For the DENV-Fluc, susceptibility to the well-characterized anti-DENV inhibitors mycophenolic acid (MPA), NITD008, and type I and III interferons is similar to that of parent DENV in Huh7 cells [49,56]. Interestingly, researchers further used this DENV-Fluc in an *in vivo* mouse model, together with substitution of a single amino acid mutation in NS4B, in order to examine the virulence of this mutation, which has been shown to enhance viral RNA synthesis in mice [57]. After infection of AG129 mice, which lack interferon-α, -β, and -γ receptors, bioluminescence imaging data showed that DENV localizes predominantly to lymphoid- and gut-associated tissues [58–60]. This observation is consistent with the results of another study on non-reporter DENV-infected AG129 mice. Furthermore, the use of DENV-Fluc has been used to demonstrate susceptibility to the anti-DENV compounds MPA and NITD008 and to neutralizing antibodies in AG129 mice. These results suggest that DENV-Fluc could provide a platform for screening and assessment of antiviral compounds and for analysis of DENV pathogenesis in living animals.

7. Conclusions and Perspectives

The development of novel biological assays is required to continue advancements in the discovery of anti-DENV agents and to improve our understanding of the mechanism of DENV replication. Conventionally, anti-DENV activity is analyzed using infectious live virus. For example, quantification of the amount of infectious virus by plaque assay, observation of cytotoxic effects, or determination of viral RNA by reverse transcription polymerase chain reaction (RT-PCR). However, these assays are low throughput and cannot be used easily to screen large compound libraries. The development of the DENV replicon cell culture system is one of the most significant advances in DENV basic research and antiviral discovery. In recent years, exploitation of replication-competent reporter-expressing transient replicon systems, replicon cells, and single-round infectious particles has led to additional advancements in the field of DENV research. Furthermore, full-length reporter viruses are also useful tools for screening of inhibitors that affect all steps of the DENV life cycle and for examination of the mechanism of DENV replication and pathogenesis *in vivo* and *in vitro*. These model systems will further expand our understanding of virus-host interactions, viral pathogenesis, and immunological responses to DENV infection, thereby facilitating the development of drugs and vaccines.

Acknowledgments: This work was supported by a Grant-in-Aid for Young Scientists (B) from the Japan Society for the Promotion of Science (15K19109) and the Research Program on Emerging and Re-emerging Infectious Diseases from Japan Agency for Medical Research and Development, AMED.

Author Contributions: Fumihiro Kato and Takayuki Hishiki wrote the manuscript.

Conflicts of Interest: The authors declare no conflict of interest.

References

1. Gubler, D.J. Dengue and dengue hemorrhagic fever. *Clin. Microbiol. Rev.* **1998**, *11*, 480–496.
2. Bhatt, S.; Gething, P.W.; Brady, O.J.; Messina, J.P.; Farlow, A.W.; Moyes, C.L.; Drake, J.M.; Brownstein, J.S.; Hoen, A.G.; Sankoh, O.; *et al.* The global distribution and burden of dengue. *Nature* **2013**, *496*, 504–507.
3. Calisher, C.H.; Karabatsos, N.; Dalrymple, J.M.; Shope, R.E.; Porterfield, J.S.; Westaway, E.G.; Brandt, W.E. Antigenic relationships between flaviviruses as determined by cross-neutralization tests with polyclonal antisera. *J. Gen. Virol.* **1989**, *70*, 37–43.
4. Chambers, T.J.; Hahn, C.S.; Galler, R.; Rice, C.M. Flavivirus genome organization, expression, and replication. *Annu. Rev. Microbiol.* **1990**, *44*, 649–688.
5. Alvarez, D.E.; Lodeiro, M.F.; Luduena, S.J.; Pietrasanta, L.I.; Gamarnik, A.V. Long-range rna-rna interactions circularize the dengue virus genome. *J. Virol.* **2005**, *79*, 6631–6643.
6. Alvarez, D.E.; Filomatori, C.V.; Gamarnik, A.V. Functional analysis of dengue virus cyclization sequences located at the 5′ and 3′utrs. *Virology* **2008**, *375*, 223–235.
7. Friebe, P.; Harris, E. Interplay of rna elements in the dengue virus 5′ and 3′ ends required for viral rna replication. *J. Virol.* **2010**, *84*, 6103–6118.
8. Liu, Z.Y.; Li, X.F.; Jiang, T.; Deng, Y.Q.; Zhao, H.; Wang, H.J.; Ye, Q.; Zhu, S.Y.; Qiu, Y.; Zhou, X.; *et al.* Novel cis-acting element within the capsid-coding region enhances flavivirus viral-rna replication by regulating genome cyclization. *J. Virol.* **2013**, *87*, 6804–6818.
9. Xiong, C.; Levis, R.; Shen, P.; Schlesinger, S.; Rice, C.M.; Huang, H.V. Sindbis virus: An efficient, broad host range vector for gene expression in animal cells. *Science* **1989**, *243*, 1188–1191.
10. Hagino-Yamagishi, K.; Nomoto, A. *In vitro* construction of poliovirus defective interfering particles. *J. Virol.* **1989**, *63*, 5386–5392.
11. Liljestrom, P.; Garoff, H. A new generation of animal cell expression vectors based on the semliki forest virus replicon. *Biotechnology (N. Y.)* **1991**, *9*, 1356–1361.
12. McKnight, K.L.; Lemon, S.M. Capsid coding sequence is required for efficient replication of human rhinovirus 14 rna. *J. Virol.* **1996**, *70*, 1941–1952.
13. Almazan, F.; Sola, I.; Zuniga, S.; Marquez-Jurado, S.; Morales, L.; Becares, M.; Enjuanes, L. Coronavirus reverse genetic systems: Infectious clones and replicons. *Virus Res.* **2014**, *189*, 262–270.
14. Lohmann, V.; Korner, F.; Koch, J.; Herian, U.; Theilmann, L.; Bartenschlager, R. Replication of subgenomic hepatitis c virus rnas in a hepatoma cell line. *Science* **1999**, *285*, 110–113.
15. Khromykh, A.A.; Westaway, E.G. Subgenomic replicons of the flavivirus kunjin: Construction and applications. *J. Virol.* **1997**, *71*, 1497–1505.

16. Pang, X.; Zhang, M.; Dayton, A.I. Development of dengue virus type 2 replicons capable of prolonged expression in host cells. *BMC Microbiol.* **2001**, *1*, 18.

17. Shi, P.Y.; Tilgner, M.; Lo, M.K. Construction and characterization of subgenomic replicons of New York strain of west nile virus. *Virology* **2002**, *296*, 219–233.

18. Corver, J.; Lenches, E.; Smith, K.; Robison, R.A.; Sando, T.; Strauss, E.G.; Strauss, J.H. Fine mapping of a cis-acting sequence element in yellow fever virus rna that is required for rna replication and cyclization. *J. Virol.* **2003**, *77*, 2265–2270.

19. Gehrke, R.; Ecker, M.; Aberle, S.W.; Allison, S.L.; Heinz, F.X.; Mandl, C.W. Incorporation of tick-borne encephalitis virus replicons into virus-like particles by a packaging cell line. *J. Virol.* **2003**, *77*, 8924–8933.

20. Cruz-Oliveira, C.; Freire, J.M.; Conceicao, T.M.; Higa, L.M.; Castanho, M.A.; da Poian, A.T. Receptors and routes of dengue virus entry into the host cells. *FEMS Microbiol. Rev.* **2015**, *39*, 155–170.

21. Teo, C.S.; Chu, J.J. Cellular vimentin regulates construction of dengue virus replication complexes through interaction with ns4a protein. *J. Virol.* **2014**, *88*, 1897–1913.

22. Welsch, S.; Miller, S.; Romero-Brey, I.; Merz, A.; Bleck, C.K.; Walther, P.; Fuller, S.D.; Antony, C.; Krijnse-Locker, J.; Bartenschlager, R. Composition and three-dimensional architecture of the dengue virus replication and assembly sites. *Cell Host Microbe* **2009**, *5*, 365–375.

23. Holden, K.L.; Stein, D.A.; Pierson, T.C.; Ahmed, A.A.; Clyde, K.; Iversen, P.L.; Harris, E. Inhibition of dengue virus translation and rna synthesis by a morpholino oligomer targeted to the top of the terminal 3′ stem-loop structure. *Virology* **2006**, *344*, 439–452.

24. Suzuki, R.; de Borba, L.; Duarte dos Santos, C.N.; Mason, P.W. Construction of an infectious cdna clone for a brazilian prototype strain of dengue virus type 1: Characterization of a temperature-sensitive mutation in ns1. *Virology* **2007**, *362*, 374–383.

25. Alcaraz-Estrada, S.L.; Manzano, M.I.; del Angel, R.M.; Levis, R.; Padmanabhan, R. Construction of a dengue virus type 4 reporter replicon and analysis of temperature-sensitive mutations in non-structural proteins 3 and 5. *J. Gen. Virol.* **2010**, *91*, 2713–2718.

26. Scaturro, P.; Trist, I.M.; Paul, D.; Kumar, A.; Acosta, E.G.; Byrd, C.M.; Jordan, R.; Brancale, A.; Bartenschlager, R. Characterization of the mode of action of a potent dengue virus capsid inhibitor. *J. Virol.* **2014**, *88*, 11540–11555.

27. Zou, G.; Xu, H.Y.; Qing, M.; Wang, Q.Y.; Shi, P.Y. Development and characterization of a stable luciferase dengue virus for high-throughput screening. *Antivir. Res.* **2011**, *91*, 11–19.

28. Alvarez, D.E.; de Lella Ezcurra, A.L.; Fucito, S.; Gamarnik, A.V. Role of rna structures present at the 3′utr of dengue virus on translation, rna synthesis, and viral replication. *Virology* **2005**, *339*, 200–212.

29. Puig-Basagoiti, F.; Tilgner, M.; Forshey, B.M.; Philpott, S.M.; Espina, N.G.; Wentworth, D.E.; Goebel, S.J.; Masters, P.S.; Falgout, B.; Ren, P.; *et al.* Triaryl pyrazoline compound inhibits flavivirus rna replication. *Antimicrob. Agents Chemother.* **2006**, *50*, 1320–1329.

30. Hsu, Y.C.; Chen, N.C.; Chen, P.C.; Wang, C.C.; Cheng, W.C.; Wu, H.N. Identification of a small-molecule inhibitor of dengue virus using a replicon system. *Arch. Virol.* **2012**, *157*, 681–688.

31. Leardkamolkarn, V.; Sirigulpanit, W.; Chotiwan, N.; Kumkate, S.; Huang, C.Y. Development of dengue type-2 virus replicons expressing gfp reporter gene in study of viral rna replication. *Virus Res.* **2012**, *163*, 552–562.

32. Kato, F.; Kobayashi, T.; Tajima, S.; Takasaki, T.; Miura, T.; Igarashi, T.; Hishiki, T. Development of a novel dengue-1 virus replicon system expressing secretory gaussia luciferase for analysis of viral replication and discovery of antiviral drugs. *Jpn. J. Infect. Dis.* **2014**, *67*, 209–212.

33. Khromykh, A.A.; Meka, H.; Guyatt, K.J.; Westaway, E.G. Essential role of cyclization sequences in flavivirus rna replication. *J. Virol.* **2001**, *75*, 6719–6728.

34. You, S.; Falgout, B.; Markoff, L.; Padmanabhan, R. *In vitro* rna synthesis from exogenous dengue viral rna templates requires long range interactions between 5′- and 3′-terminal regions that influence rna structure. *J. Biol. Chem.* **2001**, *276*, 15581–15591.

35. Ryan, M.D.; Drew, J. Foot-and-mouth disease virus 2a oligopeptide mediated cleavage of an artificial polyprotein. *EMBO J.* **1994**, *13*, 928–933.

36. Ljungberg, K.; Whitmore, A.C.; Fluet, M.E.; Moran, T.P.; Shabman, R.S.; Collier, M.L.; Kraus, A.A.; Thompson, J.M.; Montefiori, D.C.; Beard, C.; *et al.* Increased immunogenicity of a DNA-launched venezuelan equine encephalitis virus-based replicon DNA vaccine. *J. Virol.* **2007**, *81*, 13412–13423.

37. Huang, Q.; Yao, Q.; Fan, H.; Xiao, S.; Si, Y.; Chen, H. Development of a vaccine vector based on a subgenomic replicon of porcine reproductive and respiratory syndrome virus. *J. Virol. Methods* **2009**, *160*, 22–28.

38. Cao, F.; Li, X.F.; Yu, X.D.; Deng, Y.Q.; Jiang, T.; Zhu, Q.Y.; Qin, E.D.; Qin, C.F. A DNA-based west nile virus replicon elicits humoral and cellular immune responses in mice. *J. Virol. Methods* **2011**, *178*, 87–93.

39. Yang, C.C.; Tsai, M.H.; Hu, H.S.; Pu, S.Y.; Wu, R.H.; Wu, S.H.; Lin, H.M.; Song, J.S.; Chao, Y.S.; Yueh, A. Characterization of an efficient dengue virus replicon for development of assays of discovery of small molecules against dengue virus. *Antivir. Res.* **2013**, *98*, 228–241.

40. Tannous, B.A.; Kim, D.E.; Fernandez, J.L.; Weissleder, R.; Breakefield, X.O. Codon-optimized gaussia luciferase cdna for mammalian gene expression in culture and *in vivo*. *Mol. Ther.* **2005**, *11*, 435–443.

41. Masse, N.; Davidson, A.; Ferron, F.; Alvarez, K.; Jacobs, M.; Romette, J.L.; Canard, B.; Guillemot, J.C. Dengue virus replicons: Production of an interserotypic chimera and cell lines from different species, and establishment of a cell-based fluorescent assay to screen inhibitors, validated by the evaluation of ribavirin's activity. *Antivir. Res.* **2010**, *86*, 296–305.

42. Ng, C.Y.; Gu, F.; Phong, W.Y.; Chen, Y.L.; Lim, S.P.; Davidson, A.; Vasudevan, S.G. Construction and characterization of a stable subgenomic dengue virus type 2 replicon system for antiviral compound and sirna testing. *Antivir. Res.* **2007**, *76*, 222–231.

43. Yang, C.C.; Hsieh, Y.C.; Lee, S.J.; Wu, S.H.; Liao, C.L.; Tsao, C.H.; Chao, Y.S.; Chern, J.H.; Wu, C.P.; Yueh, A. Novel dengue virus-specific ns2b/ns3 protease inhibitor, bp2109, discovered by a high-throughput screening assay. *Antimicrob. Agents Chemother.* **2011**, *55*, 229–238.

44. Leardkamolkarn, V.; Sirigulpanit, W. Establishment of a stable cell line coexpressing dengue virus-2 and green fluorescent protein for screening of antiviral compounds. *J. Biomol. Screen* **2012**, *17*, 283–292.

45. Jones, C.T.; Patkar, C.G.; Kuhn, R.J. Construction and applications of yellow fever virus replicons. *Virology* **2005**, *331*, 247–259.

46. Davis, C.W.; Nguyen, H.Y.; Hanna, S.L.; Sanchez, M.D.; Doms, R.W.; Pierson, T.C. West nile virus discriminates between dc-sign and dc-signr for cellular attachment and infection. *J. Virol.* **2006**, *80*, 1290–1301.

47. Qing, M.; Liu, W.; Yuan, Z.; Gu, F.; Shi, P.Y. A high-throughput assay using dengue-1 virus-like particles for drug discovery. *Antivir. Res.* **2010**, *86*, 163–171.

48. Klasse, P.J.; Sattentau, Q.J. Mechanisms of virus neutralization by antibody. *Curr. Top. Microbiol. Immunol.* **2001**, *260*, 87–108.

49. Yin, Z.; Chen, Y.L.; Schul, W.; Wang, Q.Y.; Gu, F.; Duraiswamy, J.; Kondreddi, R.R.; Niyomrattanakit, P.; Lakshminarayana, S.B.; Goh, A.; *et al.* An adenosine nucleoside inhibitor of dengue virus. *Proc. Natl. Acad. Sci. USA* **2009**, *106*, 20435–20439.

50. Ansarah-Sobrinho, C.; Nelson, S.; Jost, C.A.; Whitehead, S.S.; Pierson, T.C. Temperature-dependent production of pseudoinfectious dengue reporter virus particles by complementation. *Virology* **2008**, *381*, 67–74.

51. Mattia, K.; Puffer, B.A.; Williams, K.L.; Gonzalez, R.; Murray, M.; Sluzas, E.; Pagano, D.; Ajith, S.; Bower, M.; Berdougo, E.; *et al.* Dengue reporter virus particles for measuring neutralizing antibodies against each of the four dengue serotypes. *PLoS ONE* **2011**, *6*, e27252.

52. Schoggins, J.W.; Dorner, M.; Feulner, M.; Imanaka, N.; Murphy, M.Y.; Ploss, A.; Rice, C.M. Dengue reporter viruses reveal viral dynamics in interferon receptor-deficient mice and sensitivity to interferon effectors *in vitro*. *Proc. Natl. Acad. Sci. USA* **2012**, *109*, 14610–14615.

53. Mondotte, J.A.; Lozach, P.Y.; Amara, A.; Gamarnik, A.V. Essential role of dengue virus envelope protein n glycosylation at asparagine-67 during viral propagation. *J. Virol.* **2007**, *81*, 7136–7148.

54. Kaptein, S.J.; de Burghgraeve, T.; Froeyen, M.; Pastorino, B.; Alen, M.M.; Mondotte, J.A.; Herdewijn, P.; Jacobs, M.; de Lamballerie, X.; Schols, D.; *et al.* A derivate of the antibiotic doxorubicin is a selective inhibitor of dengue and yellow fever virus replication *in vitro*. *Antimicrob. Agents Chemother.* **2010**, *54*, 5269–5280.

55. Deas, T.S.; Binduga-Gajewska, I.; Tilgner, M.; Ren, P.; Stein, D.A.; Moulton, H.M.; Iversen, P.L.; Kauffman, E.B.; Kramer, L.D.; Shi, P.Y. Inhibition of flavivirus infections by antisense oligomers specifically suppressing viral translation and rna replication. *J. Virol.* **2005**, *79*, 4599–4609.

56. Diamond, M.S.; Zachariah, M.; Harris, E. Mycophenolic acid inhibits dengue virus infection by preventing replication of viral rna. *Virology* **2002**, *304*, 211–221.

57. Grant, D.; Tan, G.K.; Qing, M.; Ng, J.K.; Yip, A.; Zou, G.; Xie, X.; Yuan, Z.; Schreiber, M.J.; Schul, W.; *et al.* A single amino acid in nonstructural protein ns4b confers virulence to dengue virus in ag129 mice through enhancement of viral rna synthesis. *J. Virol.* **2011**, *85*, 7775–7787.

58. Kyle, J.L.; Beatty, P.R.; Harris, E. Dengue virus infects macrophages and dendritic cells in a mouse model of infection. *J. Infect. Dis.* **2007**, *195*, 1808–1817.

59. Balsitis, S.J.; Coloma, J.; Castro, G.; Alava, A.; Flores, D.; McKerrow, J.H.; Beatty, P.R.; Harris, E. Tropism of dengue virus in mice and humans defined by viral nonstructural protein 3-specific immunostaining. *Am. J. Trop. Med. Hyg.* **2009**, *80*, 416–424.

60. Zellweger, R.M.; Prestwood, T.R.; Shresta, S. Enhanced infection of liver sinusoidal endothelial cells in a mouse model of antibody-induced severe dengue disease. *Cell Host Microbe* **2010**, *7*, 128–139.

Applications of Replicating-Competent Reporter-Expressing Viruses in Diagnostic and Molecular Virology

Yongfeng Li, Lian-Feng Li, Shaoxiong Yu, Xiao Wang, Lingkai Zhang, Jiahui Yu, Libao Xie, Weike Li, Razim Ali and Hua-Ji Qiu

Abstract: Commonly used tests based on wild-type viruses, such as immunostaining, cannot meet the demands for rapid detection of viral replication, high-throughput screening for antivirals, as well as for tracking viral proteins or virus transport in real time. Notably, the development of replicating-competent reporter-expressing viruses (RCREVs) has provided an excellent option to detect directly viral replication without the use of secondary labeling, which represents a significant advance in virology. This article reviews the applications of RCREVs in diagnostic and molecular virology, including rapid neutralization tests, high-throughput screening systems, identification of viral receptors and virus-host interactions, dynamics of viral infections *in vitro* and *in vivo*, vaccination approaches and others. However, there remain various challenges associated with RCREVs, including pathogenicity alterations due to the insertion of a reporter gene, instability or loss of the reporter gene expression, or attenuation of reporter signals *in vivo*. Despite all these limitations, RCREVs have become powerful tools for both basic and applied virology with the development of new technologies for generating RCREVs, the inventions of novel reporters and the better understanding of regulation of viral replication.

Reprinted from *Viruses*. Cite as: Li, Y.; Li, L.-F.; Yu, S.; Wang, X.; Zhang, L.; Yu, J.; Xie, L.; Li, W.; Ali, R.; Qiu, H.-J. Applications of Replicating-Competent Reporter-Expressing Viruses in Diagnostic and Molecular Virology. *Viruses* **2016**, *8*, 127.

1. Introduction

The commonly used tests based on wild-type viruses, such as immunostaining, are often time-consuming and labor-intensive. Furthermore, these methods cannot meet the demands for high-throughput screening (HTS) of antivirals, rapid, sensitive and quantitative detection of neutralizing antibodies (NAbs), visual tracking of viral proteins or viruses *in vitro* and *in vivo* and other fields of virology.

Replicating-competent reporter-expressing viruses (RCREVs) are one type of artificially modified viruses that not only retain the viral genetic characteristics but also possess the new properties of the reporter genes, which represent a useful tool for quantitative analysis of viral replication and tracking viral protein transport in both living cells and animals.

2. Technologies for the Generation of Replicating-Competent Reporter-Expressing Viruses (RCREVs)

To date, advances in technologies enable the generation of RCREVs, which have been successfully applied in diagnostic and molecular virology.

2.1. Reverse Genetics Technologies

Currently, reverse genetics systems for many viruses have been well-established [1–13], providing powerful tools for generating RCREVs. Since some viruses possess a large genome, they usually permit a large extrinsic genetic insertion without impairing viral replication. For example, vaccinia virus (VACV) contains a 192 kb genome, capable of accepting up to 25 kb insertion [14]. However, for most RNA and some DNA viruses containing a small-sized genome, a recurring difficulty in generating RCREVs is the genetic instability, especially for a larger reporter gene. For some viruses with a segmented RNA genome, the insertion of a large reporter gene into the genome is difficult or even impossible to achieve.

2.2. Reporters in RCREVs

Commonly used reporters in RCREVs include fluorescent proteins, such as enhanced green fluorescent protein (EGFP) (green), GFP mutants (enhanced cyan fluorescent protein (ECFP) (blue), mCherry (red) and Venus (yellow)), far-red fluorescent reporters (red fluorescent protein (RFP), Katushka 2, dTomato and DsRed)), near-infrared fluorescent proteins (iRFPs) and tetracysteine (TC); bioluminescent reporters, such as firefly luciferase (Fluc), *Renilla* luciferase (Rluc) and Gaussia luciferase (Gluc); in addition to other reporters, such as neomycin-resistance gene (Neo^R) and Cre recombinase. These reporters are mainly used to rapidly quantify viral replication and track viral proteins or viruses by living imaging *in vitro* and *in vivo*. However, different reporters may have different influences on the biological properties of various viruses, and the loss of the reporter gene expression is a significant concern for some RCREVs. Therefore, choosing a suitable reporter is a critical decision on designing RCREVs. For the planned applications, the reporters with a smaller size may be a promising option due to their minimum effects on the viral biology. For example, the Rluc gene (933 bp) is better than the Fluc gene (1653 bp) and has a minimal influence on the growth of the engineered classical swine fever virus (CSFV) expressing the reporters [15,16].

2.3. Reporters Expressing Strategies

Various strategies associated with the reporter gene expression have been developed. An extensively used expression strategy is to fuse the reporters to one of the viral proteins. For instance, the Rluc activities from engineered CSFV carrying the Rluc fused to the viral N^{pro} protein were detected [15]. A nonessential viral gene can be replaced with a reporter gene to generate a reporter virus. In addition, the Cre-LoxP recombination is widely used to control reporter gene expression in cell cultures or animal models. Notably, reporters can be expressed from an additional transcriptional unit (ATU), in which a reporter gene is generally flanked by highly conserved gene start-and-stop signals. For instance, GFP was expressed as a separate protein from the ATU in the recombinant peste des petits ruminants virus (PPRV) [17]. Furthermore, the reporters can be expressed separately by introduction of an internal ribosome entry site (IRES) or foot-and-mouth disease virus (FMDV) 2A self-cleaving peptide (2A) (LLNFDLLKLAGDVESNPG↓P), which is able to undergo self-cleavage allowing simultaneous expression. For example, recombinant alphaviruses expressing a separate Fluc by 2A-mediated cleavage were successfully used to screen viral receptors [18].

Since the properties of RCREVs and the stability of reporter genes may vary among different strategies, the selection of expression strategy is another principal consideration on designing RCREVs for specific applications. Notably, the same strategy might lead to different effects on the growth of the same virus due to the distinct insertion site. For example, a recombinant respiratory syncytial virus (RSV) expressing a reporter protein from an ATU upstream of NS1 displayed negligible attenuation in cell cultures [19], whereas the RSV expressing a reporter from an ATU inserted between F and G genes was significantly attenuated in cell cultures [20]. Additionally, the use of 2A peptide to achieve expression of a separate reporter might constitute a promising approach as 2A peptide is small and can readily be self-cleaving while minimizing the possibility of the loss of functions of the viral proteins.

3. Applications of RCREVs in Serum-Virus Neutralization Tests

The neutralization immunofluorescence test (NIFT) is currently a gold standard for the detection of NAbs against many noncytopathogenic viruses. However, NIFT is labor-intensive and time-consuming due to the necessary incubation and staining procedures. It would be convenient to use RCREVs to detect NAbs directly without immunostaining. There are many successful applications of RCREVs harboring EGFP, Rluc or Fluc in the rapid neutralization tests [17,21–25].

For viruses causing slight or no cytopathic effects (CPEs) in cultured cells, the EGFP reporter can be chosen to generate RCREVs for quantifying NAbs with higher specificity through direct observation of EGFP fluorescence. Due to the structural characteristics of EGFP, the fluorescence of EGFP fused to a viral protein may be attenuated or quenched. Therefore, EGFP should be separated from the viral protein by introduction of an ATU, IRES or 2A sequence when constructing the RCREVs. Owing to the simple assaying for Gluc activity compared with the Fluc, Rluc and other bioluminescent reporters, it is advantageous to determine the neutralizing antibody titers based on Gluc-expressing viruses.

Notably, attenuated RCREVs can also be applied for rapid neutralization tests due to high sensitivity and operational simplicity for detection of the reporters.

4. Application of RCREVs in Screening Systems

Antiviral compounds, interferon-stimulated genes (ISGs) or small interfering RNAs (siRNAs) have potential applications in the treatment of many diseases. The traditional screening methods of them are developed by a cell-based HTS, in which the treated cells were observed under a microscope for the inhibitory activity of the compounds for CPEs [26], enzyme-linked immunosorbent assay (ELISA) [27] or fluoresces-linked immunosorbent assay [28]. Using these approaches, the scientists have screened and identified a series of small antiviral molecules or inhibitors [29,30]. However, the traditional methods based on wild-type viruses are inefficient for antiviral screening.

To overcome this problem, RCREVs have been applied for the purpose of antiviral screening, because RCREVs can target the complete virus life cycle and offer a higher throughput of antiviral screening than traditional assays. RCREVs represent powerful screening tools for identifying antiviral compounds against various highly pathogenic viruses [31–34]. For example, a high-throughput assay for Zaire EBOV has been established using the recombinant EBOV expressing the EGFP reporter gene [31]. Interestingly, reporter viruses in combination with other approaches, such as RNAi, have been applied to screen anti-CSFV ISGs [15], which is time- and cost-effective. Importantly, RCREVs with slightly reduced growth ability compared with the wild-type viruses can also be applied for screening antiviral ISGs. In addition, RCREVs can be used for siRNAs HTS with high efficiency. For instance, a reporter CSFV expressing the Fluc gene has been used to screen antiviral siRNAs efficiently [16]. Recently, a recombinant EBOV carrying a luciferase reporter was used to screen siRNAs with higher screening efficiency than the wild-type virus [25].

However, there are some problems associated with RCREVs in HTS applications. First, the interference of compound fluorescence may occur when screening antivirals using fluorescent reporter-expressing viruses. Second, the antiviral effects of screened out antivirals by RCREVs need to be verified with the parental viruses. Furthermore, the antiviral effects may be different between RCREVs and the wild-type viruses due to the occasionally inclusive fluorescence signals for the wild-type viruses in indirect immunofluorescent assay (IFA) and higher sensitivity for RCREVs in Fluc/Rluc activity assay. Third, RCREVs are not ideal tools for screening of antivirals targeting specific step(s) of viral infection, since RCREVs can undergo a complete virus life cycle. For example, currently, a set of ISGs against hepatitis C virus (HCV), yellow fever virus (YFV), West Nile virus (WNV), Chikungunya virus (CHIKV), Venezuelan equine encephalitis virus (VEEV) and human immunodeficiency virus (HIV-1) have been documented, but their exact antiviral step(s) remain(s) unknown [35–38]. A practical challenge lies in the explanations of their antiviral mechanisms for antiviral ISGs screened by RCREVs. Despite these limitations, the following strategies may address some of the above issues. Fluc, Rluc and other bioluminescent reporters provide a viable alternative to fluorescent reporters in HTS assays for drug discovery [39]. This facilitates the development of highly sensitive, cell-based reporter assays [40], eliminates the problem of compound fluorescence [41], and possesses several advantages such as high reliability, convenience and adaptability to HTS assays. Remarkably, primary HTS followed by validation using traditional assays based on the parental viruses will greatly aid the discovery of novel antivirals against infectious diseases. Finally, the use of replicons or pseudoparticles would help to identify the step(s) of the viral life cycle as the potential targets of antivirals.

5. Applications of RCREVs in Basic Research

5.1. In Identification of Cellular Receptors/Membrane Proteins

Identification of cellular receptors facilitates understanding of the mechanisms of virus entry into host cells [42,43]. Moreover, the receptors are regarded as promising targets for development of novel antivirals [44–47]. While reporter-expressing pseudoparticles are widely used to screen viral receptors [48,49], RCREVs carrying Fluc [18,50], GFP [51] or Neo^R [52] as new useful tools have been applied for screening of viral receptors (Table 1). Since RCREVs can infect the cells with multiple life cycles in contrast to pseudoparticles, more false-positive receptors may be screened. In spite of these few limitations, RCREVs are still powerful tools to screen viral receptors in combination with unsusceptible cells and cDNA library derived from susceptible cells [51,52] or a set of siRNAs against a number of genes encoding cell membrane proteins [18,50] (Table 1).

Table 1. Cellular receptors screened by representative RCREVs.

Reporters	Viruses	Expression Strategies	Screened Cellular Receptors Proteins
Firefly luciferase (Fluc)	Classical swine fever virus (CSFV)	Fusion with a viral protein	Laminin receptor (LamR) [50]
	Alphaviruses	Introduction of foot-and-mouth disease virus 2A-enconding sequence	Fuzzy homolog (FUZ) and tetraspanin membrane protein TSPAN9 [18]
Green fluorescent protein (GFP)	Porcine reproductive and respiratory syndrome virus (PRRSV)	Fusion with a viral protein	CD163 [51]
Neomycin resistance gene (Neo^R)	Equine infectious anemia virus (EIAV)	Introduction of an additional transcriptional unit	Equine lentivirus receptor 1 (ELR1) [52]

5.2. Virus Tracking and Live Imaging in Vitro and in Vivo

With the development of reverse genetics systems, RCREVs provide an ideal tool for monitoring the dynamics of viral infection progression *in vitro* and *in vivo* due to eliminating the need for secondary labeling, which represents a significant advance in the study of the biology of viruses (Table 2).

RCREVs carrying a GFP reporter gene have been successfully used for tracking viral protein(s) or viral infection *in vitro* and *in vivo* [53–56], which indicates that the GFP reporter gene is suitable for generating RCREVs to track viral proteins either in cell cultures or animal models. Furthermore, GFPs in RCREVs can be expressed efficiently in rodent brain for a long time [57] and show lower autofluorescence in the tissue [56]. Therefore, GFP may be a promising option when RCREVs are used to study the infection of viruses replicating in the brain. Additionally, an engineered virus expressing the split-green fluorescent protein (split-GFP) in the presence of cell lines expressing the complementing GFP can facilitate the tracking of viral infection in living cells [58].

Compared with the most commonly used EGFP tag, the TC tag enables the fusion protein to fluoresce more quickly, with a minimum risk of disrupting the overall structure and function of the targeted protein [59]. The TC-labeling technology has led to successful tracking of the nonstructural or structural proteins of diverse viruses [60–65]. However, since the TC-tag technology contains a biarsenical labeling process [66,67], the engineered replication-competent TC-tagged viruses are not suitable for tracking viral protein *in vivo*.

In addition, recombinant canine distemper virus (CDV) expressing dTomato was used to investigate the routes of virus spread *in vivo* [56]. A fully functional recombinant pneumonia virus of mice (PVM) with Katushka 2 has been developed to track infection of target cells *in vivo* [68]. Compared with far-red GFP-like proteins, iRFP has a substantially higher signal-to-background ratio in a mouse model due to its infrared-shifted spectra [69,70]. Interestingly, the Cre recombinase as a reporter is used to generate RCREVs for visualizing virus infection in engineered cell lines or transgenic animals harboring a loxP-flanked fluorescent marker upstream of another otherwise silenced fluorescent reporter [71].

Recently, several influenza viruses expressing fluorescent proteins of different colors ("Color-flu" viruses expressing ECFP, EGFP, Venus or mCherry) or a toolbox of influenza A and B reporter viruses were generated to facilitate the study of viral infection in *in vivo* models. Whole-mount images of transparent lung tissues were obtained using a fluorescent stereomicroscope [72–76]. In addition, bioluminescent and fluorescent dual-reporter Marek's disease viruses are engineered to track viral replication in cell cultures or animal models [77]. In the future, "color" or dual-reporter viruses will be powerful tools to analyze viral infection at the cellular level *in vivo* to better understand the pathogenesis of various viruses.

Table 2. Applications of representative RCREVs in virus tracking and live imaging *in vitro* and *in vivo*.

Reporters	Viruses	Tracking and Live Imaging
Green fluorescent protein (GFP)	Influenza virus	Dynamics of virus infection progression in mice [53]
	Herpes simplex virus (HSV)	Compartmentalization of protein by autofluorescent particles [54]
	Borna disease virus (BDV)	In rodent brains [57]
	Canine distemper virus (CDV)	Routes of virus spread *in vivo* [56]
	Vesicular stomatitis virus (VSV)	Intracellular transport [55]
Tetracysteine (TC)	Vesicular stomatitis virus (VSV)	Dynamic imaging of M protein and virus uncoating in infected cells [60]
	Influenza A virus	Visualization of NS1 protein nuclear import in virus-infected cells in real time [61]
	Classical swine fever virus (CSFV)	Nucleus import and export [62]
	Hepatitis C virus (HCV)	Virus particle assembly [63]
	Human immunodeficiency virus (HIV)	Viral component complexes [64] *de novo* HIV production [65]
ECFP, EGFP, Venus, RFP, mCherry, NanoLuc and Gluc split-GFP, Cre recombinase	Influenza A/B virus	Viral infection *in vitro* or in lung tissues [58,71–76]
Katushka 2	Pneumonia virus of mice (PVM)	Tracking of viral infection of target cells *in vivo* [68]
iRFPs	Adenovirus	In mouse model [69]
dTomato	Canine distemper virus (CDV)	Routes of virus spread *in vivo* [56]
EGFP+Rluc/Gluc	Marek's disease virus (MDV)	Tracking of viral replication *in vitro* and *in vivo* [77]

Notably, reporters fused with viral proteins are very suitable for investigating the localization and distribution of the proteins in infected living cells. RCREVs will help advance virus-related live-imaging studies *in vitro* and *in vivo*, which allow localization of the infection and tracking of changes in the distribution of viruses in animals in real time.

5.3. In Identification of Virus-Host Interactions

Elucidating various aspects of pathogen-host interactions is essential for the comprehensive understanding of pathogenesis. Compared with the most frequently used techniques for mapping of virus-host interactions, the approaches based on RCREVs can recapitulate the virus life cycle [78]. Split-Gluc (Gluc1 and Gluc2) has been applied for identification of virus-host interactions. For example, a recombinant influenza virus carrying a Gluc1-tagged polymerase subunit is used to infect the cultured cells expressing a pool of Gluc2-tagged cellular proteins involved in nucleocytoplasmic-transporting pathways for confirming virus-host interactions [79]. In addition, split-GFP reporter has huge potential in this application. However, the reporter activity based on the interactions of RCREVs with the cellular proteins may not be detected due to the interference of the space structure.

6. Other Applications

The RCREVs are also useful in modified live vaccines containing genetic markers, which have been developed for many viruses by inserting EGFP as a positive marker [80–82]. For example, a genetically marked recombinant rinderpest vaccine expressing GFP has been developed [81]. In addition, a recombinant GFP-tagged PRRSV containing a deletion of an immunogenic epitope, in accompany with the diagnostic tests (GFP- and epitope-based ELISAs), enables serological differentiation between the marker virus-infected animals and those infected with the wild-type virus [82]. A recombinant viral hemorrhagic septicemia virus (VHSV) harboring RFP gene was utilized to evaluate VHSV-based viral-vectored vaccines [83]. More recently, the marker vaccine vSMEGFP-HCLV3′UTR in the context of the CSFV Shimen strain was generated by inserting EGFP to create a positive marker [84].

For those viruses causing CPEs, RCREVs can be used as an intermediate to generate and purify expected variants. For example, a novel gE-deleted pseudorabies virus (PRV) was obtained by gE/gI-deleted virus expressing EGFP [85]. In addition, Katushka 2 as a reporter was used to evaluate a novel reverse genetic system of RSV [19].

Interestingly, oncolytic recombinant viruses harboring reporter genes have been developed and applied for the disease progression tracking and accurate visualization of tumor burden [14]. Since oncolytic viruses selectively infect as well as replicate within cancer cells, the recombinant oncolytic viruses expressing reporter genes, particularly for far-red fluorescent proteins, will be a promising option for real-time monitoring of viral infection in cancer tissues [14].

While RCREVs harboring a reporter fused to a viral protein are the most suitable for studying the localization of the protein in infected cells, RCREVs carrying separate reporters are useful for other basic research purposes. For example, the preferential translation of viral RNAs over host RNAs during VSV infection has been demonstrated by the EGFP reporter expressed from the recombinant VSV [86]. Recently, the contribution of EBOV proteins in modulating dendritic cells (DC) maturation was investigated using the recombinant virus carrying EGFP [87]. Furthermore, unique profiles of RFP expression acquired from thousands of co-infected cells with viable and defective viruses showed how the interference of defective viruses acts at multiple steps of infection [88].

7. Limitations and Prospects

Firstly, a practical challenge for some viruses lies in not allowing the insertion of reporter genes. As we stated above, it is difficult to insert a reporter gene into the genome of influenza viruses. Despite the challenge, reporter-expressing influenza viruses have been developed and applied in basic science [58,61,71–76]. To address the question, there are three necessary considerations, including the reporter protein itself, expression strategy, and structure of the viral protein. For example, the loop/linker regions are usually chosen to insert the TC tag based on the structure of NS1 of influenza viruses [61].

Although RCREVs have been developed and applied *in vitro* and *in vivo*, one question arises regarding the expression stability of the reporter gene in RCREVs during the viral replication *in vitro* and *in vivo* [53,89]. One potential consequence of RCREVs' attenuation is the purging of the inserted reporter from the viral genome. In this regard, we need to better understand the mechanism of regulation of viral genome replication and gene expression [90,91], the association between structure and function of viral proteins, as well as the application of novel reporters such as NanoLuc due to its small size [92].

One of the biggest challenges is that RCREVs are possibly attenuated and may not accurately reflect natural infections [93,94], which partially limits the applications of the RCREVs, especially *in vivo*. Replacement of currently used expression strategies may be a promising approach to overcoming this problem. As an example, IRES or 2A peptide-encoding sequence has been used to express separately the reporter from viral protein [71,95]. Importantly, the use of split-GFP or split-luciferase may not compromise viral replication competency due to their smaller sizes [58,79]. However, whether these reporter viruses will be attenuated *in vivo* needs further investigation in the future. More recently, it has been reported that after mouse adaptation, influenza virus H5N1 expressing the Venus reporter gene became more pathogenic to mice and the Venus gene was more highly and stably expressed [96], which may be another promising avenue that maintains the pathogenicity of the reporter viruses.

Luciferase imaging uses the luciferases to catalyze reactions that produce visible light *in vivo* at body temperature, which is used to determine the sites of virus replication, monitor viral dissemination in real time [97]. However, there are many caveats in the process of obtaining accurate luciferase imaging [98]. For example, the reporter signal from RCREVs is attenuated when *in vivo* imaging in tissues. Despite these limitations, luciferases will still become major reporters for *in vivo* imaging in real time in the future as they have a number of advantages compared with the fluorescent reporters, such as no intrinsic autoluminescence. In addition, iRFPs are in high demand for *in vivo* imaging, which exhibit high brightness in mammalian cells and tissues and are suitable for long-term studies with multicolor imaging.

Finally, in view of the advantages and disadvantages of different reporters, there seems no universal reporter for various applications. Fortunately, ever-increasing novel reporters, including GFP mutants, "red-shifted" analogs of luciferase, variants of luciferase and novel luciferase NanoLuc, can be chosen to design RCREVs for specific purposes. Moreover, the dual-reporter RCREVs may be widely used to address the scientific questions. Although reporter-based assays require costly automated imaging equipment, the detection of the reporter gene expression could be also performed with inexpensive, small and simple-to-use equipment, such as a PCR device based on the development of the technologies discussed in this article.

8. Conclusions

RCREVs have proved themselves to be powerful tools for applied and basic sciences. Despite their limitations, RCREVs have many more far-reaching benefits in virus research: a genome-wide RNAi screening for host factors required for virus replication, identifying antivirals against viral infections using HTS settings, monitoring viral infections *in vitro* and *in vivo* in real time, or evaluating vaccination approaches, as well as detecting antiviral NAbs.

Acknowledgments: This work was supported by Natural Science Foundation of China (no. 31572540 and no. 31400146) and the Natural Science Foundation of Heilongjiang Province of China (no. QC2015039). We thank Lintao Liu at Lerner Research Institute, United States of America for improving the language of the manuscript.

Author Contributions: Hua-Ji Qiu and Yongfeng Li conceived, wrote and edited the manuscript. Lian-Feng Li, Shaoxiong Yu, Xiao Wang, Lingkai Zhang, Jiahui Yu and Libao Xie wrote the manuscript. Weike Li and Razim Ali edited the manuscript.

Conflicts of Interest: The authors declare no conflict of interest.

References

1.   Meyers, G.; Thiel, H.J.; Rümenapf, T. Classical swine fever virus: Recovery of infectious viruses from cDNA constructs and generation of recombinant cytopathogenic defective interfering particles. *J. Virol.* **1996**, *70*, 1588–1595.

2.   Moormann, R.J.; van Gennip, H.G.; Miedema, G.K.; Hulst, M.M.; van Rijn, P.A. Infectious RNA transcribed from an engineered full-length cDNA template of the genome of a pestivirus. *J. Virol.* **1996**, *70*, 763–770.

3. Hoffmann, E.; Neumann, G.; Kawaoka, Y.; Hobom, G.; Webster, R.G. A DNA transfection system for generation of influenza a virus from eight plasmids. *Proc. Natl. Acad. Sci. USA* **2000**, *97*, 6108–6113.

4. Martin, A.; Staeheli, P.; Schneider, U. RNA polymerase II-controlled expression of antigenomic RNA enhances the rescue efficacies of two different members of the *Mononegavirales* independently of the site of viral genome replication. *J. Virol.* **2006**, *80*, 5708–5715.

5. Ward, V.K.; McCormick, C.J.; Clarke, I.N.; Salim, O.; Wobus, C.E.; Thackray, L.B.; Virgin, H.W., IV; Lambden, P.R. Recovery of infectious murine norovirus using pol II-driven expression of full-length cDNA. *Proc. Natl. Acad. Sci. USA* **2007**, *104*, 11050–11055.

6. Ben Abdeljelil, N.; Khabouchi, N.; Mardassi, H. Efficient rescue of infectious bursal disease virus using a simplified RNA polymerase II-based reverse genetics strategy. *Arch. Virol.* **2008**, *153*, 1131–1137.

7. Li, B.Y.; Li, X.R.; Lan, X.; Yin, X.P.; Li, Z.Y.; Yang, B.; Liu, J.X. Rescue of Newcastle disease virus from cloned cDNA using an RNA polymerase II promoter. *Arch. Virol.* **2011**, *156*, 979–986.

8. Hoenen, T.; Groseth, A.; de Kok-Mercado, F.; Kuhn, J.H.; Wahl-Jensen, V. Minigenomes, transcription and replication competent virus-like particles and beyond: Reverse genetics systems for filoviruses and other negative stranded hemorrhagic fever viruses. *Antiviral Res.* **2011**, *91*, 195–208.

9. Römer-Oberdörfer, A.; Mundt, E.; Mebatsion, T.; Buchholz, U.J.; Mettenleiter, T.C. Generation of recombinant lentogenic Newcastle disease virus from cDNA. *J. Gen. Virol.* **1999**, *80*, 2987–2995.

10. Buchholz, U.J.; Finke, S.; Conzelmann, K.K. Generation of bovine respiratory syncytial virus (BRSV) from cDNA: BRSV NS2 is not essential for virus replication in tissue culture, and the human RSV leader region acts as a functional BRSV genome promoter. *J. Virol.* **1999**, *73*, 251–259.

11. Kovacs, G.R.; Parks, C.L.; Vasilakis, N.; Udem, S.A. Enhanced genetic rescue of negative-strand RNA viruses: Use of an MVA-T7 RNA polymerase vector and DNA replication inhibitors. *J. Virol. Methods* **2003**, *111*, 29–36.

12. Liu, G.; Zhang, Y.; Ni, Z.; Yun, T.; Sheng, Z.; Liang, H.; Hua, J.; Li, S.; Du, Q.; Chen, J. Recovery of infectious rabbit hemorrhagic disease virus from rabbits after direct inoculation with *in vitro*-transcribed RNA. *J. Virol.* **2006**, *80*, 6597–6602.

13. Chaudhry, Y.; Skinner, M.A.; Goodfellow, I.G. Recovery of genetically defined murine norovirus in tissue culture by using a fowlpox virus expressing T7 RNA polymerase. *J. Gen. Virol.* **2007**, *88*, 2091–2100.

14. Ady, J.W.; Johnsen, C.; Mojica, K.; Heffner, J.; Love, D.; Pugalenthi, A.; Belin, L.J.; Chen, N.G.; Yu, Y.A.; Szalay, A.A.; et al. Oncolytic gene therapy with recombinant vaccinia strain GLV-2b372 efficiently kills hepatocellular carcinoma. *Surgery* **2015**, *158*, 331–338.

15. Wang, X.; Li, Y.; Li, L.F.; Shen, L.; Zhang, L.; Yu, J.; Luo, Y.; Sun, Y.; Li, S.; Qiu, H.J. RNA interference screening of interferon-stimulated genes with antiviral activities against classical swine fever virus using a reporter virus. *Antiviral Res.* **2016**, *128*, 49–56.

16. Shen, L.; Li, Y.; Chen, J.; Li, C.; Huang, J.; Luo, Y.; Sun, Y.; Li, S.; Qiu, H.J. Generation of a recombinant classical swine fever virus stably expressing the firefly luciferase gene for quantitative antiviral assay. *Antiviral Res.* **2014**, *109*, 15–21.

17. Hu, Q.; Chen, W.; Huang, K.; Baron, M.D.; Bu, Z. Rescue of recombinant peste des petits ruminants virus: Creation of a GFP-expressing virus and application in rapid virus neutralization test. *Vet. Res.* **2012**, *43*.

18. Ooi, Y.S.; Stiles, K.M.; Liu, C.Y.; Taylor, G.M.; Kielian, M. Genome-wide RNAi screen identifies novel host proteins required for alphavirus entry. *PLoS Pathog.* **2013**, *9*, e1003835.

19. Hotard, A.L.; Shaikh, F.Y.; Lee, S.; Yan, D.; Teng, M.N.; Plemper, R.K.; Crowe, J.E., Jr.; Moore, M.L. A stabilized respiratory syncytial virus reverse genetics system amenable to recombination-mediated mutagenesis. *Virology* **2012**, *434*, 129–136.

20. Bukreyev, A.; Camargo, E.; Collins, P.L. Recovery of infectious respiratory syncytial virus expressing an additional, foreign gene. *J. Virol.* **1996**, *70*, 6634–6641.

21. Li, Y.; Shen, L.; Sun, Y.; Yuan, J.; Huang, J.; Li, C.; Li, S.; Luo, Y.; Qiu, H.J. A simplified serum-neutralization test based on enhanced green fluorescent protein-tagged classical swine fever virus. *J. Clin. Microbiol.* **2013**, *51*, 2710–2712.

22. Wang, Q.; Li, X.; Ji, X.; Wang, J.; Shen, N.; Gao, Y.; Qi, X.; Wang, Y.; Gao, H.; Zhang, S.; *et al.* A recombinant avian leukosis virus subgroup j for directly monitoring viral infection and the selection of neutralizing antibodies. *PLoS One* **2014**, *9*, e115422.

23. Xue, X.; Zheng, X.; Liang, H.; Feng, N.; Zhao, Y.; Gao, Y.; Wang, H.; Yang, S.; Xia, X. Generation of recombinant rabies Virus CVS-11 expressing eGFP applied to the rapid virus neutralization test. *Viruses* **2014**, *6*, 1578–1589.

24. Zhou, M.; Kitagawa, Y.; Yamaguchi, M.; Uchiyama, C.; Itoh, M.; Gotoh, B. Expeditious neutralization assay for human metapneumovirus based on a recombinant virus expressing Renilla luciferase. *J. Clin. Virol.* **2013**, *56*, 31–36.

25. Hoenen, T.; Groseth, A.; Callison, J.; Takada, A.; Feldmann, H. A novel Ebola virus expressing luciferase allows for rapid and quantitative testing of antivirals. *Antiviral Res.* **2013**, *99*, 207–213.

26. Lundin, A.; Bergstrom, T.; Bendrioua, L.; Kann, N.; Adamiak, B.; Trybala, E. Two novel fusion inhibitors of human respiratory syncytial virus. *Antiviral Res.* **2010**, *88*, 317–324.

27. Jiang, S.B.; Lin, K.; Zhang, L.; Debnath, A.K. A screening assay for antiviral compounds targeted to the HIV-1 gp41 core structure using a conformation-specific monoclonal antibody. *J. Virol. Methods* **1999**, *80*, 85–96.

28. Liu, S.W.; Boyer-Chatenet, L.; Lu, H.; Jiang, S.B. Rapid and automated fluorescence-linked immunosorbent assay for high-throughput screening of HIV-1 fusion inhibitors targeting gp41. *J. Biomol. Screen* **2003**, *8*, 685–693.

29. Jiang, S.B.; Lu, H.; Liu, S.W.; Zhao, Q.; He, Y.X.; Debnath, A.K. N-substituted pyrrole derivatives as novel human immunodeficiency virus type 1 entry inhibitors that interfere with the gp41 six-helix bundle formation and block virus fusion. *Antimicrob. Agents Chemother.* **2004**, *48*, 4349–4359.

30. Park, M.; Matsuura, H.; Lamb, R.A.; Barron, A.E.; Jardetzky, T.S. A fluorescence polarization assay using an engineered human respiratory syncytial virus F protein as a direct screening platform. *Anal. Biochem.* **2011**, *409*, 195–201.

31. Towner, J.S.; Paragas, J.; Dover, J.E.; Gupta, M.; Goldsmith, C.S.; Huggins, J.W.; Nichol, S.T. Generation of eGFP expressing recombinant Zaire Ebola virus for analysis of early pathogenesis events and high throughput antiviral drug screening. *Virology* **2005**, *332*, 20–27.

32. Jin, G.; Lee, S.; Choi, M.; Son, S.; Kim, G.W.; Oh, J.W.; Lee, C.; Lee, K. Chemical genetics-based discovery of indole derivatives as HCV NS5B polymerase inhibitors. *Eur. J. Med. Chem.* **2014**, *75*, 413–425.

33. Hu, Z.; Lan, K.H.; He, S.; Swaroop, M.; Hu, X.; Southall, N.; Zheng, W.; Liang, T.J. Novel cell-based hepatitis C virus infection assay for quantitative high-throughput screening of anti-hepatitis C virus compounds. *Antimicrob. Agents Chemother.* **2014**, *58*, 995–1004.

34. He, S.; Jain, P.; Lin, B.; Ferrer, M.; Hu, Z.; Southall, N.; Hu, X.; Zheng, W.; Neuenswander, B.; Cho, C.H.; *et al.* High-throughput screening, discovery, and optimization to develop a benzofuran class of hepatitis C virus inhibitors. *ACS Comb. Sci.* **2015**, *17*, 641–652.

35. Schoggins, J.W.; Wilson, S.J.; Panis, M.; Murphy, M.Y.; Jones, C.T.; Bieniasz, P.; Rice, C.M. A diverse range of gene products are effectors of the type I interferon antiviral response. *Nature* **2011**, *472*, 481–485.

36. Zhao, H.; Lin, W.; Kumthip, K.; Cheng, D.; Fusco, D.N.; Hofmann, O.; Jilg, N.; Tai, A.W.; Goto, K.; Zhang, L.; *et al.* A functional genomic screen reveals novel host genes that mediate interferon-alpha's effects against hepatitis C virus. *J. Hepatol.* **2012**, *56*, 326–333.

37. Metz, P.; Reuter, A.; Bender, S.; Bartenschlager, R. Interferon-stimulated genes and their role in controlling hepatitis C virus. *J. Hepatol.* **2013**, *59*, 1331–1341.

38. Mihm, S. Activation of type I and type III interferons in chronic hepatitis C. *J. Innate Immun.* **2015**, *7*, 251–259.

39. Miraglia, L.J.; King, F.J.; Damoiseaux, R. Seeing the light: Luminescent reporter gene assays. *Comb. Chem. High Throughput Screen* **2011**, *14*, 648–657.

40. Thorne, N.; Inglese, J.; Auld, D.S. Illuminating insights into firefly luciferase and other bioluminescent reporters used in chemical biology. *Chem. Biol.* **2010**, *17*, 646–657.

41. Simeonov, A.; Jadhav, A.; Thomas, C.J.; Wang, Y.; Huang, R.; Southall, N.T.; Shinn, P.; Smith, J.; Austin, C.P.; Auld, D.S.; *et al.* Fluorescence spectroscopic profiling of compound libraries. *J. Med. Chem.* **2008**, *51*, 2363–2371.

42. Thorley, J.A.; McKeating, J.A.; Rappoport, J.Z. Mechanisms of viral entry: Sneaking in the front door. *Protoplasma* **2010**, *244*, 5–24.

43. Grove, J.; Marsh, M. The cell biology of receptor-mediated virus entry. *J. Cell. Biol.* **2011**, *195*, 1071–1182.

44. Dorr, P.; Westby, M.; Dobbs, S.; Griffin, P.; Irvine, B.; Macartney, M.; Mori, J.; Rickett, G.; Smith-Burchnell, C.; Napier, C.; *et al.* Maraviroc (UK-427,857), a potent, orally bioavailable, and selective small-molecule inhibitor of chemokine receptor CCR5 with broad-spectrum anti-human immunodeficiency virus type 1 activity. *Antimicrob. Agents Chemother.* **2005**, *49*, 4721–4732.

45. Lu, L.; Liu, Q.; Zhu, Y.; Chan, K.H.; Qin, L.; Li, Y.; Wang, Q.; Chan, J.F.; Du, L.; Yu, F.; *et al.* Structure-based discovery of Middle East respiratory syndrome coronavirus fusion inhibitor. *Nat. Commun.* **2014**, *5*, 1661–1667.

46. Raj, V.S.; Mou, H.H.; Smits, S.L.; Dekkers, D.H.W.; Muller, M.A.; Dijkman, R.; Muth, D.; Demmers, J.A.; Zaki, A.; Fouchier, R.A.; *et al.* Dipeptidyl peptidase 4 is a functional receptor for the emerging human coronavirus-EMC. *Nature* **2013**, *495*, 251–254.

47. Yan, H.; Zhong, G.C.; Xu, G.W.; He, W.H.; Jing, Z.Y.; Gao, Z.C.; Huang, Y.; Qi, Y.H.; Peng, B.; Wang, H.M.; *et al.* Sodium taurocholate cotransporting polypeptide is a functional receptor for human hepatitis B and D virus. *eLife* **2012**, *1*, e00049.

48. Evans, M.J.; von Hahn, T.; Tscherne, D.M.; Syder, A.J.; Panis, M.; Wölk, B.; Hatziioannou, T.; McKeating, J.A.; Bieniasz, P.D.; Rice, C.M. Claudin-1 is a hepatitis c virus co-receptor required for a late step in entry. *Nature* **2007**, *446*, 801–805.

49. Ploss, A.; Evans, M.J.; Gaysinskaya, V.A.; Panis, M.; You, H.; de Jong, Y.P.; Rice, C.M. Human occludin is a hepatitis C virus entry factor required for infection of mouse cells. *Nature* **2009**, *457*, 882–886.

50. Chen, J.; He, W.R.; Shen, L.; Dong, H.; Yu, J.; Wang, X.; Yu, S.; Li, Y.; Li, S.; Luo, Y.; *et al.* The Laminin receptor is a cellular attachment receptor for classical swine fever virus. *J. Virol.* **2015**, *89*, 4894–4906.

51. Calvert, J.G.; Slade, D.E.; Shields, S.L.; Jolie, R.; Mannan, R.M.; Ankenbauer, R.G.; Welch, K. CD163 expression confers susceptibility to porcine reproductive and respiratory syndrome viruses. *J. Virol.* **2007**, *81*, 7371–7379.

52. Zhang, B.S.; Jin, S.; Jin, J.; Li, F.; Montelaro, R.C. A tumor necrosis factor receptor family protein serves as a cellular receptor for the macrophage-tropic equine lentivirus. *Proc. Natl. Acad. Sci. USA* **2005**, *102*, 9918–9923.

53. Manicassamy, B.; Manicassamy, S.; Belicha-Villanueva, A.; Pisanelli, G.; Pulendran, B.; García-Sastre, A. Analysis of *in vivo* dynamics of influenza virus infection in mice using a GFP reporter virus. *Proc. Natl. Acad. Sci. USA* **2010**, *107*, 11531–11536.

54. La Boissière, S.; Izeta, A.; Malcomber, S.; O'Hare, P. Compartmentalization of VP16 in cells infected with recombinant herpes simplex virus expressing VP16-green fluorescent protein fusion proteins. *J. Virol.* **2004**, *78*, 8002–8014.

55. Das, S.C.; Nayak, D.; Zhou, Y.; Pattnaik, A.K. Visualization of intracellular transport of vesicular stomatitis virus nucleocapsids in living cells. *J. Virol.* **2006**, *80*, 6368–6377.

56. Ludlow, M.; Nguyen, D.T.; Silin, D.; Lyubomska, O.; de Vries, R.D.; von Messling, V.; McQuaid, S.; de Swart, R.L.; Duprex, W.P. Recombinant canine distemper virus strain Snyder Hill expressing green or red fluorescent proteins causes meningoencephalitis in the ferret. *J. Virol.* **2012**, *86*, 7508–7519.

57. Daito, T.; Fujino, K.; Honda, T.; Matsumoto, Y.; Watanabe, Y.; Tomonaga, K. A novel borna disease virus vector system that stably expresses foreign proteins from an intercistronic noncoding region. *J. Virol.* **2011**, *85*, 12170–12178.

58. Avilov, S.V.; Moisy, D.; Munier, S.; Schraidt, O.; Naffakh, N.; Cusack, S. Replication-competent influenza A virus that encodes a split-green fluorescent protein-tagged PB2 polymerase subunit allows live-cell imaging of the virus life cycle. *J. Virol.* **2012**, *86*, 1433–1448.

59. Tsien, R.Y. The green fluorescent protein. *Annu. Rev. Biochem.* **1998**, *67*, 509–544.

60. Das, S.C.; Panda, D.; Nayak, D.; Pattnaik, A.K. Biarsenical labeling of vesicular stomatitis virus encoding tetracysteine-tagged M protein allows dynamic imaging of M protein and virus uncoating in infected cells. *J. Virol.* **2009**, *83*, 2611–2622.

61. Li, Y.; Lu, X.; Li, J.; Berube, N.; Giest, K.L.; Liu, Q.; Anderson, D.H.; Zhou, Y. Genetically engineered, biarsenically labeled influenza virus allows visualization of viral NS1 protein in living cells. *J. Virol.* **2010**, *84*, 7204–7213.

62. Li, Y.; Shen, L.; Li, C.; Huang, J.; Zhao, B.; Sun, Y.; Li, S.; Luo, Y.; Qiu, H.J. Visualization of the N^pro protein in living cells using biarsenically labeling tetracysteine-tagged classical swine fever virus. *Virus Res.* **2014**, *189*, 67–74.

63. Counihan, N.A.; Rawlinson, S.M.; Lindenbach, B.D. Trafficking of hepatitis C virus core protein during virus particle assembly. *PLoS Pathog.* **2011**, *7*, e1002302.

64. Arhel, N.; Genovesio, A.; Kim, K.A.; Miko, S.; Perret, E.; Olivo-Marin, J.C.; Shorte, S.; Charneau, P. Quantitative four-dimensional tracking of cytoplasmic and nuclear HIV-1 complexes. *Nat. Methods* **2006**, *3*, 817–824.

65. Turville, S.G.; Aravantinou, M.; Stossel, H.; Romani, N.; Robbiani, M. Resolution of *de novo* HIV production and trafficking in immature dendritic cells. *Nat. Methods* **2008**, *5*, 75–85.

66. Griffin, B.A.; Adams, S.R.; Tsien, R.Y. Specific covalent labeling of recombinant protein molecules inside live cells. *Science* **1998**, *281*, 269–272.

67. Martin, B.R.; Giepmans, B.N.; Adams, S.R.; Tsien, R.Y. Mammalian cell-based optimization of the biarsenical-binding tetracysteine motif for improved fluorescence and affinity. *Nat. Biotechnol.* **2005**, *23*, 1308–1314.

68. Dyer, K.D.; Drummond, R.A.; Rice, T.A.; Percopo, C.M.; Brenner, T.A.; Barisas, D.A.; Karpe, K.A.; Moore, M.L.; Rosenberg, H.F. Priming of the respiratory tract with immunobiotic *Lactobacillus plantarum* limits infection of alveolar macrophages with recombinant pneumonia virus of mice (rK2-PVM). *J. Virol.* **2015**, *90*, 979–991.

69. Filonov, G.S.; Piatkevich, K.D.; Ting, L.M.; Zhang, J.; Kim, K.; Verkhusha, V.V. Bright and stable near-infrared fluorescent protein for *in vivo* imaging. *Nat. Biotechnol.* **2011**, *29*, 757–761.

70. Shcherbakova, D.M.; Verkhusha, V.V. Near-infrared fluorescent proteins for multicolor *in vivo* imaging. *Nat. Methods* **2013**, *10*, 751–754.

71. Reuther, P.; Göpfert, K.; Dudek, A.H.; Heiner, M.; Herold, S.; Schwemmle, M. Generation of a variety of stable influenza A reporter viruses by genetic engineering of the NS gene segment. *Sci. Rep.* **2015**, *5*, 11346.

72. Fukuyama, S.; Katsura, H.; Zhao, D.; Ozawa, M.; Ando, T.; Shoemaker, J.E.; Ishikawa, I.; Yamada, S.; Neumann, G.; Watanabe, S.; *et al.* Multi-spectral fluorescent reporter influenza viruses (Color-flu) as powerful tools for *in vivo* studies. *Nat. Commun.* **2015**, *6*, 6600.

73. Eckert, N.; Wrensch, F.; Gartner, S.; Palanisamy, N.; Goedecke, U.; Jager, N.; Pohlmann, S.; Winkler, M. Influenza A virus encoding secreted Gaussia luciferase as useful tool to analyze viral replication and its inhibition by antiviral compounds and cellular proteins. *PLoS One* **2014**, *9*, e97695.

74. Nogales, A.; Rodríguez-Sánchez, I.; Monte, K.; Lenschow, D.J.; Perez, D.R.; Martínez-Sobrido, L. Replication-competent fluorescent-expressing influenza B virus. *Virus Res.* **2015**, *213*, 69–81.

75. Nogales, A.; Baker, S.F.; Martinez-Sobrido, L. Replication-competent influenza A viruses expressing a red fluorescent protein. *Virology* **2015**, *476*, 206–216.

76. Tran, V.; Poole, D.S.; Jeffery, J.J.; Sheahan, T.P.; Creech, D.; Yevtodiyenko, A.; Peat, A.J.; Francis, K.P.; You, S.; Mehle, A. Multi-modal imaging with a toolbox of influenza A reporter viruses. *Viruses* **2015**, *7*, 5319–5327.

77. Harmache, A. A virulent bioluminescent and fluorescent dual-reporter Marek's disease virus unveils an alternative spreading pathway in addition to cell-to-cell contact. *J. Virol.* **2014**, *88*, 11617–11623.

78. Komarova, A.V.; Combredet, C.; Meyniel-Schicklin, L.; Chapelle, M.; Caignard, G.; Camadro, J.M.; Lotteau, V.; Vidalain, P.O.; Tangy, F. Proteomic analysis of virus-host interactions in an infectious context using recombinant viruses. *Mol. Cell. Proteom.* **2011**, *10*, M110.007443.

79. Munier, S.; Rolland, T.; Diot, C.; Jacob, Y.; Naffakh, N. Exploration of binary virus-host interactions using an infectious protein complementation assay. *Mol. Cell. Proteom.* **2013**, *12*, 2845–2855.

80. Dong, X.N.; Chen, Y.H. Marker vaccine strategies and candidate CSFV marker vaccines. *Vaccine* **2007**, *25*, 205–230.

81. Walsh, E.P.; Baron, M.D.; Rennie, L.F.; Anderson, J.; Barrett, T. Development of a genetically marked recombinant rinderpest vaccine expressing green fluorescent protein. *J. Gen. Virol.* **2000**, *81*, 709–718.

82. Fang, Y.; Christopher-Hennings, J.; Brown, E.; Liu, H.X.; Chen, Z.H.; Lawson, S.R.; Breen, R.; Clement, T.; Gao, X.F.; Bao, J.J.; *et al.* Development of genetic markers in the nonstructural protein 2 region of a US type 1 porcine reproductive and respiratory syndrome virus: Implications for future recombinant marker vaccine development. *J. Gen. Virol.* **2008**, *89*, 3086–3096.

83. Kim, M.S.; Park, J.S.; Kim, K.H. Optimal place of a foreign gene in the genome of viral haemorrhagic septicaemia virus (VHSV) for development of VHSV-based viral-vectored vaccines. *J. Appl. Microbiol.* **2013**, *114*, 1866–1873.

84. Li, Y.; Wang, X.; Sun, Y.; Li, L.F.; Zhang, L.; Li, S.; Luo, Y.; Qiu, H.J. Generation and evaluation of a chimeric classical swine fever virus expressing a visible marker gene. *Arch. Virol.* **2016**, *161*, 563–571.

85. Wang, C.H.; Yuan, J.; Qin, H.Y.; Luo, Y.; Cong, X.; Li, Y.; Chen, J.; Li, S.; Sun, Y.; Qiu, H.J. A novel gE-deleted pseudorabies virus (PRV) provides rapid and complete protection from lethal challenge with the PRV variant emerging in Bartha-K61-vaccinated swine population in China. *Vaccine* **2014**, *32*, 3379–3385.

86. Whitlow, Z.W.; Connor, J.H.; Lyles, D.S. Preferential translation of vesicular stomatitis virus mRNAs is conferred by transcription from the viral genome. *J. Virol.* **2006**, *80*, 11733–11742.

87. Lubaki, N.M.; Ilinykh, P.; Pietzsch, C.; Tigabu, B.; Freiberg, A.N.; Koup, R.A.; Bukreyev, A. The lack of maturation of Ebola virus-infected dendritic cells results from the cooperative effect of at least two viral domains. *J. Virol.* **2013**, *87*, 7471–7485.

88. Akpinar, F.; Timm, A.; Yin, J. High-throughput single-cell kinetics of virus infections in the presence of defective interfering particles. *J. Virol.* **2015**, *90*, 1599–1612.

89. Dinh, P.X.; Panda, D.; Das, P.B.; Das, S.C.; Das, A.; Pattnaik, A.K. A single amino acid change resulting in loss of fluorescence of eGFP in a viral fusion protein confers fitness and growth advantage to the recombinant vesicular stomatitis virus. *Virology* **2012**, *432*, 460–469.

90. Gasanov, N.B.; Toshchakov, S.V.; Georgiev, P.G.; Maksimenko, O.G. The use of transcription terminators to generate transgenic lines of Chinese hamster ovary cells (CHO) with stable and high level of reporter gene expression. *Acta Nat.* **2015**, *7*, 74–80.

91. Shin, Y.C.; Bischof, G.F.; Lauer, W.A.; Desrosiers, R.C. Importance of codon usage for the temporal regulation of viral gene expression. *Proc. Natl. Acad. Sci. USA* **2015**, *112*, 14030–14035.

92. Hall, M.P.; Unch, J.; Binkowski, B.F.; Valley, M.P.; Butler, B.L.; Wood, M.G.; Otto, P.; Zimmerman, K.; Vidugiris, G.; Machleidt, T.; *et al.* Engineered luciferase reporter from a deep sea shrimp utilizing a novel imidazopyrazinone substrate. *ACS Chem. Biol.* **2012**, *7*, 1848–1857.

93. Kittel, C.; Sereinig, S.; Ferko, B.; Stasakova, J.; Romanova, J.; Wolkerstorfer, A.; Katinger, H.; Egorov, A. Rescue of influenza virus expressing GFP from the NS1 reading frame. *Virology* **2004**, *324*, 67–73.

94. Shinya, K.; Fujii, Y.; Ito, H.; Ito, T.; Kawaoka, Y. Characterization of a neuraminidase-deficient influenza A virus as a potential gene delivery vector and a live vaccine. *J. Virol.* **2004**, *78*, 3083–3088.

95. De Felipe, P.; Luke, G.A.; Hughes, L.E.; Gani, D.; Halpin, C.; Ryan, M.D. E unum pluribus: Multiple proteins from a self-processing polyprotein. *Trends Biotechnol.* **2006**, *24*, 68–75.

96. Zhao, D.; Fukuyama, S.; Yamada, S.; Lopes, T.J.; Maemura, T.; Katsura, H.; Ozawa, M.; Watanabe, S.; Neumann, G.; Kawaoka, Y. Molecular determinants of virulence and stability of a reporter-expressing H5N1 influenza A virus. *J. Virol.* **2015**, *89*, 11337–11346.

97. Luker, K.E.; Luker, G.D. Applications of bioluminescence imaging to antiviral research and therapy: Multiple luciferase enzymes and quantitation. *Antiviral Res.* **2008**, *78*, 179–187.

98. Barry, M.A.; May, S.; Weaver, E.A. Imaging luciferase-expressing viruses. *Methods Mol. Biol.* **2012**, *797*, 79–87.

Engineering Hepadnaviruses as Reporter-Expressing Vectors: Recent Progress and Future Perspectives

Weiya Bai, Xiaoxian Cui, Youhua Xie and Jing Liu

Abstract: The *Hepadnaviridae* family of small, enveloped DNA viruses are characterized by a strict host range and hepatocyte tropism. The prototype hepatitis B virus (HBV) is a major human pathogen and constitutes a public health problem, especially in high-incidence areas. Reporter-expressing recombinant viruses are powerful tools in both studies of basic virology and development of antiviral therapeutics. In addition, the highly restricted tropism of HBV for human hepatocytes makes it an ideal tool for hepatocyte-targeting in vivo applications such as liver-specific gene delivery. However, compact genome organization and complex replication mechanisms of hepadnaviruses have made it difficult to engineer replication-competent recombinant viruses that express biologically-relevant cargo genes. This review analyzes difficulties associated with recombinant hepadnavirus vector development, summarizes and compares the progress made in this field both historically and recently, and discusses future perspectives regarding both vector design and application.

Reprinted from *Viruses*. Cite as: Bai, W.; Cui, X.; Xie, Y.; Liu, J. Engineering Hepadnaviruses as Reporter-Expressing Vectors: Recent Progress and Future Perspectives. *Viruses* **2016**, *8*, 125.

1. Introduction

Hepadnaviridae is a family of small, enveloped DNA viruses with notable hepatic tropism, especially in mammals, and transmission is achieved predominantly through parenteral routes [1,2]. The viral genome consists of partially double-stranded, relaxed circular DNA (rcDNA) that is produced through a process involving a reverse transcription step similar to retroviruses [2,3]. These features led to the classification of hepadnaviruses under group VII (dsDNA(RT) or pararetrovirus) in the Baltimore system, along with certain similar DNA viruses infecting plants.

Hepadnaviruses usually have highly-restricted host ranges and have traditionally been classified into two genera based on host specificity [4,5]. Orthohepadnaviruses infect mammals, with members including the prototype hepatitis B virus (HBV) of humans, woolly monkey hepatitis B virus (WMHBV), woodchuck hepatitis virus (WHV), and ground squirrel hepatitis virus (GSHV), *etc.* Avihepadnaviruses infect various domesticated and wild birds, with members including the prototype duck hepatitis B virus (DHBV), as well as heron hepatitis B virus (HHBV), *etc*. In recent years, the advent and advances of next-generation sequencing and other metagenomics technologies have enabled the discovery of new HBV-like viruses that infect hosts previously not known to be affected by hepadnaviruses, such as bats [6] and fish [7]. In addition, analyses of whole genome sequencing data have also led to the discovery of endogenous hepadnaviral sequences in genomes of avian [8–10] and reptilian [11] species, suggesting a family history spanning millions of years. In light of these recent discoveries, hepadnaviruses, including extant and now extinct ones, are obviously far more diverse than previously understood and the taxonomy may well be expanded and modified in the future.

Among extant hepadnaviruses, orthohepadnaviruses productively infect only hepatocytes of the liver, whereas DHBV has been shown to additionally infect certain other cell types of the liver and non-liver organs [3]. Hepato-tropism has been considered the result of tissue-specific distribution of both receptor(s) required for viral entry and transcription factors required for viral expression [12,13]. Accordingly, liver pathologies including hepatitis are major manifestations of symptomatic hepadnaviral infections in both human and animals [1,3]. However, as hepadnavirus infection is neither cytopathic nor cytolytic, hepatitis is generally considered a consequence of the activated host immune response against infected hepatocytes.

HBV is a major human pathogen and constitutes a severe public health problem in high-incidence areas [5,14]. HBV infection of adults is usually asymptomatic or manifests as self-resolving acute hepatitis, while a small percentage of patients fail to clear the virus and become infected for life. Vertical transmission of HBV from infected mothers to neonates typically results in asymptomatic chronic infection accompanied by immunotolerance towards HBV, which would be broken in later life leading to active hepatitis. Chronic HBV infection is associated with higher risks of cirrhosis and hepatocellular carcinoma (HCC) [1]. Although extensive adoption of preventive HBV vaccine has drastically reduced incidence of new infections, the World Health Organization estimated that HBV chronically infects ~240 million people worldwide and causes about 600,000 related deaths annually [14].

Duck/DHBV and woodchuck/WHV have been used as model systems of HBV infections for decades, and have helped significantly in understanding hepadnavirus virology and developing anti-HBV therapeutics [15]. However, chronic DHBV infection is not associated with liver cirrhosis or HCC in ducks, while WHV-related HCC is mechanistically much more homogenous than HBV-associated human HCC [3], underlining the fact that HBV and human or humanized animal systems are required for studying various important aspects of HBV pathogenesis.

Reverse genetic systems were established for HBV as well as DHBV decades ago and have become standard tools in studies of viral functions as well as virus-host interactions. In contrast, owing to certain characteristics in genome organization and life cycle (see next section), development of reporter-expressing virus systems based on hepadnaviruses has met with far less successes than many other virus families. Nevertheless, recombinant hepadnavirus vector systems could serve as powerful tools for both studying fundamental questions in hepadnavirus virology and evaluating clinical interventions for chronic HBV infection, as have been demonstrated by other recombinant virus systems. In addition, the highly restricted host range and hepato-tropism of HBV makes it a uniquely ideal tool for hepatocyte-targeting applications, such as liver-specific therapeutic gene delivery. This review attempts to summarize difficulties associated with recombinant hepadnavirus vector development and progress made historically and recently, and discuss future perspectives in this field regarding both vector design and application.

2. Genome Organization and Life Cycle of Hepadnaviruses

This section will briefly describe aspects of hepadnavirus virology that most pertains to recombinant virus development. For a more detailed discussion of current understandings in this field, readers are referred to comprehensive reviews published recently [3,12,16].

All extant hepadnaviruses have a genome length of ~3.0–3.4 kb and nearly identical, highly-compact genome organization (Figure 1). The orthohepadnavirus genome contains four overlapping ORFs that encompass the entire genome: preC/C ORF encodes the nucleocapsid protein C (core, HBcAg) and an N-terminally extended secreted form called e antigen (HBeAg) using two distinct start codons; P ORF encodes the viral polymerase; preS1/preS2/S ORF encodes three co-terminal forms of viral envelope protein named small (S), middle (M), and large (L) surface antigen (HBsAg) that are translated from three distinct start codons; X ORF encodes the X protein (HBx) that plays multiple functions in the viral life cycle and virus-host interactions. Compared to orthohepadnaviruses, avihepadnaviruses harbor a preS/S ORF, instead of preS1/preS2/S, that only encodes two forms of co-terminal envelope proteins, and avihepadnaviral P and preC/C ORFs overlap each other, lacking a conventional X ORF in between (Figure 1A). In addition to overlapping ORFs, hepadnaviral genomes also contain multiple cis-acting elements essential for various steps of the viral life cycle: promoters (Cp, Sp1, Sp2, and Xp) and enhancers (EnI and EnII) for transcription, epsilon (encapsidation) signal for initiation of reverse transcription and capsid packaging, direct repeats 1 and 2 (DR1 and DR2) for polymerase translocation during genome replication, etc.

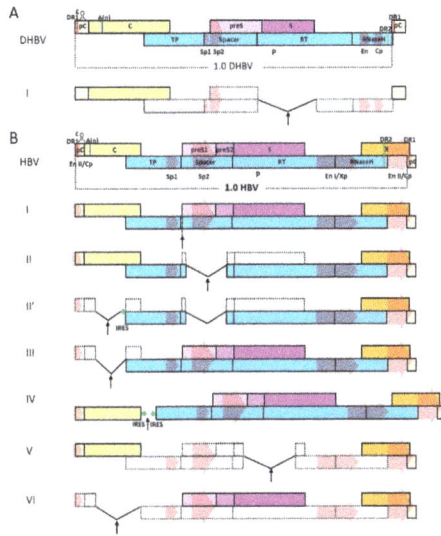

Figure 1. Schematic representation of genome organization of HBV and DHBV and major recombinant (D)HBV vector designs. Terminally-redundant wild-type genomes are shown to reflect the circularity and 1.0 copy of each genome is marked out beginning at the start codon of preC(pC)/C ORF. ORFs are represented by boxes and ORFs destroyed in recombinant vectors are depicted using dotted lines. Promoters (Cp, Sp1, Sp2, and Xp) and enhancers (En, EnI, EnII) are represented by arrows pointing in the direction of transcription. *Cis* elements required for replication and encapsidation are also depicted. ε, encapsidation signal. A(n), polyadenylation signal. DR, direct repeats. (**A**) Wild-type DHBV genome and recombinant DHBV design: I, cargo gene replaces S ORF in-frame and destroys the overlapping P ORF. (**B**) wild-type HBV genome and recombinant HBV designs: I, cargo gene is inserted in-frame with P ORF within spacer region, upstream of preS1 ORF. P ORF may or may not be terminated depending on cargo. II, cargo gene is inserted in a deletion in polymerase spacer in-frame with preS1/preS2 ORF, without terminating the overlapping P ORF. II', derivative of II. The cargo gene replaces the central part of prematurely-terminated C ORF, followed by IRES upstream of P ORF with a maximized deletion in the spacer region that retains polymerase activity but destroys preS1/preS2 ORF. III, the cargo gene replaces central part of C ORF and may or may not be expressly fused to the remaining N-terminal of preC/C, depending on the design. IV, IRES units are introduced to separate de-overlapped C and P ORFs. No viral ORF is obliterated. V, the cargo gene replaces S ORF in-frame and destroys the overlapping P ORF. VI, the cargo gene replaces the central and C-terminal of C ORF, destroying the overlapping P ORF. Arrows indicate insertion sites. Vector designs with obliterated C and/or P ORFs require *trans*-complementation of the obliterated proteins for genome replication. Vector designs with obliterated envelope ORFs require *trans*-complementation of envelope proteins for production of progeny virus.

107

Mature, infectious hepadnavirus virions contain the rcDNA genome and enter hepatocytes through interactions between large surface antigens and specific cellular receptor(s). Once inside, the viral capsid releases its genome contents into nucleus, where rcDNA is converted into covalently-closed circular DNA (cccDNA) that then serves as a transcription template for all viral RNA species. Both core and viral polymerase are translated from the same transcript, termed pregenomic RNA (pgRNA), which also serves as a template for genome replication. Most notably, for wild-type viruses, nascent polymerase co-translationally binds to the epsilon signal at the 5′ terminal of its own pgRNA *in cis*, and initiates reverse transcription using an internal, conserved tyrosine residue as a primer for negative-strand DNA synthesis. A vital translocation step then follows, with the polymerase-primer "jumping" from the 5′ terminal to the 3′ terminal of pgRNA, which involves the DR1/DR2 direct repeat sequences at both termini. Packaging of the pgRNA/polymerase complex by viral core proteins is essential for productive genome replication and eventual formation of progeny rcDNA. Capsids containing newly-formed rcDNA then obtain host-derived membranes containing viral envelope proteins to produce progeny viruses and exit through budding at the ER. Alternatively, capsids are rerouted to the nucleus and the released rcDNA are converted into cccDNA to increase the level of transcription templates.

3. Hepadnavirus-Specific Difficulties for the Design and Development of Viral Vectors

Characteristic peculiarities in the genome organization and life cycle of hepadnaviruses pose specific difficulties and problems for efforts aimed at engineering recombinant hepadnavirus vectors.

First of all, genome sizes of wild-type viruses very likely represent the maximum limit that could be tolerated by the viral replication and packaging mechanisms. This has been demonstrated by *in vitro* experiments using artificially-created longer-than-wild-type genomes [17], and is possibly also reflected in a general lack of natural over-size insertion mutants reported in the literature. The viral capsid can also be expected to impose an intrinsic restriction on the size of genome that could be packaged within, along with the associated polymerase. Furthermore, studies of hepadnaviral reverse transcription mechanisms have revealed that efficiency of proper polymerase translocation giving rise to correct replication products deteriorates when the distance between 5′ and 3′ DRs exceeds the wild-type genome length [17,18].

Secondly, although initiation of reverse transcription of pgRNA could be mediated *in trans* by polymerase translated from other mRNA, the efficiency has been shown to be far inferior to that initiated *in cis* by polymerase translated from pgRNA [19]. Such tight coupling of polymerase translation with genome replication dictates that recombinant genome needs to encode a functional polymerase for it to be able to replicate with a wild-type level of efficiency and subsequently produce applicable quantities of recombinant virions. Given that coding sequences of wild-type polymerase make up ~75% of the hepadnavirus genome, very limited space is available for engineering purposes if polymerase ORF is to be left untouched, heavily restricting cargo capacity. On the other hand, obliterating polymerase ORF would markedly free up more space for cargo sequences, but at the expense of replication competence.

Thirdly, orthohepadnaviruses encode an X protein (HBx) that has been shown to be capable of playing a myriad of functions in virus-host interactions *in vitro* and/or *in vivo* [20]. Most notably, HBx is apparently a key stimulator of both cccDNA-directed transcription and, indirectly via the regulation of pgRNA synthesis, genome replication [21]. HBx does not appear to be packaged into HBV virions and experiments *in vitro* have demonstrated that HBV mutants lacking functional HBx ORF could not establish or sustain productive infection in susceptible cells [22,23]. Therefore, for a recombinant HBV to be optimally active in infected cells, intact HBx ORF is most likely required. Fortunately, since the entire length of HBx is overlapping with other vital elements of HBV genome (Figure 1), HBx ORF can be easily left untouched in recombinant HBV vector design.

Intrinsic peculiarities of hepadnavirus virology not only impose the above challenges for the design and engineering of recombinant virus vectors, but they also make testing of these vectors technically difficult. For example, the envelopment of hepadnaviral capsids appears to be a highly-specific process that, in distinct contrast to some other enveloped viruses like poxviruses and lentiviruses, does not normally incorporate membrane proteins other than the corresponding hepadnaviral envelope proteins. This makes it highly difficult to create pseudotyped hepadnaviruses that could be tested on cells of non-liver origin. Consequently, testing of recombinant hepadnavirus vectors have to rely on hepatocytes that support infection by wild type hepadnavirus.

For the prototype avihepadnavirus DHBV, effective infection *in vivo* of ducklings and *in vitro* of either primary duck hepatocytes or the chicken hepatoma-derived LMH cell line have been commonly used [3,15]. For HBV, however, stably reproducible infection *in vivo* has only been demonstrated for certain higher primates, especially chimpanzees [15,24], which are economically and ethically prohibitive for routine experiments. Alternatively, chimeric immunodeficient mice harboring human primary hepatocytes can be used to simulate HBV infection *in vivo* [25,26], but in the absence of normal immune functions, virus-host interactions are not fully reflected. *In vitro*, primary human and tree shrew hepatocytes [27], hepatoma-derived HepaRG cells [28] and, recently, liver cell lines stably transfected with the HBV receptor NTCP (Na$^+$-taurocholate cotransporting polypeptide) [29], have been used. Although these systems support HBV infection with varying efficiencies, so far only primary infection can be achieved, without demonstrable secondary infection by progeny viruses. For WHV, although woodchucks can be routinely infected using viruses prepared from naturally- or experimentally-infected animals, efficient production of infectious virus *in vitro* has yet to be achieved.

4. Hepadnavirus-Derived Viral Vectors

It was more than 25 years ago when the first work dealing with recombinant hepadnaviruses was published [30]. Since then, attempts at engineering usable hepadnavirus vectors have been made repeatedly, with varying degrees of success. Due to its obvious clinical relevance, most efforts have targeted HBV. This section will summarize historical and recent progresses by describing and comparing vector designs validated by sufficient experimental data (Table 1).

As discussed in the previous section, reverse transcription of hepadnaviral pgRNA is far more efficient if it is initiated *in cis* by polymerase translated from the same pgRNA. Most researchers, therefore, have opted to retain polymerase expression in their vector design in the hope of achieving higher replicative competence. Recombinant hepadnavirus designs are, therefore, divided into two categories according to whether functional polymerase is encoded by the vector or needs to be complemented *in trans*, and discussed respectively.

4.1. Vectors that Encode Functional Polymerase

Characterization of hepadnavirus polymerase functions revealed early on that it is made up of four structurally and functionally distinct domains: an N-terminal TP (terminal priming) domain that is responsible for catalyzing the priming of negative strand DNA synthesis after binding to pgRNA epsilon *in cis*, a spacer or tether domain that joins the neighboring domains, an RT (reverse transcription) domain that synthesizes both negative and positive strands of the progeny viral genome, and an RH (RNaseH) domain that degrades pgRNA after negative-strand DNA is synthesized [3]. Among the polymerase domains, the spacer is the least conserved and predicted to be the least structurally ordered. Most likely, it only functions as a physical linker or hinge.

Table 1. Listing and comparison of recombinant hepadnavirus vector design.

Publication	Virus Base	Required trans-Complementation [1]	Cargo Insertion Site and Insertion Strategy	Tested Cargo(s) (Length)	Replication Evidence	Replication Efficiency	Virion Formation Evidence [2]	Virion Formation Efficiency	Virion Infectivity Evidence
Recombinant vectors that express functional polymerase									
Chang et al., 1990 [30]	DHBV	preS/S	Cargo ORF inserted in-frame in Pol spacer between preS and S with own ATG but no stop codon	Protein A (369)	EPA, Southern blot	Comparable	N.D.	N.D.	N.D.
Chaisomchit et al., 1997 [31]	HBV	None	Cargo ORF inserted in-frame in Pol spacer between Sp1 and preS1 start codon with own start but no stop codon	HIV Tat (267)	EPA	Severely reduced	N.D.	Severely reduced	N.D.
Wang et al., 2002 [32]	HBV	C	Cargo ORF replaced C ORF between ε/A(n) and Pol start codon, fused to remaining N-terminal of C	Flag (48)	Southern blot	Severely reduced	Capture by S antibodies followed by Southern blot	N.D.	PHH
Yoo et al., 2002 [33]	HBV	C	Cargo ORF replaced C ORF between ε/A(n) and Pol start codon	GFP	EPA, Southern blot	Severely reduced	Density gradient ultracentrifuge followed by Southern blot and PCR	N.D.	PHH, HepG2 [3]
Deng et al., 2009 [34] Wang et al., 2014 [35]	HBV	C	Cargo ORF replaced C ORF between ε/A(n) and Pol start codon, fused to remaining N-terminal of C. Kozak sequences of Pol were optimized.	Peptide (180)	Southern blot	Increased	Density gradient ultracentrifuge and capture by S antibodies followed by Southern blot	N.D.	PTH
Wang et al., 2013 [36]	HBV	None	Cargo ORF inserted between separated C and Pol ORF with intervening short IRES	BsdR (399) GFP (720)	EPA, Southern blot	Comparable to severely reduced depending on cargo	Density gradient ultracentrifuge followed by Southern blot	Reduced	PTH
Hong et al., 2013 [37]	HBV	preS1/preS2/S	Cargo sequences replaced 384 bp of Pol spacer (preS1/preS2) in-frame with preS1, without terminating Pol ORF. Start codon of preS1 mutated.	HIV Tat (207) DsRed (678) shRNA cassette (294)	Southern blot	Comparable to reduced depending on cargo	Density gradient ultracentrifuge followed by Southern blot	Comparable to severely reduced depending on cargo	HepaRG
Bai et al., (submitted)	HBV	C preS1/preS2/S	Cargo sequences replaced C ORF between ε/A(n) and Pol start codon. Short IRES placed before Pol start codon and 384 bp of Pol spacer (preS1/preS2), same as above, deleted. C ORF prematurely terminated.	ZeoR (375) NanoLuc (522/606) [4] DsRed (678) GFP (747) shRNA cassette (294)	Southern blot	Comparable to reduced depending on cargo	Capture by preS1 mAb followed by Southern blot	Comparable to reduced depending on cargo	PTH

Table 1. *Cont.*

Recombinant vectors that do not express functional polymerase

Publication	Virus Base	Required *trans*-Complementation [1]	Cargo Insertion Site and Insertion Strategy	Tested Cargo(s) (Length)	Replication Evidence	Replication Efficiency	Virion Formation Evidence [2]	Virion Formation Efficiency	Virion Infectivity Evidence
Chaisomchit *et al.*, 1997 [31]	HBV	Pol	Cargo ORF inserted in-frame in Pol spacer between Sp1 and preS1 start codon with own start and stop codon	ZeoR (372)	EPA	Severely reduced	N.D.	N.D.	N.D.
Protzer *et al.*, 1999 [17]	DHBV	Pol preS/S	Cargo ORF replaced 558 bp of S ORF in-frame	GFP (733) Duck IFN (591)	N.D.	N.D.	Density gradient ultracentrifuge followed by dot blot	N.D.	PDH
Ibid.	HBV	Pol preS1/preS2/S	Cargo ORF replaced 939 bp of S ORF in-frame	GFP (733)	N.D.	N.D.	*Ibid.*	N.D.	PHH
Untergasser *et al.*, 2004 [38]	HBV	All HBV ORFs	Cargo ORF replaced 939 nt of S ORF in-frame. All other HBV ORFs are prematurely terminated by mutation. Some constructs replaced ~311 nt of SP2 with exogenous promoters 366 nt or 575 nt long.	GFP (733) RLuc (942)	Southern blot	N.D.	*Ibid.*	Comparable	PHH
Liu *et al.* 2013 [39]	HBV	All HBV ORFs	Cargo ORF replaced S ORF in-frame. All other HBV ORFs are prematurely terminated or nulled by mutation.	GFP RFP	Southern blot	Reduced	N.D.	N.D.	HepaRG
Nishitsuji *et al.*, 2015 [40]	HBV	Pol C	Cargo ORF with own start and stop codons replaced 562 nt of C ORF downstream of ε/A(n) and the N-terminal of P ORF	NanoLuc (513)	N.D.	N.D.	Density gradient ultracentrifuge followed by Southern blot	N.D.	PXB NTCP cell lines

[1] core and polymerase are absolutely required for genome replication; envelope proteins (DHBV preS/S and HBV preS1/preS2/S) are only required for virion formation. [2] due to secretion of non-enveloped capsids by transfected cells [35–37,40], detection of viral DNA in transfection supernatants without virion-specific separation or enrichment step(s) is considered only evidence of replication. [3] despite early controversies, human hepatoma cells lines, such as HepG2 and Huh-7, are currently generally accepted as not susceptible to HBV infection. [4] two forms, one intracellular and one secreted, were tested. ε, signal on pre-genomic RNA. A(n), polyadenylation signal. EPA, endogenous polymerase activity assay, which measures polymerase-catalyzed incorporation of isotope-labelled nucleotides into progeny genomes within viral capsids. PHH, PTH, and PDH refer to primary human, tupaia, and duck hepatocytes, respectively. PXB, hepatocytes prepared from chimeric mouse harboring human primary hepatocytes. N.D., not done or not shown.

4.1.1. Vectors that Use Polymerase Spacer Region for Cargo Insertion

While examining the ability of DHBV polymerase to tolerate mutations and insertions at various locations, Chang *et al.* found that insertion of coding sequences for bacterial protein A (369 nt) without termination codon in-frame into the spacer region, between preS and S in the overlapping ORF, did not abolish polymerase activity and self-sufficient replication of the resultant recombinant genome [30]. The cargo ORF carried its own start codon, making it theoretically possible for preS/S mRNA and probably also pgRNA to translate protein A fused to the C-terminal half of polymerase, in addition to full-length polymerase with protein A sequences embedded in the spacer region, but this former type of fusion protein was apparently not translated to detectable levels in transfected cells. Since the preS/S ORF is interrupted by cargo insertion, the recombinant genome would require *trans*-complemented envelope proteins to form mature progeny virions. Although this early work was not a study devoted to recombinant HBV and recombinant virion production was not tested, the results showed that polymerase spacer is a viable cargo insertion site for engineering self-replicating recombinant hepadnavirus.

The first comprehensive study of recombinant hepadnavirus was published by Chaisomchit *et al.* in 1997 [31] with a design scheme similar to the work by Chang *et al.* The authors proposed recombinant HBV as "more efficient means for gene delivery to the liver" compared to retroviral and adenoviral vectors. They first tested the possibility by inserting coding sequences for HIV Tat protein (267 nt) without a stop codon in-frame into the polymerase spacer, between the Sp1 promoter and preS1 start codon in the overlapping ORF (Figure 1B, design I). This insertion site was located more upstream and closer to the TP domain compared to the protein A insertion site in the previous work. HBV polymerase with Tat sequences inserted in spacer was functional and replicated recombinant genome at efficiencies that were 1.5%–4% of wild-type HBV. As only P ORF is affected, the recombinant genome does not require *trans*-complementation of viral structural proteins and enveloped recombinant virions could be detected in transfection supernatants, but only at very low levels. Infectivity of recombinant virions was not tested. Tat-induced transcription activation of promoters could be detected in transfected cells, and it was shown that Tat fused to C-terminal part of polymerase could be translated from Sp1 transcribed mRNA. This pioneering study demonstrated with compelling evidence that polymerase spacers could tolerate fairly long insertions at the cost of replication efficiency.

Both of these two early studies used similar design that inserted cargo genes into polymerase spacer of wild-type genomes. Our lab also made an attempt to harness the polymerase spacer as an insertion site to develop recombinant HBV vectors for hepatocyte-specific delivery of reporter and functional genes. However, instead of wild-type HBV, we based our design on a clinically-isolated, highly-replicative

HBV mutant that harbors a large in-frame deletion of 207 nt in the polymerase spacer [37]. The mutant does not encode functional envelope proteins due to the partial loss of preS1 ORF and non-sense mutations in S ORF, and the polymerase contains a 69 amino acid deletion in the spacer region. However, the mutant replicates more efficiently than the wild-type, and when *trans*-complemented with functional envelope proteins, produces mature enveloped progeny viruses also more efficiently than the wild-type. We inserted terminated ORF encoding the N-terminal activation domain of HIV Tat (207 nt) into the deletion in the polymerase spacer, but unlike the previous two studies, the inserted ORF was in-frame with preS1 so as not to be expressed fused to polymerase, and multiple synonymous mutations were used to avoid terminating the overlapping P ORF (Figure 1B, design II). The recombinant HBV replicated and produced progeny virions with efficiencies comparable to the wild-type, and Tat expression driven by Sp1 could be detected using reporter assay. The polymerase spacer deletion in the mutant was then maximized to increase cargo capacity and we obtained a vector with a 384 nt in-frame deletion that replicated as efficiently as the parental mutant. The vector could tolerate insertions of up to 675 nt and still retain wild-type-level replicative competence, as long as P ORF is not interrupted. We demonstrated that recombinant HBV carrying synonymously-mutated sequences encoding DsRed infected PTH with high efficiency. Moreover, we showed that the vector could carry and express functional RNA in infected PTH, which was the first report of recombinant HBV delivering non-protein cargo. Naturally, the major limitation of this design is that cargo sequences must not introduce stop codons in P ORF, which is often difficult and sometimes impossible.

4.1.2. Vectors that Use Core Region for Cargo Insertion

Inserting cargo sequences into polymerase spacer while keeping the resultant polymerase active is inherently difficult. Consequently, other published designs in this category chose to avoid changing P ORF. On the hepadnaviral genome, all *cis*-acting sequence elements required for genome replication and packaging are clustered closely together between the C-terminal part of polymerase and N-terminal of core, with the remaining part of polymerase taking up almost all of the rest of genome space (Figure 1). The only segment of the genome that is apparently replaceable without affecting P ORF or *cis* elements is the middle part of C ORF (~350 nt) between epsilon packaging/polyadenylation signals and the start codon of P ORF.

In an attempt to test recombinant HBV as a potential liver-targeting delivery vector, Wang *et al.* examined a series of HBV deletion mutants, one of which allowed the replacement of the central part of C ORF with a short terminated ORF encoding Flag tag (48 nt) that was inserted in-frame with the preceding N-terminal of

C ORF (Figure 1B, design III) [32]. The recombinant genome could replicate at levels much lower than wild type when *trans*-complemented with core, and produced recombinant virions that infected PHH with low efficiency. The Flag tag is expected to be expressed as a C-terminal fusion to the remaining N-terminal of e and c antigens. However, extending the Flag tag with C-terminal addition of full-length or truncated GFP sequences resulted in a loss of replicative competence. In a similar study roughly coinciding with this work, Yoo *et al.* replaced the same part of C ORF with GFP-encoding sequences (~720 nt), without specifying whether the insertion was in-frame with C or whether it carried its own start and stop codons. Replication efficiency of this recombinant HBV was about 3% of wild-type HBV in the presence of *trans*-complemented core expression [33]. However, no marked increase in genome size was observed in Southern blot as should be expected. Enveloped virions were demonstrated in co-transfection supernatants at significantly reduced levels compared to the wild-type. Recombinant virions infected primary human hepatocytes and questionably, also HepG2, and resulted in detectable fluorescence in infected cells.

These two studies demonstrated that the central part of C ORF is also a viable choice as cargo insertion site, but the reported insertions severely affected replication competence. In wild-type HBV, both core and polymerase are translated from pgRNA. Multiple mechanisms have been shown to be involved in translation initiation of the downstream P ORF, including leaky scanning and ribosome re-initiation [41], and sequences upstream of polymerase start codon significantly affect the translation efficiency of polymerase and consequently, replication efficiency of the genome. In light of this, inefficient expression of polymerase might be responsible, at least partially, for the low replicative competence observed in the two studies discussed above. In a study aimed at liver-targeting delivery of immunogenic peptides, Deng *et al.* attempted to alleviate this problem by optimizing sequences surrounding HBV polymerase start codon according to Kozak's rules [34], while replacing the upstream central part of C ORF with sequences encoding a polyepitope peptide (180 nt) in a fashion similar to the design of Wang *et al.* above. The recombinant HBV apparently replicated better than wild-type HBV in the presence of a *trans*-complemented core, and in a follow-up study, mature virions infectious for primary tupaia hepatocytes (PTH) were demonstrated to be produced at levels lower than the wild-type [35]. Expression of the cargo peptide, however, was only shown using plasmid or recombinant adenovirus as delivery vectors [34,35].

An alternative approach to enhancing polymerase translation is to make it independent of upstream sequences by inserting an internal ribosome entry site (IRES) before its start codon. In the work by Wang *et al.* [35], the overlapping part of C/P ORFs was duplicated to create non-overlapping C and P ORFs. ORF encoding blastidicin resistance protein (BsdR, 399 nt) or GFP (720 nt) with a termination codon at the 3′ end was then inserted between C and P ORFs, and short IRES units were used to separate the three (Figure 1B, design IV). Since all viral ORFs are still present, such a design requires no *trans*-complementation of wild-type HBV proteins. Recombinant HBV harboring the shorter BsdR replicated and produced progeny virions with efficiencies generally comparable to wild-type HBV, but insertion of the longer GFP resulted in severely reduced replication and nearly undetectable progeny virus secretion. As duplication of part of C ORF increases the genome size beyond that of the wild-type, even without cargo insertion, inability to harbor long insertions is not surprising. Infectivity of the recombinant virus was then demonstrated using HepaRG cells. An analogous strategy has been used in our lab (submitted) to improve the previously-described vector with a maximized deletion in polymerase spacer [37]. A short artificial IRES was placed before the start codon of P ORF containing the maximized deletion in spacer, and cargo sequences replaced the central part of C ORF (Figure 1B, design II′). Unlike the above studies, cargo protein genes carried own start and stop codons and upstream remaining C ORF was terminated to avoid expressing cargo genes as fusions to the remaining N-terminal of core. Recombinant HBV harboring fluorescent and bioluminescent reporters of 375–747 nt replicated, in the presence of *trans*-complemented core, with varying efficiencies that were mostly comparable to the wild-type, depending on the length and type of the insertion. Enveloped progeny viruses were obtained by providing wild-type core and envelope proteins *in trans* to recombinant genomes and infectivity of recombinant viruses harboring protein or RNA genes was demonstrated using PTH. Compared to its parental vector, the switch to C ORF for insertion allowed freer choice of cargo sequences, while the inheritance of replication-enhancing deletion in polymerase ORF and isolation of polymerase translation through introduction of IRES provided acceptable replication efficiencies for most of the tested cargos.

4.2. Vectors that Do Not Encode Functional Polymerase

Theoretically, since all *cis*-acting sequence elements required for genome replication and packaging are located between the C-terminal part of the polymerase and N-terminal of the core (Figure 1), cargo sequences can be inserted anywhere, or replace any segment(s), on the rest of hepadnavirus genome, if polymerase expression does not need to be retained by the recombinant vector. Such vectors would have maximal capacity for harboring cargo sequences. However, all except one of the few reports on vectors belonging to this category chose to use the S or P ORF for cargo insertion, fairly distant from those *cis*-acting elements, which might be beneficial by minimizing interference of their functions by cargo sequences.

In their pioneering work on recombinant HBV vectors, Chaisomchit *et al.* also tested a non-replicative version of their design by replacing the un-terminated in-frame Tat insertion in polymerase spacer with a terminated in-frame insertion of ZeoR (372 nt)

(Figure 1B, design I) [31]. Self-sufficient replication of the recombinant genome was expectedly obliterated, and even with *trans*-complemented polymerase, replication was only 1.5%–3% of wild-type HBV, which is similar to the replicative vector with non-terminated Tat insertion. Virion formation and infection were not examined.

Later, in the hope of achieving delivery of therapeutic genes to hepatocytes, Protzer *et al.* developed DHBV and HBV vectors that have most of the S ORF replaced with GFP or duck interferon coding sequences in-frame (Figure 1A, design I and Figure 1B, design V) [17]. The overlapping P ORF is prematurely terminated by the cargo sequences. Recombinant viruses infectious for primary duck or human hepatocytes (PDH and PHH) could be obtained from transfection supernatants, if *trans*-complemented with both polymerase and envelope proteins. Notably, the authors showed that a recombinant DHBV-expressing duck interferon was able to inhibit co-infecting wild-type DHBV in PDH infection assay, which constituted the first demonstration of the therapeutic value of recombinant hepadnavirus. Following up on this work, Untergasser *et al.* attempted to improve the vector from the safety perspective of potential gene therapy applications by prematurely terminating all viral ORFs in the vector so that the recombinant HBV would express only the cargo gene product [38]. Since there was no significant change to the vector design, recombinant HBV harboring GFP or renilla luciferase (942 nt) were obtained when *trans*-complemented wild-type HBV proteins, and the virions infected PHH and expressed the cargo genes. In addition, stronger, exogenous promoters were tested in replacement of Sp2 in the hope of enhancing cargo gene expression, but although recombinant viruses infectious for PHH were produced, enhanced expression was not demonstrated using infection assay. Similar design was used by Liu *et al.* in a later study also aimed at liver-targeting delivery of therapeutic genes [39]. The authors replaced S ORF in-frame with sequences encoding GFP or RFP, and observed reduced replication compared to wild type HBV in the presence of *trans*-complemented HBV proteins. GFP expression by recombinant virions in infected HepaRG was demonstrated.

Recently, Nishitsuji *et al.* [40] reported a system allowing quantitative detection of HBV infection *in vitro* based on a polymerase-negative design of recombinant HBV that uses C ORF for cargo insertion, instead of polymerase or S ORF, as done in the previous designs in this category. However, in contrast to vectors that use C ORF as insertion site in the first category, they replaced the entire C ORF downstream of epsilon packaging/polyadenylation signals (562 nt), including the part overlapping with the N-terminal of P ORF, thus destroying polymerase expression (Figure 1B, design VI). Recombinant HBV harboring the NanoLuc reporter (513 nt) could be produced when *trans*-complementation of core and polymerase were provided and infection of human hepatocytes and NTCP-transfected hepatoma cell lines was demonstrated.

4.3. Comparison of Vector Designs from the Perspective of Potential Applications

Potential *in vitro* applications of recombinant hepadnavirus vectors include the study of fundamentals of hepadnavirus life cycle, most notably entry into target cells and formation of initial cccDNA from the incoming virus genome. At the same time, reporter-expressing recombinant hepadnaviruses would facilitate the development of drug screening and evaluating systems. *In vivo*, recombinant hepadnaviruses could form the basis of hepatocyte-targeting therapeutic interventions for liver-afflicting conditions, including HBV-related and non-HBV-related hepatitis, cirrhosis, and HCC, as well as life-threatening diseases not directly involving liver, such as type I diabetes. Reporter-expressing recombinant HBV would be useful for characterizing hepatotropism, sustenance, and biosafety of such therapeutics before *in vivo* applications could be attempted.

Different applications of recombinant virus vectors sometimes have different requirements regarding the vectors, in addition to functions of the cargo gene. For instance, for therapeutic interventions targeting chronic hepatitis B patients using recombinant HBV expressing interferon or HBV-targeting siRNA precursors, continuous high activity of a recombinant virus is only desirable while wild-type HBV is active in co-infected hepatocytes, but not thereafter. In contrast, sustained activity of recombinant HBV capable of surviving hepatocyte propagation and turnover, regardless of the presence or absence of wild-type HBV, is desired when, for example, using insulin-expressing recombinant HBV for treating type I diabetes.

Features of the recombinant hepadnavirus vector designs reviewed above are summarized in Table 2 with a focus on their capabilities of self-sufficient replication and progeny virus production. Recombinant hepadnaviruses encoding both functional core and polymerase proteins [31,36,37] are expected to be able to replicate their genomes and expand intranuclear cccDNA pools resulting in sustained high expression of cargo genes in infected cells, irrespective of wild-type co-infection. Consequently, such vectors would persist in infected hepatocytes and are, therefore, ideal for *in vivo* applications targeting non-HBV-related life-long diseases, and for studies where highly-sensitive detection of infection is desired. Progeny recombinant viruses will also be produced if core- and polymerase-producing recombinant viruses also encode functional envelope proteins [31,36]. This would theoretically allow secondary infection by progeny recombinant viruses of surrounding susceptible cells, further enhancing the level and continuity of cargo gene expression.

Loss of functional core and/or polymerase expression makes the majority of recombinant vectors incapable of replication, and consequently, progeny production, without co-infecting wild-type virus (Table 2). These vectors would likely persist with low activity and no expansion in mono-infected cells, but would start replication and progeny virus production, giving rise to enhanced cargo gene expression and expanded infection by recombinant virus, if the infected cells are to be

118

super-infected with wild-type virus. Activity of recombinant virus will recede along with wild-type virus, if and when the latter is under control. Such activation by the wild-type makes these vectors ideal for therapeutic applications targeting infection by wild-type viruses.

Table 2. Comparison of recombinant hepadnavirus vector designs.

Representative Publication(s)	Obliterated ORF(s)	cccDNA Pool Expansion [1]	Progeny Virus Production [2]
Chaisomchit et al., 1997 [31] Wang et al., 2013 [36]	None	Self-sufficient	Self-sufficient
Hong et al., 2013 [37]	S	Self-sufficient	Requires help
Wang et al., 2002 [32] Yoo et al., 2002 [33] Deng et al., 2009 [34] Wang et al., 2014 [35]	C	Requires help	Requires help
Bai et al. (submitted)	C/S	Requires help	Requires help
Chaisomchit et al., 1997 [31]	P	Requires help	Requires help
Protzer et al., 1999 [17] Chang et al., 1990 [30]	P/S	Requires help	Requires help
Nishitsuji et al., 2015 [40]	P/C	Requires help	Requires help
Untergasser et al., 2004 [38] Liu et al., 2013 [39]	All	Requires help	Requires help

[1] vectors with functional C and P ORFs are expected to be able to replicate self-sufficiently and form additional cccDNA in infected cells which would, in turn, result in higher cargo gene expression. [2] vectors retaining all functional viral ORFs are expected to be able to replicate self-sufficiently and produce infectious progeny recombinant viruses, which would in turn result in infection of additional susceptible cells. Requires help: *trans*-complementation of obliterated proteins by a co-infecting wild-type virus is required for indicated functions.

5. Future Perspectives on Recombinant Hepadnavirus Vector Design

Minimal requirements for hepadnavirus genome replications include *cis* elements on pgRNA (Figure 1), functional core *in cis* or *in trans*, and functional polymerase preferably expressed *in cis*. Designs that do not retain functional polymerase on the recombinant genome theoretically would allow much freer choice of cargo length and insertion site. However, published studies in this category have not demonstrated such potential, and are generally limited to using fairly short cargo genes to replace viral ORF (Table 1). It is possible that future work on such vectors will liberate additional viral genome space and sites for cargo insertions, and enable the design of recombinant viruses harboring larger genes or multiple small genes to provide better or more complex functionalities.

Recombinant hepadnavirus designs that retain functional polymerase have inherently limited capacity for cargo insertions. Even with non-inactivating deletions in polymerase spacer [37], 600–700 nt is very likely the maximum cargo length that is practical for HBV. The capacity will be further reduced if functional core is also to be retained [31,36]. However, the relative ease of recombinant virus production and expected higher expression of cargo genes in infected cells compared to polymerase-negative designs make such designs a favored choice in most cases. Finding or engineering small-sized cargo genes with experimentally- or clinically-important functions will be the key to making these vector designs more relevant to the field.

6. Future Perspectives on Applications of Recombinant Hepadnavirus Vector

Work on developing recombinant hepadnavirus vectors has been going on for two decades and there have been more than ten designs with varying degrees of similarity and innovation (Table 1). Some of the studies characterized the cargo capacity, replication, progeny virus production and infectivity in fairly great detail and convincingly demonstrated the usability of the corresponding vector, at least when used with the tested cargo. A couple of studies went further and showed the huge potential of such vectors for possible therapeutic applications *in vivo* [17,37].

Admittedly, compared to other more commonly used recombinant viral vector systems, such as retrovirus/lentivirus, poxvirus, adenovirus, and adeno-associated virus vectors, progress in the development and application of recombinant hepadnaviruses both *in vivo* and *in vitro* has been much slower and less fruitful. Nevertheless, lack of strict hepatotropism makes the other recombinant virus vectors intrinsically inferior to hepadnavirus vectors for hepatocyte-targeting applications. Moreover, for potential applications in chronic HBV-infected patients, recombinant HBV is minimally affected by vector-targeting immune reactions, in distinct contrast to other virus vectors. Needless to say, for studying basic virology of hepadnaviruses, only recombinant hepadnaviruses are irreplaceable tools for obtaining meaningful results. So far, however, studies that actually take advantage of the reported recombinant hepadnavirus vector systems to address unanswered virological questions and unmet clinical demands have been rare and mostly restricted to labs that originally developed the systems. The reasons behind such apparent lack of application are manifold.

Firstly, large-scale preparation of recombinant viruses is a very cumbersome process, especially for poorly replicative constructs. The use of stronger promoter, such as CMV promoter, instead of native Cp promoter to drive pgRNA transcription could enhance, to a limited degree, the production of progeny viruses. In future, identifying and counteracting cellular mechanisms restricting hepadnaviral replication might give rise to novel production systems. The necessity of providing *trans*-complementation of viral proteins for most designs further complicates production and may also incur the risk of wild-type contamination through homologous recombination [38], the mitigation of which would require extensive synonymous mutations. Engineering stably-transfected cell lines that continuously produce recombinant viruses at acceptable levels in the supernatants will significantly boost adoption and application of recombinant hepadnavirus vectors.

Secondly, infection systems for HBV used to be tedious and costly to establish and use, especially for *in vivo* applications. The identification of NTCP receptor for HBV and demonstration that NTCP-overexpressing hepatoma cell lines supporting wild-type and recombinant HBV infection no doubt represent a significant advance in this respect [29]. With the advent and general availability of easier-to-handle *in vitro* HBV infection systems, like HepG2/NTCP [29], interest in and applications of viable recombinant hepadnavirus vectors can be expected to grow. It is also possible that with further understanding of HBV infection mechanisms, which could be significantly promoted by the use of reporter-expressing recombinant HBV, transgenic mice supporting HBV infection might be eventually obtained. In addition, advances in WHV reverse genetics might also enable the development of recombinant WHV vectors usable in this important model of HBV.

Thirdly, most reported recombinant hepadnavirus vectors have been demonstrated using only one or two cargo genes (Table 1) and only our lab's work has used non-protein cargos [37]. Such limited demonstration of recombinant hepadnaviruses' capability does not help in attracting potentially interested researchers. There are, of course, inherent restrictions on possible choices of cargo genes (see previous sections), but extensive and in-depth characterization of existing vectors for their ability to deliver commonly used, as well as novel cargo sequences with relevant functions, will surely encourage and facilitate wider applications of recombinant hepadnaviruses in both laboratory and clinical settings.

Last, but not least, integration of hepadnavirus sequences into hepatocyte genomes is often detected in chronically-infected subjects and has been linked, at least in some studies, to HCC development [1–3]. Unlike retroviruses and lentiviruses, integration of viral genome into a host chromosome is not a necessary step in hepadnavirus life cycle, and probably represents an opportunistic event during the long-term presence and activity of hepadnaviruses in hepatocytes. Unfortunately, fairly limited information is available on the integration mechanisms, as well as preferred integration sites or lack thereof. Similarly, a potential link to HCC has also been proposed for HBx protein without detailed understanding of the underlying molecular details [1,20]. These constitute a major safety concern for any potential *in vivo* applications of recombinant HBV for subjects not chronically infected with HBV. Further understanding of hepadnavirus integration and HBx functions, which could be aided by studies using recombinant hepadnaviruses *in vivo* and *in vitro*, might eventually enable more realistic evaluation of the associated risks and allow recombinant HBV-mediated gene delivery to be applicable to more patients.

Acknowledgments: This work was supported by the National Key Project for Infectious Diseases of China (2012ZX10002-006, 2012ZX10004-503, 2012ZX10002012-003), National Basic Research Program of China (2012CB519002), National High-Tech Program of China (2012AA02A407), Natural Science Foundation of China (31071143, 31170148), Shanghai Municipal R&D Program (11DZ2291900, GWDTR201216), and MingDao Project of Fudan University.

Conflicts of Interest: The authors declare no conflict of interest.

References

1. Liang, T.J. Hepatitis B: The virus and disease. *Hepatology* **2009**, *49*, S13–S21.
2. Seeger, C.; Mason, W.S. Hepatitis B virus biology. *Microbiol. Mol. Biol. Rev.* **2000**, *64*, 51–68.
3. Seeger, C.; Mason, W.S. Molecular biology of hepatitis B virus infection. *Virology* **2015**, *479*, 672–686.
4. Littlejohn, M.; Locarnini, S.; Yuen, L. Origins and evolution of hepatitis B virus and hepatitis D virus. *Cold Spring Harb. Perspect. Med.* **2016**, *6*.
5. MacLachlan, J.H.; Cowie, B.C. Hepatitis B virus epidemiology. *Cold Spring Harb. Perspect. Med.* **2015**, *5*.
6. Drexler, J.F.; Geipel, A.; Konig, A.; Corman, V.M.; van Riel, D.; Leijten, L.M.; Bremer, C.M.; Rasche, A.; Cottontail, V.M.; Maganga, G.D.; *et al.* Bats carry pathogenic hepadnaviruses antigenically related to hepatitis B virus and capable of infecting human hepatocytes. *Proc. Natl. Acad. Sci. USA* **2013**, *110*, 16151–16156.
7. Hahn, C.M.; Iwanowicz, L.R.; Cornman, R.S.; Conway, C.M.; Winton, J.R.; Blazer, V.S. Characterization of a novel hepadnavirus in the White Sucker (*Catostomus commersonii*) from the Great Lakes Region of the United States. *J. Virol.* **2015**, *89*, 11801–11811.
8. Suh, A.; Brosius, J.; Schmitz, J.; Kriegs, J.O. The genome of a Mesozoic paleovirus reveals the evolution of hepatitis B viruses. *Nat. Commun.* **2013**, *4*.
9. Liu, W.; Pan, S.; Yang, H.; Bai, W.; Shen, Z.; Liu, J.; Xie, Y. The first full-length endogenous hepadnaviruses: Identification and analysis. *J. Virol.* **2012**, *86*, 9510–9513.
10. Gilbert, C.; Feschotte, C. Genomic fossils calibrate the long-term evolution of hepadnaviruses. *PLoS Biol.* **2010**, *8*.
11. Gilbert, C.; Meik, J.M.; Dashevsky, D.; Card, D.C.; Castoe, T.A.; Schaack, S. Endogenous hepadnaviruses, bornaviruses and circoviruses in snakes. *Proc. R. Soc. London B Bio. Sci.* **2014**, *281*.
12. Watashi, K.; Wakita, T. Hepatitis B virus and hepatitis D virus entry, species specificity, and tissue tropism. *Cold Spring Harb. Perspect. Med.* **2015**, *5*.
13. Winer, B.Y.; Ploss, A. Determinants of hepatitis B and delta virus host tropism. *Curr. Opin. Virol.* **2015**, *13*, 109–116.
14. World Health Organization Hepatitis B. World Health Organization Fact Sheet 204 (Revised July 2013). Available online: http://www.who.int/mediacentre/factsheets/fs204/en/ (accessed on 10 December 2013).

15. Mason, W.S. Animal models and the molecular biology of hepadnavirus infection. *Cold Spring Harb. Perspect. Med.* **2015**, *5*.

16. Hu, J.M.; Seeger, C. Hepadnavirus genome replication and persistence. *Cold Spring Harb. Perspect. Med.* **2015**, *5*.

17. Protzer, U.; Nassal, M.; Chiang, P.W.; Kirschfink, M.; Schaller, H. Interferon gene transfer by a hepatitis B virus vector efficiently suppresses wild-type virus infection. *Proc. Natl. Acad. Sci. USA* **1999**, *96*, 10818–10823.

18. Ho, T.C.; Jeng, K.S.; Hu, C.P.; Chang, C. Effects of genomic length on translocation of hepatitis B virus polymerase-linked oligomer. *J. Virol.* **2000**, *74*, 9010–9018.

19. Bartenschlager, R.; Schaller, H. Hepadnaviral assembly is initiated by polymerase binding to the encapsidation signal in the viral RNA genome. *EMBO J.* **1992**, *11*, 3413–3420.

20. Feitelson, M.A.; Bonamassa, B.; Arzumanyan, A. The roles of hepatitis B virus-encoded X protein in virus replication and the pathogenesis of chronic liver disease. *Expert Opin. Ther. Targets* **2014**, *18*, 293–306.

21. Belloni, L.; Pollicino, T.; de Nicola, F.; Guerrieri, F.; Raffa, G.; Fanciulli, M.; Raimondo, G.; Levrero, M. Nuclear HBx binds the HBV minichromosome and modifies the epigenetic regulation of cccDNA function. *Proc. Natl. Acad. Sci. USA* **2009**, *106*, 19975–19979.

22. Lucifora, J.; Arzberger, S.; Durantel, D.; Belloni, L.; Strubin, M.; Levrero, M.; Zoulim, F.; Hantz, O.; Protzer, U. Hepatitis B virus X protein is essential to initiate and maintain virus replication after infection. *J. Hepatol.* **2011**, *55*, 996–1003.

23. Tsuge, M.; Hiraga, N.; Akiyama, R.; Tanaka, S.; Matsushita, M.; Mitsui, F.; Abe, H.; Kitamura, S.; Hatakeyama, T.; Kimura, T.; *et al.* HBx protein is indispensable for development of viraemia in human hepatocyte chimeric mice. *J. Gen. Virol.* **2010**, *91*, 1854–1864.

24. Thung, S.N.; Gerber, M.A.; Purcell, R.H.; London, W.T.; Mihalik, K.B.; Popper, H. Animal model of human disease. Chimpanzee carriers of hepatitis B virus. Chimpanzee hepatitis B carriers. *Am. J. Pathol.* **1981**, *105*, 328–332.

25. Dandri, M.; Burda, M.R.; Zuckerman, D.M.; Wursthorn, K.; Matschl, U.; Pollok, J.M.; Rogiers, X.; Gocht, A.; Kock, J.; Blum, H.E.; *et al.* Chronic infection with hepatitis B viruses and antiviral drug evaluation in uPA mice after liver repopulation with tupaia hepatocytes. *J. Hepatol.* **2005**, *42*, 54–60.

26. Dandri, M.; Burda, M.R.; Torok, E.; Pollok, J.M.; Iwanska, A.; Sommer, G.; Rogiers, X.; Rogler, C.E.; Gupta, S.; Will, H.; *et al.* Repopulation of mouse liver with human hepatocytes and *in vivo* infection with hepatitis B virus. *Hepatology* **2001**, *33*, 981–988.

27. Kock, J.; Nassal, M.; MacNelly, S.; Baumert, T.F.; Blum, H.E.; von Weizsacker, F. Efficient infection of primary tupaia hepatocytes with purified human and woolly monkey hepatitis B virus. *J. Virol.* **2001**, *75*, 5084–5089.

28. Gripon, P.; Rumin, S.; Urban, S.; le Seyec, J.; Glaise, D.; Cannie, I.; Guyomard, C.; Lucas, J.; Trepo, C.; Guguen-Guillouzo, C. Infection of a human hepatoma cell line by hepatitis B virus. *Proc. Natl. Acad. Sci. USA* **2002**, *99*, 15655–15660.

29. Yan, H.; Zhong, G.; Xu, G.; He, W.; Jing, Z.; Gao, Z.; Huang, Y.; Qi, Y.; Peng, B.; Wang, H.; *et al.* Sodium taurocholate cotransporting polypeptide is a functional receptor for human hepatitis B and D virus. *eLife* **2012**, *1*.

30. Chang, L.J.; Hirsch, R.C.; Ganem, D.; Varmus, H.E. Effects of insertional and point mutations on the functions of the duck hepatitis B virus polymerase. *J. Virol.* **1990**, *64*, 5553–5558.

31. Chaisomchit, S.; Tyrrell, D.L.; Chang, L.J. Development of replicative and nonreplicative hepatitis B virus vectors. *Gene Ther.* **1997**, *4*, 1330–1340.

32. Wang, L.; Kaneko, S.; Honda, M.; Kobayashi, K. Approach to establishing a liver targeting gene therapeutic vector using naturally occurring defective hepatitis B viruses devoid of immunogenic T cell epitope. *Virus Res.* **2002**, *85*, 187–197.

33. Yoo, J.; Rho, J.; Lee, D.; Shin, S.; Jung, G. Hepatitis B virus vector carries a foreign gene into liver cells *in vitro*. *Virus Genes* **2002**, *24*, 215–224.

34. Deng, Q.; Mancini-Bourgine, M.; Zhang, X.; Cumont, M.C.; Zhu, R.; Lone, Y.C.; Michel, M.L. Hepatitis B virus as a gene delivery vector activating foreign antigenic T cell response that abrogates viral expression in mouse models. *Hepatology* **2009**, *50*, 1380–1391.

35. Wang, Z.; Zhu, K.; Bai, W.; Jia, B.; Hu, H.; Zhou, D.; Zhang, X.; Zhang, X.; Xie, Y.; Bourgine, M.M.; *et al.* Adenoviral delivery of recombinant hepatitis B virus expressing foreign antigenic epitopes for immunotherapy of persistent viral infection. *J. Virol.* **2014**, *88*, 3004–3015.

36. Wang, Z.; Wu, L.; Cheng, X.; Liu, S.; Li, B.; Li, H.; Kang, F.; Wang, J.; Xia, H.; Ping, C.; *et al.* Replication-competent infectious hepatitis B virus vectors carrying substantially sized transgenes by redesigned viral polymerase translation. *PLoS ONE* **2013**, *8*, e60306.

37. Hong, R.; Bai, W.; Zhai, J.; Liu, W.; Li, X.; Zhang, J.; Cui, X.; Zhao, X.; Ye, X.; Deng, Q.; *et al.* Novel recombinant hepatitis B virus vectors efficiently deliver protein and RNA encoding genes into primary hepatocytes. *J. Virol.* **2013**, *87*, 6615–6624.

38. Untergasser, A.; Protzer, U. Hepatitis B virus-based vectors allow the elimination of viral gene expression and the insertion of foreign promoters. *Hum. Gene Ther.* **2004**, *15*, 203–210.

39. Liu, J.; Cheng, X.; Guo, Z.; Wang, Z.; Li, D.; Kang, F.; Li, H.; Li, B.; Cao, Z.; Nassal, M.; *et al.* Truncated active human matrix metalloproteinase-8 delivered by a chimeric adenovirus-hepatitis B virus vector ameliorates rat liver cirrhosis. *PLoS ONE* **2013**, *8*, e53392.

40. Nishitsuji, H.; Ujino, S.; Shimizu, Y.; Harada, K.; Zhang, J.; Sugiyama, M.; Mizokami, M.; Shimotohno, K. Novel reporter system to monitor early stages of the hepatitis B virus life cycle. *Cancer Sci.* **2015**, *106*, 1616–1624.

41. Chen, A.; Kao, Y.F.; Brown, C.M. Translation of the first upstream ORF in the hepatitis B virus pregenomic RNA modulates translation at the core and polymerase initiation codons. *Nucleic Acids Res.* **2005**, *33*, 1169–1181.

Use of Reporter Genes in the Generation of Vaccinia Virus-Derived Vectors

Sally Al Ali, Sara Baldanta, Mercedes Fernández-Escobar and Susana Guerra

Abstract: Vaccinia virus (VACV) is one of the most extensively-studied viruses of the *Poxviridae* family. It is easy to genetically modify, so it has become a key tool for many applications. In this context, reporter genes facilitate the study of the role of foreign genes introduced into the genome of VACV. In this review, we describe the type of reporter genes that have been used to generate reporter-expressing VACV and the applications of the recombinant viruses obtained. Reporter-expressing VACV are currently employed in basic and immunology research, in the development of vaccines and cancer treatment.

Reprinted from *Viruses*. Cite as: Ali, S.A.; Baldanta, S.; Fernández-Escobar, M.; Guerra, S. Use of Reporter Genes in the Generation of Vaccinia Virus-Derived Vectors. *Viruses* **2016**, *8*, 134.

1. Introduction

Since the first description of recombinant DNA techniques, many advances have been achieved in the field of molecular biology and genetic modification. Currently, there is a wide variety of tools that allow the genetic modification of animals, plants, bacteria and viruses [1–4]. The genetic modification of viruses has become one of the best strategies for introducing nucleic acids into different cells, tissues or even in *in vivo* models, given the high transfection efficiency and ease of carrying it out, compared to chemical or physiological methods [5,6].

After the description of recombination events in cells infected with vaccinia virus (VACV) and through recombinant DNA technology [7,8], VACV has become a suitable model for the generation of recombinant virus vectors [9]. At first, the main purpose for introducing foreign genes into virus genomes was basic research about the biology of the viruses both *in vitro* and *in vivo*. However, with the latest technical advances and the higher understanding of the VACV viral cycle, virus genetic modification is getting a wider spectrum purpose. Thus, they can also be used for the development of vaccines or as oncolytic agents. This review aims to highlight the main aspects of the genetic modification of VACV and the generation and application of reporter-expressing virus in this model.

2. Biology of VACV

VACV is the prototype member of the *Poxviridae* family, so most research of poxvirus has been focused on its use [10]. VACV is a large DNA double-stranded virus, with a complex envelope. It was the live vaccine used to eradicate smallpox and nowadays is also used as a viral vector for recombinant vaccines and cancer therapy [9,11]. The VACV genome is one of the largest of all DNA viruses, with a size of 190 kbp and about 250 encoding genes [12]. The genome has a high genetic compaction, with a few intergenic and small non-coding regions. The coding regions are continuous, thereby not given to splicing [13,14].

VACV have a complete replicating cycle inside the cytoplasm of the host cell, even though it is a DNA virus (Figure 1) [10]. This fact determines the genetic characteristics of the virus, being completely independent of the replication and transcription machinery of the host cell. Once the virion infects the host cell, the viral core is uncoated, and nearly 100 early viral genes are transcribed [15,16]. Early genes produce the required enzymes for catalyzing the viral core breakdown, viral DNA replication and the modulation of the host antiviral response [17]. Viral DNA begins to replicate inside the infected host cell using viral enzymes at 3 h post-infection. As soon as the viral replication starts, transcription of downstream genes encoding for regulatory proteins that induce the expression of the late genes occurs. Late genes encode for proteins and enzymes required for the assembly of new viral particles. After DNA and all viral proteins are synthesized, the process known as morphogenesis begins, which results in the formation of the new virions [18,19]. These can be retained inside the cell until cellular lysis or released to the environment by other mechanism [10,18].

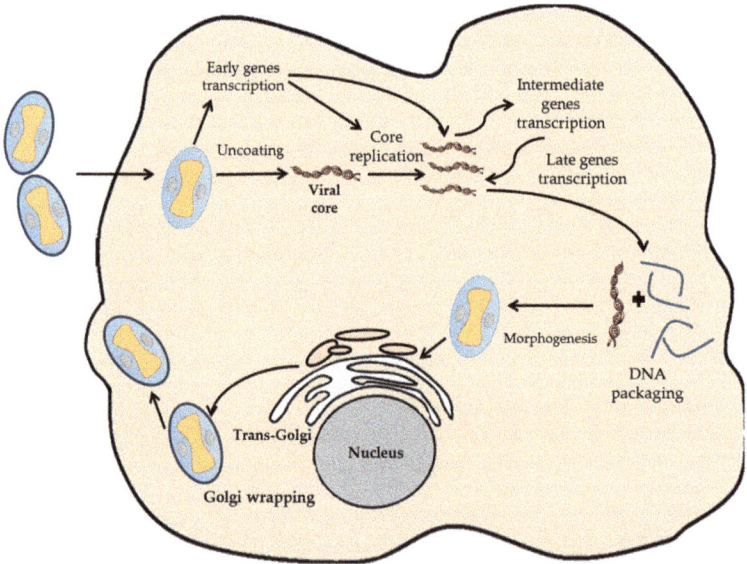

Figure 1. Diagram representative of the VACV infection cycle. The different steps of the VACV cycle are indicated.

3. VACV as a Vector

Several features of the biology of VACV make it suitable for its use as a vector in biological experiments, vaccine design or cancer therapy. The complete cytoplasmic replication of VACV facilitates the expression of foreign genes inserted in the viral genome and its detection or isolation [20,21]. Usually, bacterial or non-mammalian viral vectors fail to make the expressed proteins to perform its full activity as antigens. However, VACV has the ability to transcribe its genes using its own transcription factors and enzymes. That means that if a foreign gene is inserted directly to a VACV promoter element, it will be transcribed with foreign proteins reaching high levels of expression in the infected cell. Moreover, this replication cycle is an appropriate model for molecular and genetic investigations of *cis* and *trans* factors that are mainly required for gene expression [12,22]. Furthermore, since VACV remains in the cytoplasm, the risk of insertional mutagenesis and oncogenesis, the main problems encountered in gene therapy using integrative viruses, disappears. In some cases, patients treated with retroviral vectors have developed cancer years after they have been treated [5,23].

VACV can replicate in different cell lines, primary cell cultures, and also grows in several animal species, such as mice, guinea pigs, rabbits, *etc.* [10]. This broad host range allows infection of cell lines with recombinant viruses for large-scale expression of heterologous proteins, which reduces its cost in comparison to other production systems [21,24]. Additionally, VACV enables high production titers, so it is an advantage in the manufacturing of a large amount of vaccines [6].

Although the VACV genome is large and compact, it can tolerate the deletion of certain viral sequences and the insertion of exogenous genetic material [25]. A VACV vector has a transgene capacity of approximately 25–30 kb, higher than other viral vectors, including adeno-associated virus (4.5 kb), adenovirus (8–10 kb) and retrovirus (7–8 kb) [4]. Thus, VACV is an excellent candidate vector in the design of polyvalent vaccines with antigens from several pathogens or different antigens from the same pathogen [9,26].

Finally, as far as its use as a vaccine vector is concerned, VACV is clearly immunogenic effective, strong evidence being the eradication of smallpox in 1980 [11]. VACV is also safe and easy to inoculate, since it can be administrated intradermally or with an air gun without medical training. In some organisms, it has been found that it can cause problems by preexisting immunity, but the probability of having post-vaccination complications, such as progressive VACV infection or encephalitis, is significantly low[27]. Nowadays, due to the better knowledge of the VACV biology and the immune response generated after vaccination, vaccines based on this virus are becoming safer [9]. In addition, it is important to remark that VACV vectors are very stable and can be lyophilized and kept frozen for several years, facilitating its transport and storage [23].

4. Design Considerations in the Generation of VACV Vectors

To get recombinant VACV expressing foreign genes, the main method used is homologous recombination (Figure 2) [28]. First, it is necessary to construct a plasmid that contains the gene or transgene of interest. After that, the cells have to be infected with the virus and subsequently transfected with the plasmid that contains the transgene. An alternative method could be used, employing two viruses, one defective for some genes and one wild-type acting as a helper [4,29]. For both methods, the recombinant viruses are produced by homologous recombination inside the infected cell.

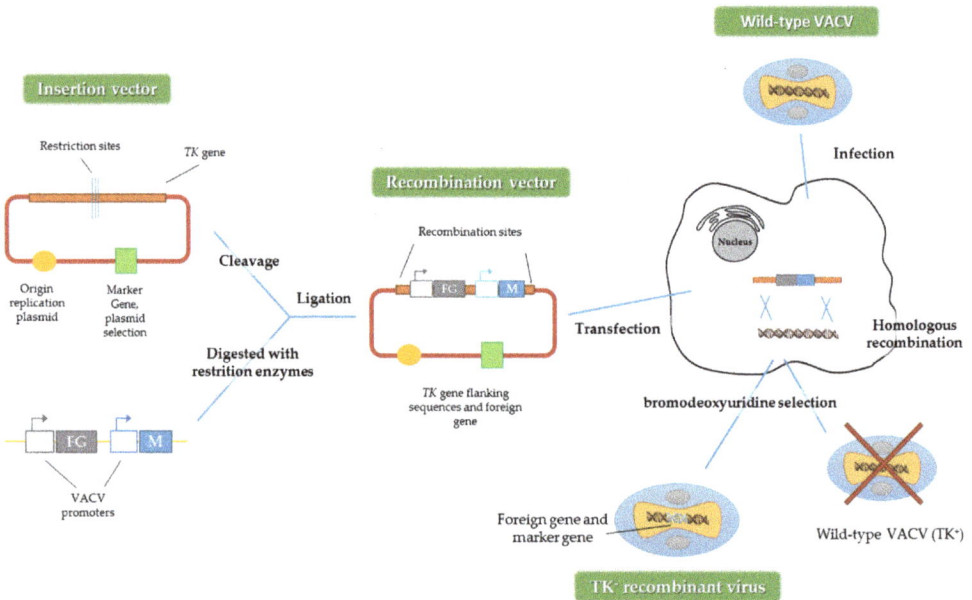

Figure 2. Construction of recombinant VACV vectors by homologous recombination. FG represents the foreign gene and M represents the marker gene and *TK: thymidine kinase* gene. Adapted from [28].

Another way to generate recombinant viruses is the method described by Falkner and Moss [30], denominated transient dominant selection (TDS), which allows the introduction of site-directed mutations into the VACV genome. Generally, the recombinant viruses obtained by this method are rescued by metabolic selection, using the *guanine phosphoribosyltransferase* gene (*gpt*) from *Escherichia coli* as a marker. The presence of the protein encoded by *gpt* allows the recombinant viruses to grow in the presence of mycophenolic acid, xanthine and hypoxanthine [31]. Subsequently, after this first metabolic selection, a second recombination event must occur to eliminate the selection marker, maintaining the mutation introduced into the VACV genome (Figure 3) [32]. In contrast to the method described above, in the TDS technique the marker should not be flanked by homologous regions of the VACV genome [30]. Alternatively, puromycin resistance could be used as a selection marker in TDS, increasing the recombinant viruses' generation efficiency [33].

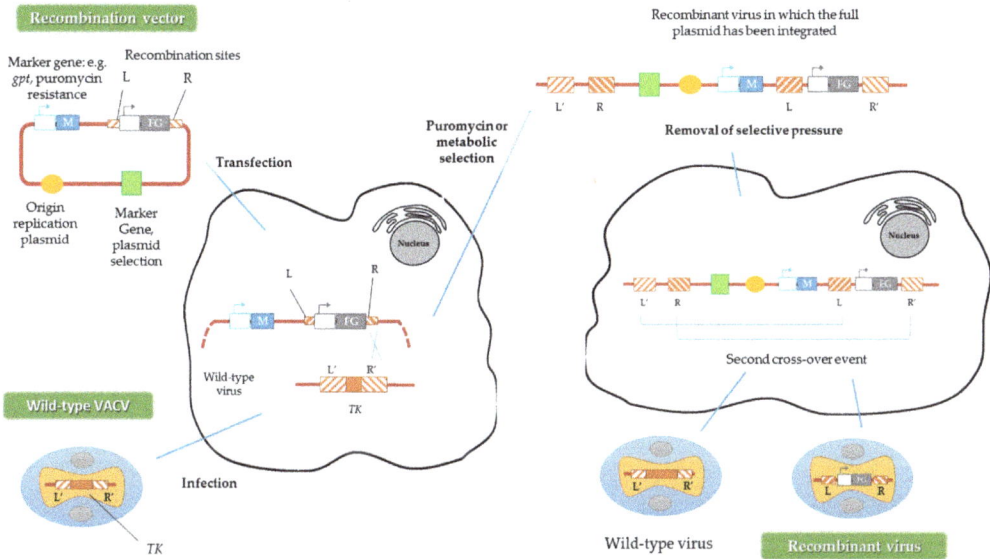

Figure 3. Construction of recombinant VACV vectors by the transient dominant selection (TDS) technique. FG represents the foreign gene and M represents the marker gene. R and L represent the right and left flanking regions of the *TK* gene in the plasmid, and R' and L' represent the same regions of the *TK* gene in the VACV genome. Adapted from [33].

Two important aspects to be considered when obtaining recombinant poxvirus are the VACV genome insertion sites and the reporter genes introduced.

4.1. VACV Genome Insertion Sites

The VACV genome has about seven known insertion sites where foreign genes can be inserted (Figure 4) [13]. The insertion site choice depends mainly on the future application of the recombinant viruses. It may also be important in the later selection of the recombinant viruses obtained. For instance, inserting the gene of interest in the *thymidine kinase (TK)* locus confers a detectable phenotype (TK-): the recombinant viruses are able to grow in the presence of 5-bromo-2'-deoxyuridine (BrdU), a synthetic analog of thymidine [28,34]. Another important site of insertion that allows a subsequent selection is the VACV *hemagglutinin (HA)* gene as the recombinant viruses can be easily recognized by their disability to bind erythrocytes in a hemagglutination test [35–37].

VACV has five more places of insertion: the *Bam*HI site of the *Hind*III-F DNA fragment [38]; the *VACV growth factor* gene (*VGF*), located in both inverted terminal repeats (ITRs) [39]; the *N2* and *M1* genes located on the left side of the VACV genome [40]; the M1 subunit of the *ribonucleotide reductase* (*RR*) gene in the *Hind*III-I DNA fragment [41]; and the *A27L* gene encoding the 14 kDa fusion protein, in the large *Hind*III-A DNA fragment [42]. It is noteworthy that some strains of VACV have only one copy of *VFG*, such as VACV Lister variants [33]. Recombinant production using these insertion sites, although successfully occurring, requires the use of a marker gene or other strategies for later selection of the recombinant viruses. Due to these limitations, the *TK* gene is the most common site of insertion in the VACV genome [5]. Some authors have used temperature-sensitive VACV strains, allowing the recombinant viruses to be selected in culture at 40 °C [43]. However, the most common way for an easy identification of recombinant viruses is the use of reporter genes as selectable markers, which will be discussed in Section 4.2 [44].

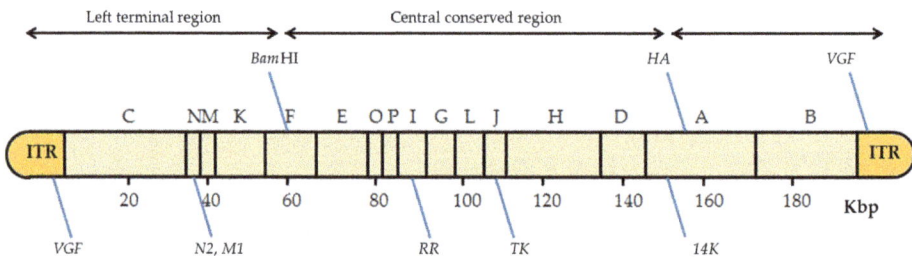

Figure 4. Scheme of the insertion sites in the VACV genome. The diagram of the VACV genome with the *Hind*III restriction sites is shown, including the location of the different insertion sites. *Bam*HI: *Bam*HI site of the *Hind*III F fragment; *HA*: hemagglutinin gene; *VGF*: *VACV growth factor* gene; *N2*: *N2* gene; *M1*: *M1* gene encodes the large subunit of ribonucleotide reductase (RR); *TK*: *thymidine kinase* gene; *A27L*: gene that encodes the 14 kDa fusion protein; ITRs: inverted terminal repeats. Adapted from [13].

In spite of the promoter or the regions between the promoter and coding region, the insertion site also influences foreign gene expression and virus virulence [13,23,25]. Insertion into the *TK, VGF, RR* or *A27L* genes has an impact on viral replication *in vivo*, but not *in vitro* [25,45]. Moreover, the method described in Figure 2 requires the use of special cell lines or mutagenic selective agents, such as TK-/- cell lines and BrdU [30]. For this reason, different strategies and new insertion sites are being studied to ensure the correct expression of the transgenes *in vitro* and *in vivo* [25,33,46,47].

4.2. Reporter-Expressing Viruses

Reporter-expressing viruses are recombinant viruses expressing a reporter gene [48]. In some cases, the reporter gene is located downstream of a viral promoter, to study biological pathways or, fused with other viral or foreign genes. As reporter genes are expected to be easily detected, they are the best indicators for screening successfully recombinant viruses. The reporter gene should be chosen considering the non-endogenous activity in the cell type, tissue or organism used to culture the viruses [44]. Reporter genes have additional applications *in vitro* and *in vivo*, as the reporter gene acts as a substitute of the gene of interest. Moreover, reporter genes facilitate the use of tissue-specific and pathway-specific promoters, as well as regulatory promoter elements as biomarkers for a particular event route. Furthermore, it is important that the existence of the reporter gene should not affect the normal physiology and general characteristics of the transfected cells [48–50]. Table 1 presents an overview of the reporter genes commonly used in the generation of recombinant VACV.

Table 1. Reporter genes commonly used in the generation of recombinant vaccinia virus (VACV).

Reporter Gene	Origin	Product	Detection	Reference
CAT	*Escherichia coli*	Chloramphenicol acetyltransferase	Thin-layer chromatography autoradiography, ELISA	[51,52]
LacZ	*Escherichia coli*	β-galactosidase	Colorimetry	[53]
GUS	*Escherichia coli*	β-glucuronidase	Colorimetry or fluorescence	[54]
GFP	*Aequorea victoria* (jellyfish)	Green fluorescent protein	Fluorescence	[50,55]
LUC or *lux*CDABE	*Photinus pyralis* (firefly) and bacteria	Luciferase	Luminescence	[56,57]

ELISA: enzyme-linked immunosorbent assay.

4.2.1. Chloramphenicol Acetyltransferase

CAT was the first reporter gene used in transcriptional assays in mammalian cells. CAT is an enzyme from *Escherichia coli* that detoxifies the antibiotic chloramphenicol, which inhibits protein synthesis in bacteria [58]. Particularly, CAT links acetyl-coenzyme A (acetyl-CoA) groups to chloramphenicol, preventing it from blocking the 50 S ribosomal subunit. This gene is not found in eukaryotes, so eukaryotic cells do not present any basal CAT activity [44]. The reaction catalyzed by CAT can be quantified using fluorogenic or radiolabeled substrates, such as ^{3}H-labeled acetyl-CoA and ^{14}C-labeled chloramphenicol. CAT can be detected either by thin-layer chromatography, autoradiography or enzyme-linked immunosorbent assay (ELISA) [51].

131

There is a strong link between *CAT* gene transcript levels and enzymatic activity, which is easy to quantify. Thus, *CAT* has become a suitable reporter gene for investigating transcriptional elements in a wide variety of experiments implicating animal and plant cells, as well as viruses [51]. There are some disadvantages of using the CAT system, such as the higher amount of cells required when compared to other assays, like the luciferase assay (detailed in Section 4.2.5). In addition, the CAT system is not suitable for use with weakly-expressed genes and *CAT* promoter activity quantification takes longer than other reporter systems [52]. Finally, this reporter gene has another important limitation due to the use of radioisotopes [44].

4.2.2. β-Galactosidase

The first study using *lacZ* as a reporter gene was published in 1980, and since then, it has become one of the most commonly-used reporter genes in molecular biology [49]. Although β-galactosidase catalyzes the cleavage of the disaccharide lactose to form glucose and galactose, it recognizes several artificial substrates, which has promoted its use as a reporter gene [58]. Thus, β-galactosidase can hydrolyze substrates such as ortho-nitrophenyl beta-galactoside (ONPG), 5-bromo-4-chloro-3-indolyl beta-D-galactopyranoside (X-Gal) and 3,4-cyclohexenoesculetin beta-D-galactopyranoside (S-Gal), resulting in a yellow, blue or black product precipitate, respectively [53,59]. Furthermore, expression of the *lacZ* gene can be stimulated with isopropyl beta-D-thiogalactopyranoside (IPTG), which is a highly stable synthetic and non-hydrolyzable analog of lactose [49].

One of the applications of the *lacZ* reporter gene is the selection of transformed bacterial colonies. The recombinant (white) and non-recombinant (blue) bacteria are discriminated based on the interruption of the *lacZ* gene by the insert DNA or gene of interest using X-Gal as a substrate [53]. Other uses are the visualization of the β-galactosidase expression in transfected eukaryotic cells or the selection of the recombinant virus by viral plaque screening [60]. Finally, *lacZ* is used to detect β-galactosidase activity in immunological and histochemical experiments [44]. One of the main advantages of using this reporter gene system is its low cost, since it does not require specific devices to detect the colorimetric reaction or to identify its expression.

4.2.3. β-Glucuronidase

Another *Escherichia coli*-derived hydrolyzing enzyme gene that lends a reporter assay is *GUS*. The β-glucuronidase protein catalyzes the breakdown of complex carbohydrates, such as glycosaminoglycans. This reporter gene system has been widely used in transgenic plants, and it has also been successfully used in mammalian cells for VACV recombinant virus selection [54]. For the β-glucuronidase (GUS) assay 4-methylumbelliferyl beta-D-glucuronide (MUG) or 5-bromo-4-chloro-3-indolyl beta-D-glucuronide (X-Gluc) can be used as substrates. They respectively lead to a fluorogenic or a blue product after cleavage [61,62]. Monitoring β-glucuronidase activity through a GUS assay allows the determination of the spatial and temporal expression of the gene of interest [63].

4.2.4. Florescent Proteins

The most known fluorescent protein is green fluorescent protein (GFP), which was cloned from the species of jellyfish *Aequorea victoria*. Because of the great impact of fluorescent proteins in molecular biology applications, the Nobel Prize in Chemistry 2008 was awarded to Osamu Shimomura, Martin Chalfie and Roger Y. Tsien for the discovery and development of GFP [64,65]. *GFP* is the most used reporter gene; however, genetic engineering has developed a wide variety of color mutants, such as red fluorescent protein (RFP) or yellow fluorescent protein (YFP) among others [49].

Fluorescent proteins tolerate N- and C-terminal fusions to a wide-range of proteins, have been expressed in most known cell types and are used as a non-harmful fluorescent marker in living cells and organisms. The use of fluorescent proteins allows a variety of applications: cell lineage tracker , reporter for gene expression assays or measure of protein-protein interactions. Additionally, cell-fixation is not needed to examine its expression, and the probability of artifacts is quite small compared to immunocytochemical methods which require cell fixation [44]. One of the disadvantages of these proteins is their size. Therefore, in some cases, they can affect the *in vivo* function of fused proteins or genes of interest. Nevertheless, one limitation of using GFP is its low sensitivity [66], another is that its signal cannot be exogenously amplified [50].

4.2.5. Luciferases

The first luciferase (LUC), from the firefly *Photinus pyralis*, was cloned in 1980 and *LUC* has been widely used as a reporter gene. Later, it was also described in bacteria and dinoflagellates [44]. Luciferases are enzymes that catalyze a chemical reaction resulting in the production of light. Firefly luciferase oxidize the D-luciferin, in the presence of oxygen and adenosine triphosphate (ATP) as the energy source. As in β-Galactosidase assays, an exogenous substrate is needed, and it may be a disadvantage [49]. In other systems, such as the luciferase identified in bacteria (*lux*CDABE operon), the enzyme catalyzes the oxidation of long-chain aldehydes and flavin mononucleotides (FMNH$_2$) in the presence of oxygen to yield green-blue light [67]. Although in bacteria this operon encodes all components necessary for light emission, it is limited in mammalian cells. Therefore, the exogenous substrate has to be added to improve the reaction [56]. Besides the different substrates required, each luciferase system is categorized by having specific kinetics, with a particular detection and sensitivities that require adjusting the experimental design [58,67].

The use of luciferase is extremely widespread in biological systems studies and includes cell proliferation assays, protein folding/secretion analyses, *in vivo* imaging and control of *in vivo* viral spreading [57,67–69]. The main advantage of using this system is its high sensitivity when compared to other systems, such as CAT. Additionally, the LUC system is more direct, rapid and suitable when it comes to weakly-expressed genes, and it can be used to quantify gene activity. One disadvantage of the LUC system is the requirement of ultrasensitive charge-coupled device (CCD) cameras to detect gene expression [56].

5. Applications of Reporter-Expressing Viruses

5.1. In Vitro Applications

Reporter-gene assays have helped the pox virologists in basic research, for example for the study of the location, structure and function of many VACV proteins during the infection cycle and their interaction with proteins of the host cell [44,70]. As shown in Dvoracek and Shors [63], the *GUS* reporter gene was used for deleting the D9R viral protein and selecting the recombinant viruses, with the aim to understand the role of this protein in the viral life cycle. In addition, the *lacZ* gene has typically been used mainly for the selection of recombinants [71]. Moreover, several studies have reported the different transgenes' insertion points and VACV promoters in which the recombinant virus production was enhanced. These studies are essential for improvement of the development of vaccines based on recombinant VACV [62,72].

On the other hand, fluorescent markers such as the GFP, YFP or luciferase are also useful for labelling VACV replicative strains. These viruses have allowed the study of processes like the input and output morphogenesis in virus-infected cells [68,69,73–76]. In these studies, fluorescence of certain viral proteins allows us to study their interaction with other viral or cellular proteins [77]. Furthermore, VACV is a clear example of how viruses have developed strategies to evade the immune response [78]. In this field, the generation of recombinant VACV with reporter genes is also useful to discern the molecular mechanism by which VACV proteins manipulate the immune system of the host. Thus, in Unterholzner *et al.* [79], the generation of a GFP-labeled recombinant VACV revealed that the C6 viral protein acted as an immunomodulatory agent, blocking the expression of type I interferon.

Another major application of reporter-expressing VACVs is the design of high-throughput assays. The generation of *lacZ* or *GFP* expressing recombinant virus can be used to optimize antibody neutralization assays [71,80].

Lastly, VACV and reporter genes have been used to study proteins from other viruses, particularly RNA viruses, such as influenza or severe acute respiratory syndrome-associated coronavirus (SARS-CoV) [81]. To genetically modify these viruses, RNA must be reverse transcribed to cDNA, since this is particularly unstable in plasmids, making VACV a good tool for functional studies of proteins from such viruses [82].

5.2. In Vivo Applications

There are several *in vivo* applications for recombinant reporter-expressing viruses. For example, in virulence studies, the use of labeled viruses allows us to follow the viral pathogenicity and detect in which organs the viral replication and dissemination occur [70,76]. For example, Zaitseva *et al.* [69] used the recombinant VACV Western Reverse strain (WR)-LUC to analyze the viral spread *in vivo* for several days reducing the number of mice used. Moreover, VACV is an effective enhancer for both humoral and cell-mediated immunity; it is used as a vector to study the immune system and the expression of proteins' antigenicity of other pathogens. Furthermore, VACV is used to explore the immunopathological mechanisms, to know which epitopes or antigens presented by a pathogen have the ability to induce the host-immune response, and to demonstrate the specific role of a particular antigen during the pathogenic process [13,83,84].

Despite the examples mentioned above, the most common uses of recombinant VACV *in vivo* are the production of prophylactic vaccines and treatments against cancer [4,85]. Table 2 shows some of the vaccines based on VACV, with the reporter gene and the insertion site employed indicated in each case. In these vaccines, VACV acts as a vector capable of delivering antigens from other organisms [23]. While in many recombinant vaccines a viral antigen has been inserted, some of them have also been developed against bacteria [86] or protists [34,87,88]. These vaccines simulate infection by the pathogen from which the antigens are and elicit the immune response, by producing antigens for different pathogens. In several vaccines, mainly against human immunodeficiency virus (HIV) or influenza, genes of immunomodulatory cytokines are added for coexpression with the antigen, improving the immunogenicity of the vaccines [23,89,90]. As summarized in Table 2, most of the transgene insertion sites are within the *TK* or the *HA* genes, making the selection of recombinants easier, as explained above. However, in several vaccines, besides using this strategy, a reporter gene is used as well. The use of reporter genes facilitates the preliminary tests of the vaccine on animal models. Moreover, especially in vaccines used in animals, the reporter gene makes it possible to distinguish between vaccinated and infected animals [48]. For example, VACV has been used for nearly twenty years to eradicate rabies from wildlife as an oral-based vaccine. In this case, the recombinant VACV expresses the rabies virus glycoprotein and has been used to vaccinate raccoons, red foxes, skunks and coyotes in the United States and

Europe. This battle has successfully purged rabies in some parts of Europe and the United States [91].

Table 2. VACV-derived vaccines.

Pathogenic Agent		Antigen	Features		Reference
			Site of Insertion	Reporter Gene	
Viral	HIV	Env	*TK* or *HA* gene	*LacZ*, *LUC*	[90,92]
		Env (TAB 13)	*HA* gene	*LacZ*	[35]
		RT	Not mentioned	*LacZ*	[93]
	Hepatitis B virus	HBsAg	*TK* gene or *Bam*HI site	Not mentioned	[38,94,95]
		PreS2-S	*TK* gene	Not mentioned	[96]
		LS	*TK* gene	Not mentioned	[97]
		MS	*TK* gene	Not mentioned	[98]
	Herpes simplex virus 1	gD	*TK* gene or *Bam*HI site	Not mentioned	[38,99,100]
		gB	Not mentioned	Not mentioned	[101,102]
		gG	Not mentioned	Not mentioned	[103]
	Influenza	HA	*TK* gene	Not mentioned	[104]
		M1, NS1, NP, PB1, PA	*TK* gene	Not mentioned	[105]
Protist	*Plasmodium yoelii*	Circumsporozoite	*TK* gene	*LacZ*	[34]
	Plasmodium knowlesi	Sporozoite antigen	*TK* gene	Not mentioned	[88]
	Plasmodium falciparum	S antigen	*TK* gene	Not mentioned	[87]
	Leishmania infantum	LACK	*TK* and *HA* gene	*LacZ* and *GUS*	[106]
Animal	*Echinococcus granulosus*	E95 antigen	*TK* gene	*LacZ*	[107]
Bacterial	*Brucella abortus*	18-kDa antigen	*TK* gene	*LacZ*	[86]
	Streptococcus pyogenes	M protein	*TK* gene	Not mentioned	[108]

Another application for VACV vectors is in cancer treatment, known as oncolytic virotherapy [26]. This is the use of replication-competent viruses to selectively attack and destroy cancer cells, without harming healthy host cells [109]. Examples of recombinant VACV used are summarized in Table 3. A promising study is the use of oncolytic VACV as a vector for the *human sodium iodide symporter* (*hNIS*) gene in prostate cancer therapy, which has been demonstrated to restrict tumor growth and to increase survival in mice [110]. VACV is also a promising therapeutic agent for pancreatic cancer [85], cholangiocarcinoma [111] and colorectal cancer [112]. It is worth mentioning that many of the viral vectors developed to treat tumors have several common characteristics. Generally, VACV oncolytic vectors have a deletion in the *TK* gene, essential for the pyrimidine synthesis pathway, which forces the virus to replicate in cells displaying a high amount of nucleotide pools, enhancing the viral tropism to cancer cells. Others have a deletion in the *VGF* gene, preventing non-infected cells from proliferation [109]. Furthermore, as in the development of vaccines, viral vectors are "armed" with genes that enhance the antitumor activity, the virus tropism or the immunoreactivity, to promote better tumor destruction, such as *granulocyte-macrophage colony-stimulating factor* (*GM-CSF*) or *erythropoietin* genes (enhanced virus; Table 3). Another particular feature is that many of these recombinants carry reporter genes, and thus viral replication can be monitored by non-invasive imaging methods [68,69,76,113].

Table 3. Oncolytic vaccinia virus (VACV) developed for cancer treatment.

Virus	Target Cancer	Features		Reference
		Inactive Genes	Additional Genes	
Initial virus				
JX-594	Melanoma, hepatocellular carcinoma, colorectal cancer	*TK* inactive,	*LacZ* and *GM-CSF*	[112,114]
GLV-1h68	Colorectal cancer, prostate cancer, salivary gland carcinoma	*TK, HA* and *F14.5L* inactive	*GFP, LacZ* and *GUS*	[111]
vvDD	Sarcomas, neuroblastoma	*TK* and *VGF* inactive	*CD*	[115,116]
Enhanced virus				
GLV-1h153	Pancreatic cancer	GLV-1h68 expressing *hNIS*		[110]
GLV-1h210	Lung cancer	GLV-1h68 expressing *hEPO*		[117]
vvDD-SR-RFP	Sarcomas, neuroblastoma	*TK* and *VGF* inactive	*CD, RFP, SR*	[118]

6. Limitations of VACV Vectors

The main limitation of using VACV as a vector is the short-term gene expression, since it is a lytic virus killing the infected cells. Thus, gene expression will not last for more than 12–24 h post-infection [13,109]. Additionally, although for some applications it is an advantage, since VACV replicates completely in the infected cell cytoplasm, it is hard to use VACV to engineer nuclear gene replacement [23]. The other main disadvantage is the limited immunogenicity in individuals vaccinated against smallpox. This pre-existing immunity reduces the effectiveness of vaccines based on VACV, although some trials have overcome this problem by mucosal vaccination with vaccinia vectors [5]. The VACV safety profile should be considered because it has progressive complications especially with immunocompromised individuals [11]. These limitations primarily affect *in vivo* applications of VACV recombinants in vaccine development, so several attenuated strains of VACV are being generated [9].

7. New Perspectives

As for other viruses, the development of vaccines or oncolytic therapies based on VACV requires the understanding of its pathogenesis and biology. Despite improvements in the vectors' design, such as the use of different promoters or insertion sites, homologous recombination has been almost exclusively the way to obtain VACV recombinants [45]. Homologous recombination requires the use of markers or reporter genes for selecting recombinants, which offers many disadvantages. Apart from the physical space needed for the marker gene, which is limited in therapeutic virus, the use of certain markers can introduce mutations or generate artifacts that are only found after an analysis of the generated virus. Sometimes, these problems cannot be detected *in vitro*, but are very important to overcome when these vectors are used *in vivo* on animal models [46,48].

In recent years, some strategies have been developed to avoid these risks using markers, or at least to remove them from the final recombinant VACV. Rice and colleagues [45] described a double selection method to improve the selection of recombinant VACV, so that a reporter or marker gene is not necessary. A helper virus is used to rescue a recombinant VACV and is subsequently grown in non-permissive cells to the helper virus; allowing the selection of a large percentage of recombinant virions. However, the method that has certainly had an enormous importance in the modification of genomes is the clustered regularly interspaced short palindromic repeats (CRISPR)/CRISPR-associated protein 9 (Cas9) system. Briefly, the CRISPR/Cas9 system consists of an endonuclease (Cas9) employing a guide RNA to generate a break in a target place of the genome, later to be repaired, either randomly or precisely using a specifically designed "restful" template [119]. The effectiveness of this system has been proven in different organisms, including viruses, such as herpes simplex virus (HSV) [120], hepatitis B virus (HBV) [121] and HIV [76]. Currently, this technique is starting to be used also in VACV [47]. For example, this system has achieved the deletion of VACV virulence genes, such as *A46L* and *N1L*. A46L and N1L are VACV intracellular proteins that inhibit nuclear factor-kappa B (NF-kB) activation, so it is undesirable that they were present in VACV vectors with therapeutic purposes [78]. Furthermore, given the efficiency of the method, "reparative" vectors with excisable marker genes have been designed. Therefore, recombinant viruses are effectively isolated, but eventually, the marker gene is eliminated [46]. Given the simplicity of recombinant VACV by the CRISPR/Cas9 system generation, an exponential increase of applications with better markers for basic research or without selectable markers for clinical application is expected [119,120].

8. Concluding Remarks

In conclusion, the development of recombinant viruses is a promising therapeutic advance in the biomedical field. In this sense, the use of reporter-expressing VACVs has become a fundamental tool for a number of applications, in basic research, vaccine design and cancer therapy. As many of these trials are still experimental, more information is required regarding the side effects of the viral treatment. Continuing efforts are necessary to develop new reporter-expressing VACVs that are safer and more effective for future therapies.

Acknowledgments: We thank all of the pox virologist who contributed to this study. This work is supported by Grant FIS2011-00127 and Reference SAF2014-54623-R to SG.

Author Contributions: Sara Baldanta wrote the paper and created the figures, Saly al Ali and the rest of the authors also wrote the paper.

Conflicts of Interest: The authors declare no conflict of interest.

Abbreviations

acetyl-CoA	acetyl-coenzyme A
ATP	adenosine triphosphate
CAT	chloramphenicol acetyltransferase
CCD	charge-coupled device
CD	cytosine deaminase
CRISPR/Cas9	clustered regularly interspaced short palindromic repeats/CRISPR-associated protein9
Env	envelope
FMNH$_2$	flavin mononucleotide
gB	glycoprotein B
gD	glycoprotein D
GFP	green fluorescent protein
gG	glycoprotein G
GM-CSF	granulocyte-macrophage colony-stimulating factor
gpt	guanine phosphoribosyltransferase gene
GUS	β-glucuronidase
HA	hemagglutinin
HBsAg	hepatitis B virus surface antigen
HBV	hepatitis B virus
hEPO	human erythropoietin
HIV	human immunodeficiency virus
hNIS	sodium iodide symporter

HSV	herpes simplex virus
IPTG	isopropyl beta-D-thiogalactopyranoside
ITRs	inverted terminal repeats
LACK	Leishmania homolog of activated C kinase
lacZ	β-galactosidase gene
LS	large surface protein
LUC	luciferase
M1	matrix 1
MS	middle surface protein
MUG	4-methylumbelliferyl beta-D-glucuronide
NF-kB	nuclear factor-kappa B
NP	nucleoprotein
NS1	non-structural protein 1
PA	polymerase acidic
PB1	polymerase basic 1
RT	retrotranscriptase
ONPG	ortho-nitrophenyl beta-galactoside
RFP	red fluorescent protein
RR	ribonucleotide reductase
SARS-CoV	severe acute respiratory syndrome-associated coronavirus
S-Gal	3,4-cyclohexenoesculetin beta-D-galactopyranoside
SR	somatostatin receptor
TDS	transient dominant selection
TK	thymidine kinase
TK-	TK-defective phenotype
VACV	vaccinia virus
VGF	vaccinia growth factor
WR	Western Reserve strain
X-Gal	5-bromo-4-chloro-3-indolyl beta-D-galactopyranoside
X-Gluc	5-bromo-4-chloro-3-indolyl beta-D-glucuronide
YFP	yellow fluorescent protein.

References

1. Wells, K.D. Genetic engineering of mammals. *Cell. Tissue Res.* **2016**, *363*, 289–294.
2. Gill, S.S.; Gill, R.; Tuteja, R.; Tuteja, N. Genetic engineering of crops: A ray of hope for enhanced food security. *Plant. Signal. Behav.* **2014**, *9*, e28545.
3. Van Pijkeren, J.P.; Britton, R.A. Precision genome engineering in lactic acid bacteria. *Microb. Cell. Fact.* **2014**, *13*, S10.

4. Ura, T.; Okuda, K.; Shimada, M. Developments in viral vector-based vaccines. *Vaccines* **2014**, *2*, 624–641.

5. Souza, A.P.; Haut, L.; Reyes-Sandoval, A.; Pinto, A.R. Recombinant viruses as vaccines against viral diseases. *Braz. J. Med. Biol. Res.* **2005**, *38*, 509–522.

6. Kim, T.K.; Eberwine, J.H. Mammalian cell transfection: The present and the future. *Anal. Bioanal. Chem.* **2010**, *397*, 3173–3178.

7. Nakano, E.; Panicali, D.; Paoletti, E. Molecular genetics of vaccinia virus: Demonstration of marker rescue. *Proc. Natl. Acad. Sci. USA* **1982**, *79*, 1593–1596.

8. Panicali, D.; Paoletti, E. Construction of poxviruses as cloning vectors: Insertion of the *thymidine kinase* gene from herpes simplex virus into the DNA of infectious vaccinia virus. *Proc. Natl. Acad. Sci. USA* **1982**, *79*, 4927–4931.

9. Sanchez-Sampedro, L.; Perdiguero, B.; Mejias-Perez, E.; Garcia-Arriaza, J.; Di Pilato, M.; Esteban, M. The evolution of poxvirus vaccines. *Viruses* **2015**, *7*, 1726–1803.

10. McFadden, G. Poxvirus tropism. *Nat. Rev. Microbiol.* **2005**, *3*, 201–213.

11. Bhattacharya, S. The world health organization and global smallpox eradication. *J. Epidemiol. Community Health* **2008**, *62*, 909–912.

12. Broyles, S.S. Vaccinia virus transcription. *J. Gen. Virol.* **2003**, *84*, 2293–2303.

13. Hruby, D.E. Vaccinia virus vectors: New strategies for producing recombinant vaccines. *Clin. Microbiol. Rev.* **1990**, *3*, 153–170.

14. Hughes, A.L.; Friedman, R. Poxvirus genome evolution by gene gain and loss. *Mol. Phylogenet. Evol.* **2005**, *35*, 186–195.

15. Moss, B. Poxvirus cell entry: How many proteins does it take? *Viruses* **2012**, *4*, 688–707.

16. Schmidt, F.I.; Bleck, C.K.; Mercer, J. Poxvirus host cell entry. *Curr. Opin. Virol.* **2012**, *2*, 20–27.

17. Rubins, K.H.; Hensley, L.E.; Bell, G.W.; Wang, C.; Lefkowitz, E.J.; Brown, P.O.; Relman, D.A. Comparative analysis of viral gene expression programs during poxvirus infection: A transcriptional map of the vaccinia and monkeypox genomes. *PLoS ONE* **2008**, *3*, e2628.

18. Roberts, K.L.; Smith, G.L. Vaccinia virus morphogenesis and dissemination. *Trends Microbiol.* **2008**, *16*, 472–479.

19. Yen, J.; Golan, R.; Rubins, K. Vaccinia virus infection & temporal analysis of virus gene expression: Part 1. *J. Vis. Exp.* **2009**, *26*, 1168.

20. Bleckwenn, N.A.; Bentley, W.E.; Shiloach, J. Exploring vaccinia virus as a tool for large-scale recombinant protein expression. *Biotechnol. Prog.* **2003**, *19*, 130–136.

21. Bleckwenn, N.A.; Bentley, W.E.; Shiloach, J. Evaluation of production parameters with the vaccinia virus expression system using microcarrier attached HeLa cells. *Biotechnol. Prog.* **2005**, *21*, 554–561.

22. Masternak, K.; Wittek, R. *Cis*- and *trans*-acting elements involved in reactivation of vaccinia virus early transcription. *J. Virol.* **1996**, *70*, 8737–8746.

23. Nascimento, I.P.; Leite, L.C. Recombinant vaccines and the development of new vaccine strategies. *Braz. J. Med. Biol. Res.* **2012**, *45*, 1102–1111.

24. Bleckwenn, N.A.; Golding, H.; Bentley, W.E.; Shiloach, J. Production of recombinant proteins by vaccinia virus in a microcarrier based mammalian cell perfusion bioreactor. *Biotechnol. Bioeng.* **2005**, *90*, 663–674.

25. Coupar, B.E.; Oke, P.G.; Andrew, M.E. Insertion sites for recombinant vaccinia virus construction: Effects on expression of a foreign protein. *J. Gen. Virol.* **2000**, *81*, 431–439.

26. Kim, M. Replicating poxviruses for human cancer therapy. *J. Microbiol.* **2015**, *53*, 209–218.

27. Garcel, A.; Fauquette, W.; Dehouck, M.P.; Crance, J.M.; Favier, A.L. Vaccinia virus-induced smallpox postvaccinal encephalitis in case of blood-brain barrier damage. *Vaccine* **2012**, *30*, 1397–1405.

28. Henderson, D.A.; Moss, B. *Recombinant vaccinia virus vaccines*, 3rd ed.; Saunders: Philadelphia, 1999.

29. Moore, A.R.; Dong, B.; Chen, L.; Xiao, W. Vaccinia virus as a subhelper for AAV replication and packaging. *Mol. Ther. Methods Clin. Dev.* **2015**, *2*, 15044.

30. Falkner, F.G.; Moss, B. Transient dominant selection of recombinant vaccinia viruses. *J. Virol.* **1990**, *64*, 3108–3111.

31. Falkner, F.G.; Moss, B. *Escherichia coli gpt* gene provides dominant selection for vaccinia virus open reading frame expression vectors. *J. Virol.* **1988**, *62*, 1849–1854.

32. Marzook, N.B.; Procter, D.J.; Lynn, H.; Yamamoto, Y.; Horsington, J.; Newsome, T.P. Methodology for the efficient generation of fluorescently tagged vaccinia virus proteins. *J. Vis. Exp.* **2014**, e51151.

33. Kochneva, G.; Zonov, E.; Grazhdantseva, A.; Yunusova, A.; Sibolobova, G.; Popov, E.; Taranov, O.; Netesov, S.; Chumakov, P.; Ryabchikova, E. Apoptin enhances the oncolytic properties of vaccinia virus and modifies mechanisms of tumor regression. *Oncotarget* **2014**, *5*, 11269–11282.

34. Rodriguez, D.; Gonzalez-Aseguinolaza, G.; Rodriguez, J.R.; Vijayan, A.; Gherardi, M.; Rueda, P.; Casal, J.I.; Esteban, M. Vaccine efficacy against malaria by the combination of porcine parvovirus-like particles and vaccinia virus vectors expressing CS of *Plasmodium*. *PLoS ONE* **2012**, *7*, e34445.

35. Gomez, C.E.; Rodriguez, D.; Rodriguez, J.R.; Abaitua, F.; Duarte, C.; Esteban, M. Enhanced CD8+ T cell immune response against a V3 loop multi-epitope polypeptide (TAB13) of HIV-1 Env after priming with purified fusion protein and booster with modified vaccinia virus ankara (MVA-TAB) recombinant: A comparison of humoral and cellular immune responses with the vaccinia virus western reserve (WR) vector. *Vaccine* **2001**, *20*, 961–971.

36. Brown, C.K.; Turner, P.C.; Moyer, R.W. Molecular characterization of the vaccinia virus hemagglutinin gene. *J. Virol.* **1991**, *65*, 3598–3606.

37. O'Brien, T.C.; Tauraso, N.M. Vaccinia virus: Kinetics of the hemagglutination-inhibition test and preparation of hemagglutinin. *Arch. Gesamte. Virusforsch.* **1972**, *36*, 158–165.

38. Paoletti, E.; Lipinskas, B.R.; Samsonoff, C.; Mercer, S.; Panicali, D. Construction of live vaccines using genetically engineered poxviruses: Biological activity of vaccinia virus recombinants expressing the hepatitis B virus surface antigen and the herpes simplex virus glycoprotein D. *Proc. Natl. Acad. Sci. USA* **1984**, *81*, 193–197.

39. Buller, R.M.; Chakrabarti, S.; Cooper, J.A.; Twardzik, D.R.; Moss, B. Deletion of the vaccinia virus growth factor gene reduces virus virulence. *J. Virol.* **1988**, *62*, 866–874.

40. Buller, R.M.; Smith, G.L.; Cremer, K.; Notkins, A.L.; Moss, B. Decreased virulence of recombinant vaccinia virus expression vectors is associated with a thymidine kinase-negative phenotype. *Nature* **1985**, *317*, 813–815.

41. Child, S.J.; Palumbo, G.J.; Buller, R.M.; Hruby, D.E. Insertional inactivation of the large subunit of ribonucleotide reductase encoded by vaccinia virus is associated with reduced virulence *in vivo*. *Virology* **1990**, *174*, 625–629.

42. Rodriguez, D.; Rodriguez, J.R.; Rodriguez, J.F.; Trauber, D.; Esteban, M. Highly attenuated vaccinia virus mutants for the generation of safe recombinant viruses. *Proc. Natl. Acad. Sci. USA* **1989**, *86*, 1287–1291.

43. Condit, R.C.; Motyczka, A.; Spizz, G. Isolation, characterization, and physical mapping of temperature-sensitive mutants of vaccinia virus. *Virology* **1983**, *128*, 429–443.

44. Jiang, T.; Xing, B.; Rao, J. Recent developments of biological reporter technology for detecting gene expression. *Biotechnol. Genet. Eng. Rev.* **2008**, *25*, 41–75.

45. Rice, A.D.; Gray, S.A.; Li, Y.; Damon, I.; Moyer, R.W. An efficient method for generating poxvirus recombinants in the absence of selection. *Viruses* **2011**, *3*, 217–232.

46. Yuan, M.; Gao, X.; Chard, L.S.; Ali, Z.; Ahmed, J.; Li, Y.; Liu, P.; Lemoine, N.R.; Wang, Y. A marker-free system for highly efficient construction of vaccinia virus vectors using CRISPR Cas9. *Mol. Ther. Methods Clin. Dev.* **2015**, *2*, 15035.

47. Yuan, M.; Zhang, W.; Wang, J.; Al Yaghchi, C.; Ahmed, J.; Chard, L.; Lemoine, N.R.; Wang, Y. Efficiently editing the vaccinia virus genome by using the CRISPR-Cas9 system. *J. Virol.* **2015**, *89*, 5176–5179.

48. Falzarano, D.; Groseth, A.; Hoenen, T. Development and application of reporter-expressing mononegaviruses: Current challenges and perspectives. *Antiviral Res.* **2014**, *103*, 78–87.

49. Ghim, C.M.; Lee, S.K.; Takayama, S.; Mitchell, R.J. The art of reporter proteins in science: Past, present and future applications. *BMB Rep.* **2010**, *43*, 451–460.

50. Swenson, E.S.; Price, J.G.; Brazelton, T.; Krause, D.S. Limitations of green fluorescent protein as a cell lineage marker. *Stem Cells* **2007**, *25*, 2593–2600.

51. Gorman, C.M.; Moffat, L.F.; Howard, B.H. Recombinant genomes which express chloramphenicol acetyltransferase in mammalian cells. *Mol. Cell. Biol.* **1982**, *2*, 1044–1051.

52. Overbeek, P.A.; Lai, S.P.; Van Quill, K.R.; Westphal, H. Tissue-specific expression in transgenic mice of a fused gene containing RSV terminal sequences. *Science* **1986**, *231*, 1574–1577.

53. Juers, D.H.; Matthews, B.W.; Huber, R.E. *LacZ* beta-galactosidase: Structure and function of an enzyme of historical and molecular biological importance. *Protein Sci.* **2012**, *21*, 1792–1807.

54. Carroll, M.W.; Moss, B. *E. coli* beta-glucuronidase (*GUS*) as a marker for recombinant vaccinia viruses. *BioTechniques* **1995**, *19*, 352–354, 356.

55. Bleckwenn, N.A.; Bentley, W.E.; Shiloach, J. Vaccinia virus-based expression of gp120 and eGFP: Survey of mammalian host cell lines. *Biotechnol. Prog.* **2005**, *21*, 186–191.

56. Gahan, C.G. The bacterial lux reporter system: Applications in bacterial localisation studies. *Curr. Gene Ther.* **2012**, *12*, 12–19.

57. Gould, S.J.; Subramani, S. Firefly luciferase as a tool in molecular and cell biology. *Anal. Biochem.* **1988**, *175*, 5–13.

58. Kirby, J.; Heath, P.R.; Shaw, P.J.; Hamdy, F.C. Gene expression assays. *Adv. Clin. Chem.* **2007**, *44*, 247–292.

59. Cui, W.; Liu, L.; Kodibagkar, V.D.; Mason, R.P. S-Gal, a novel 1h MRI reporter for beta-galactosidase. *Magn. Reson. Med.* **2010**, *64*, 65–71.

60. Chakrabarti, S.; Brechling, K.; Moss, B. Vaccinia virus expression vector: Coexpression of beta-galactosidase provides visual screening of recombinant virus plaques. *Mol. Cell. Biol.* **1985**, *5*, 3403–3409.

61. Villari, P.; Iannuzzo, M.; Torre, I. An evaluation of the use of 4-methylumbelliferyl-beta-d-glucuronide (MUG) in different solid media for the detection and enumeration of escherichia coli in foods. *Lett. Appl. Microbiol.* **1997**, *24*, 286–290.

62. Howley, P.M.; Spehner, D.; Drillien, R. A vaccinia virus transfer vector using a *GUS* reporter gene inserted into the I4L locus. *Gene* **1996**, *172*, 233–237.

63. Dvoracek, B.; Shors, T. Construction of a novel set of transfer vectors to study vaccinia virus replication and foreign gene expression. *Plasmid* **2003**, *49*, 9–17.

64. Tsuji, F.I. Early history, discovery, and expression of aequorea green fluorescent protein, with a note on an unfinished experiment. *Microsc. Res. Tech.* **2010**, *73*, 785–796.

65. The Nobel Prize in Chemistry 2008. Available online: http://www.nobelprize.org/nobel_prizes/chemistry/laureates/2008/ (accessed on 19 May 2016).

66. Coralli, C.; Cemazar, M.; Kanthou, C.; Tozer, G.M.; Dachs, G.U. Limitations of the reporter green fluorescent protein under simulated tumor conditions. *Cancer Res.* **2001**, *61*, 4784–4790.

67. Prescher, J.A.; Contag, C.H. Guided by the light: Visualizing biomolecular processes in living animals with bioluminescence. *Curr. Opin. Chem. Biol.* **2010**, *14*, 80–89.

68. Zhu, R.; Liu, Q.; Huang, W.; Yu, Y.; Wang, Y. Comparison of the replication characteristics of vaccinia virus strains Guang 9 and Tian Tan *in vivo* and *in vitro*. *Arch. Virol.* **2014**, *159*, 2587–2596.

69. Zaitseva, M.; Kapnick, S.M.; Meseda, C.A.; Shotwell, E.; King, L.R.; Manischewitz, J.; Scott, J.; Kodihalli, S.; Merchlinsky, M.; Nielsen, H.; *et al.* Passive immunotherapies protect WRvFire and IHD-J-Luc vaccinia virus-infected mice from lethality by reducing viral loads in the upper respiratory tract and internal organs. *J. Virol.* **2011**, *85*, 9147–9158.

70. Tsoneva, D.; Stritzker, J.; Bedenk, K.; Zhang, Q.; Frentzen, A.; Cappello, J.; Fischer, U.; Szalay, A.A. Drug-encoded biomarkers for monitoring biological therapies. *PLoS ONE* **2015**, *10*, e0137573.

71. Manischewitz, J.; King, L.R.; Bleckwenn, N.A.; Shiloach, J.; Taffs, R.; Merchlinsky, M.; Eller, N.; Mikolajczyk, M.G.; Clanton, D.J.; Monath, T.; *et al.* Development of a novel vaccinia-neutralization assay based on reporter-gene expression. *J. Infect. Dis.* **2003**, *188*, 440–448.

72. Li, Y.; Sheng, Y.; Chu, Y.; Ji, H.; Jiang, S.; Lan, T.; Li, M.; Chen, S.; Fan, Y.; Li, W.; *et al.* Seven major genomic deletions of vaccinia virus Tiantan strain are sufficient to decrease pathogenicity. *Antiviral Res.* **2016**, *129*, 1–12.

73. Turner, P.C.; Moyer, R.W. The vaccinia virus fusion inhibitor proteins SPI-3 (K2) and HA (A56) expressed by infected cells reduce the entry of superinfecting virus. *Virology* **2008**, *380*, 226–233.

74. Ward, B.M. Visualization and characterization of the intracellular movement of vaccinia virus intracellular mature virions. *J. Virol.* **2005**, *79*, 4755–4763.

75. Warren, R.D.; Cotter, C.A.; Moss, B. Reverse genetics analysis of poxvirus intermediate transcription factors. *J. Virol.* **2012**, *86*, 9514–9519.

76. Zhu, W.; Lei, R.; Le Duff, Y.; Li, J.; Guo, F.; Wainberg, M.A.; Liang, C. The CRISPR/Cas9 system inactivates latent HIV-1 proviral DNA. *Retrovirology* **2015**, *12*, 22.

77. Chan, W.M.; Ward, B.M. The A33-dependent incorporation of B5 into extracellular enveloped vaccinia virions is mediated through an interaction between their lumenal domains. *J. Virol.* **2012**, *86*, 8210–8220.

78. Smith, G.L.; Benfield, C.T.; Maluquer de Motes, C.; Mazzon, M.; Ember, S.W.; Ferguson, B.J.; Sumner, R.P. Vaccinia virus immune evasion: Mechanisms, virulence and immunogenicity. *J. Gen. Virol.* **2013**, *94*, 2367–2392.

79. Unterholzner, L.; Sumner, R.P.; Baran, M.; Ren, H.; Mansur, D.S.; Bourke, N.M.; Randow, F.; Smith, G.L.; Bowie, A.G. Vaccinia virus protein C6 is a virulence factor that binds tbk-1 adaptor proteins and inhibits activation of IRF3 and IRF7. *PLoS Pathog.* **2011**, *7*, e1002247.

80. Johnson, M.C.; Damon, I.K.; Karem, K.L. A rapid, high-throughput vaccinia virus neutralization assay for testing smallpox vaccine efficacy based on detection of green fluorescent protein. *J. Virol. Methods* **2008**, *150*, 14–20.

81. Wang, W.; Li, R.; Deng, Y.; Lu, N.; Chen, H.; Meng, X.; Wang, W.; Wang, X.; Yan, K.; Qi, X.; *et al.* Protective efficacy of the conserved NP, PB1, and M1 proteins as immunogens in DNA- and vaccinia virus-based universal influenza a virus vaccines in mice. *Clin. Vaccine Immunol.* **2015**, *22*, 618–630.

82. Van den Worm, S.H.; Eriksson, K.K.; Zevenhoven, J.C.; Weber, F.; Zust, R.; Kuri, T.; Dijkman, R.; Chang, G.; Siddell, S.G.; Snijder, E.J.; *et al.* Reverse genetics of SARS-related coronavirus using vaccinia virus-based recombination. *PLoS ONE* **2012**, *7*, e32857.

83. Harrington, L.E.; Most Rv, R.; Whitton, J.L.; Ahmed, R. Recombinant vaccinia virus-induced T-cell immunity: Quantitation of the response to the virus vector and the foreign epitope. *J. Virol.* **2002**, *76*, 3329–3337.

84. Qiu, S.; Ren, X.; Ben, Y.; Ren, Y.; Wang, J.; Zhang, X.; Wan, Y.; Xu, J. Fusion-expressed ctb improves both systemic and mucosal T-cell responses elicited by an intranasal DNA priming/intramuscular recombinant vaccinia boosting regimen. *J. Immunol. Res.* **2014**, *2014*, 308732.

85. Yaghchi, C.A.; Zhang, Z.; Alusi, G.; Lemoine, N.R.; Wang, Y. Vaccinia virus, a promising new therapeutic agent for pancreatic cancer. *Immunotherapy* **2015**, *7*, 1249–1258.

86. Vemulapalli, R.; Cravero, S.; Calvert, C.L.; Toth, T.E.; Sriranganathan, N.; Boyle, S.M.; Rossetti, O.L.; Schurig, G.G. Characterization of specific immune responses of mice inoculated with recombinant vaccinia virus expressing an 18-kilodalton outer membrane protein of brucella abortus. *Clin. Diagn. Lab. Immunol.* **2000**, *7*, 114–118.

87. Langford, C.J.; Edwards, S.J.; Smith, G.L.; Mitchell, G.F.; Moss, B.; Kemp, D.J.; Anders, R.F. Anchoring a secreted plasmodium antigen on the surface of recombinant vaccinia virus-infected cells increases its immunogenicity. *Mol. Cell. Biol.* **1986**, *6*, 3191–3199.

88. Smith, G.L.; Godson, G.N.; Nussenzweig, V.; Nussenzweig, R.S.; Barnwell, J.; Moss, B. *Plasmodium knowlesi* sporozoite antigen: Expression by infectious recombinant vaccinia virus. *Science* **1984**, *224*, 397–399.

89. Gherardi, M.M.; Ramirez, J.C.; Rodriguez, D.; Rodriguez, J.R.; Sano, G.; Zavala, F.; Esteban, M. IL-12 delivery from recombinant vaccinia virus attenuates the vector and enhances the cellular immune response against HIV-1 Env in a dose-dependent manner. *J. Immunol.* **1999**, *162*, 6724–6733.

90. Valkenburg, S.A.; Li, O.T.; Mak, P.W.; Mok, C.K.; Nicholls, J.M.; Guan, Y.; Waldmann, T.A.; Peiris, J.S.; Perera, L.P.; Poon, L.L. IL-15 adjuvanted multivalent vaccinia-based universal influenza vaccine requires CD4+ T cells for heterosubtypic protection. *Proc. Natl. Acad. Sci. USA* **2014**, *111*, 5676–5681.

91. Brochier, B.; Aubert, M.F.; Pastoret, P.P.; Masson, E.; Schon, J.; Lombard, M.; Chappuis, G.; Languet, B.; Desmettre, P. Field use of a vaccinia-rabies recombinant vaccine for the control of sylvatic rabies in europe and North America. *Rev. Sci. Tech.* **1996**, *15*, 947–970.

92. Gherardi, M.M.; Najera, J.L.; Perez-Jimenez, E.; Guerra, S.; Garcia-Sastre, A.; Esteban, M. Prime-boost immunization schedules based on influenza virus and vaccinia virus vectors potentiate cellular immune responses against human immunodeficiency virus Env protein systemically and in the genitorectal draining lymph nodes. *J. Virol.* **2003**, *77*, 7048–7057.

93. Walker, B.D.; Flexner, C.; Paradis, T.J.; Fuller, T.C.; Hirsch, M.S.; Schooley, R.T.; Moss, B. HIV-1 reverse transcriptase is a target for cytotoxic T lymphocytes in infected individuals. *Science* **1988**, *240*, 64–66.

94. Cheliapov, N.V.; Chernos, V.I.; Andzhaparidze, O.G. Analysis of antibody formation to the vaccinia virus in human subjects and rabbits in response to the administration of a recombinant vaccinia-hepatitis B vaccine. *Vopr. Virusol.* **1988**, *33*, 175–179.

95. Grigorieva, I.M.; Grigoriev, V.G.; Zakharova, L.G.; Pashvykina, G.V.; Shevlyagin, V.Y.; Altstein, A.D. Immunogenicity of recombinant vaccinia viruses expressing hepatitis B virus surface antigen in mice. *Immunol. Lett.* **1993**, *36*, 267–271.

96. Kutinova, L.; Ludvikova, V.; Krystofova, J.; Otavova, M.; Simonova, V.; Nemeckova, S.; Hainz, P.; Vonka, V. Influence of the parental virus strain on the virulence and immunogenicity of recombinant vaccinia viruses expressing HBV Pres2-S protein or VZV glycoprotein I. *Vaccine* **1996**, *14*, 1045–1052.

97. Cheng, K.C.; Smith, G.L.; Moss, B. Hepatitis B virus large surface protein is not secreted but is immunogenic when selectively expressed by recombinant vaccinia virus. *J. Virol.* **1986**, *60*, 337–344.

98. Cheng, K.C.; Moss, B. Selective synthesis and secretion of particles composed of the hepatitis B virus middle surface protein directed by a recombinant vaccinia virus: Induction of antibodies to pre-S and S epitopes. *J. Virol.* **1987**, *61*, 1286–1290.

99. Rooney, J.F.; Wohlenberg, C.; Cremer, K.J.; Moss, B.; Notkins, A.L. Immunization with a vaccinia virus recombinant expressing herpes simplex virus type 1 glycoprotein D: Long-term protection and effect of revaccination. *J. Virol.* **1988**, *62*, 1530–1534.

100. Martin, S.; Moss, B.; Berman, P.W.; Laskey, L.A.; Rouse, B.T. Mechanisms of antiviral immunity induced by a vaccinia virus recombinant expressing herpes simplex virus type 1 glycoprotein D: Cytotoxic t cells. *J. Virol.* **1987**, *61*, 726–734.

101. Cantin, E.M.; Eberle, R.; Baldick, J.L.; Moss, B.; Willey, D.E.; Notkins, A.L.; Openshaw, H. Expression of herpes simplex virus 1 glycoprotein B by a recombinant vaccinia virus and protection of mice against lethal herpes simplex virus 1 infection. *Proc. Natl. Acad. Sci. USA* **1987**, *84*, 5908–5912.

102. McLaughlin-Taylor, E.; Willey, D.E.; Cantin, E.M.; Eberle, R.; Moss, B.; Openshaw, H. A recombinant vaccinia virus expressing herpes simplex virus type 1 glycoprotein B induces cytotoxic T lymphocytes in mice. *J. Gen. Virol.* **1988**, *69*(Pt. 7), 1731–1734.

103. Sullivan, V.; Smith, G.L. Expression and characterization of herpes simplex virus type 1 (HSV-1) glycoprotein G (gG) by recombinant vaccinia virus: Neutralization of HSV-1 infectivity with anti-gG antibody. *J. Gen. Virol.* **1987**, *68*(Pt. 10), 2587–2598.

104. Smith, G.L.; Murphy, B.R.; Moss, B. Construction and characterization of an infectious vaccinia virus recombinant that expresses the influenza hemagglutinin gene and induces resistance to influenza virus infection in hamsters. *Proc. Natl. Acad. Sci. USA* **1983**, *80*, 7155–7159.

105. Goodman, A.G.; Heinen, P.P.; Guerra, S.; Vijayan, A.; Sorzano, C.O.; Gomez, C.E.; Esteban, M. A human multi-epitope recombinant vaccinia virus as a universal T cell vaccine candidate against influenza virus. *PLoS ONE* **2011**, *6*, e25938.

106. Ramos, I.; Alonso, A.; Marcen, J.M.; Peris, A.; Castillo, J.A.; Colmenares, M.; Larraga, V. Heterologous prime-boost vaccination with a non-replicative vaccinia recombinant vector expressing lack confers protection against canine visceral leishmaniasis with a predominant Th1-specific immune response. *Vaccine* **2008**, *26*, 333–344.

107. Cross, M.L.; Fleming, S.B.; Cowan, P.E.; Scobie, S.; Whelan, E.; Prada, D.; Mercer, A.A.; Duckworth, J.A. Vaccinia virus as a vaccine delivery system for marsupial wildlife. *Vaccine* **2011**, *29*, 4537–4543.

108. Hruby, D.E.; Hodges, W.M.; Wilson, E.M.; Franke, C.A.; Fischetti, V.A. Expression of streptococcal M protein in mammalian cells. *Proc. Natl. Acad. Sci. USA* **1988**, *85*, 5714–5717.

109. Chan, W.M.; McFadden, G. Oncolytic poxviruses. *Annu. Rev. Virol.* **2014**, *1*, 119–141.

110. Mansfield, D.C.; Kyula, J.N.; Rosenfelder, N.; Chao-Chu, J.; Kramer-Marek, G.; Khan, A.A.; Roulstone, V.; McLaughlin, M.; Melcher, A.A.; Vile, R.G.; *et al.* Oncolytic vaccinia virus as a vector for therapeutic sodium iodide symporter gene therapy in prostate cancer. *Gene Ther.* **2016**, *23*, 357–368.

111. Pugalenthi, A.; Mojica, K.; Ady, J.W.; Johnsen, C.; Love, D.; Chen, N.G.; Aguilar, R.J.; Szalay, A.A.; Fong, Y. Recombinant vaccinia virus GLV-1h68 is a promising oncolytic vector in the treatment of cholangiocarcinoma. *Cancer Gene Ther.* **2015**, *22*, 591–596.

112. Park, S.H.; Breitbach, C.J.; Lee, J.; Park, J.O.; Lim, H.Y.; Kang, W.K.; Moon, A.; Mun, J.H.; Sommermann, E.M.; Maruri Avidal, L.; *et al.* Phase 1b trial of biweekly intravenous Pexa-Vec (JX-594), an oncolytic and immunotherapeutic vaccinia virus in colorectal cancer. *Mol. Ther.* **2015**, *23*, 1532–1540.

113. Bauzon, M.; Hermiston, T. Armed therapeutic viruses - a disruptive therapy on the horizon of cancer immunotherapy. *Front. Immunol.* **2014**, *5*, 74.

114. Merrick, A.E.; Ilett, E.J.; Melcher, A.A. Jx-594, a targeted oncolytic poxvirus for the treatment of cancer. *Curr. Opin. Investig. Drugs.* **2009**, *10*, 1372–1382.

115. McCart, J.A.; Ward, J.M.; Lee, J.; Hu, Y.; Alexander, H.R.; Libutti, S.K.; Moss, B.; Bartlett, D.L. Systemic cancer therapy with a tumor-selective vaccinia virus mutant lacking thymidine kinase and vaccinia growth factor genes. *Cancer Res.* **2001**, *61*, 8751–8757.

116. Lun, X.; Ruan, Y.; Jayanthan, A.; Liu, D.J.; Singh, A.; Trippett, T.; Bell, J.; Forsyth, P.; Johnston, R.N.; Narendran, A. Double-deleted vaccinia virus in virotherapy for refractory and metastatic pediatric solid tumors. *Mol. Oncol.* **2013**, *7*, 944–954.

117. Nguyen, D.H.; Chen, N.G.; Zhang, Q.; Le, H.T.; Aguilar, R.J.; Yu, Y.A.; Cappello, J.; Szalay, A.A. Vaccinia virus-mediated expression of human erythropoietin in tumors enhances virotherapy and alleviates cancer-related anemia in mice. *Mol. Ther.* **2013**, *21*, 2054–2062.

118. McCart, J.A.; Mehta, N.; Scollard, D.; Reilly, R.M.; Carrasquillo, J.A.; Tang, N.; Deng, H.; Miller, M.; Xu, H.; Libutti, S.K.; *et al.* Oncolytic vaccinia virus expressing the human somatostatin receptor SSTR2: Molecular imaging after systemic delivery using 111in-pentetreotide. *Mol. Ther.* **2004**, *10*, 553–561.

119. Mei, Y.; Wang, Y.; Chen, H.; Sun, Z.S.; Ju, X.D. Recent progress in CRISPR/Cas9 technology. *J. Genet. Genomics.* **2016**, *43*, 63–75.

120. Suenaga, T.; Kohyama, M.; Hirayasu, K.; Arase, H. Engineering large viral DNA genomes using the CRISPR-Cas9 system. *Microbiol. Immunol.* **2014**, *58*, 513–522.

121. Lin, G.; Zhang, K.; Li, J. Application of CRISPR/Cas9 technology to HBV. *Int. J. Mol. Sci.* **2015**, *16*, 26077–26086.

Marburg Virus Reverse Genetics Systems

Kristina Maria Schmidt and Elke Mühlberger

Abstract: The highly pathogenic Marburg virus (MARV) is a member of the *Filoviridae* family and belongs to the group of nonsegmented negative-strand RNA viruses. Reverse genetics systems established for MARV have been used to study various aspects of the viral replication cycle, analyze host responses, image viral infection, and screen for antivirals. This article provides an overview of the currently established MARV reverse genetic systems based on minigenomes, infectious virus-like particles and full-length clones, and the research that has been conducted using these systems.

Reprinted from *Viruses*. Cite as: Schmidt, K.M.; Mühlberger, E. Marburg Virus Reverse Genetics Systems. *Viruses* **2016**, *8*, 178.

1. Introduction

1.1. Epidemiology

The *Filoviridae* family is subdivided into three distinct genera, *Marburgvirus*, *Ebolavirus* and *Cuevavirus*. The genus *Marburgvirus* includes a single species, *Marburg marburgvirus*, which is represented by two distinct viruses, Marburg virus (MARV) and Ravn virus (RAVV) [1]. Both MARV and RAVV cause a severe hemorrhagic disease in humans and susceptible animals [2]. The first reported MARV outbreak took place in Marburg and Frankfurt, Germany and Belgrade, Yugoslavia (now Serbia) in 1967, nine years prior to the first emergence of the better-known ebolaviruses, and was caused by infected African green monkeys imported from Uganda. Seven (22%) of the 31 infected patients succumbed to the disease (reviewed in [3]). All consecutive marburgvirus episodes and outbreaks were traced back to sub-Saharan Africa. The so far largest MARV outbreak took place in Uíge, Angola with 252 affected patients, many of them children, and a devastating case fatality rate of 90% [4,5]. This striking difference in survival rates between the 1967 MARV outbreak in resource-rich Europe, where the patients received aggressive medical treatment, and resource-poor Angola with suboptimal treatment options supports the observation during the most recent Ebola virus (EBOV) outbreak in West Africa that high-resource intensive care measurements significantly improve disease outcome [6]. However, other parameters, including transmission route and differences in the virulence of the viral strains might also have accounted for the observed differences in case fatality rates.

Similar to ebolaviruses, human-to-human transmission of MARV mainly occurs via body fluids and requires close contact to infected patients or deceased. Sexual transmission was reported during the 1967 MARV outbreak (for review see [3]). Due to the severity of the disease and the lack of treatment and vaccination options, work with MARV and RAVV is restricted to biosafety level (BSL) 4 facilities.

Filoviruses are zoonotic viruses. Strikingly, almost all primary marburgvirus infections in humans were traced back to caves inhabited by bats, and it was possible to isolate live MARV and RAVV viruses from the common Egyptian fruit bat (*Rousettus aegyptiacus* (*R. aegyptiacus*)) [7–13]. The marburgvirus bat isolates mirror the genetic diversity found in human isolates [10]. *R. aegyptiacus* bats experimentally infected with MARV did not show obvious signs of disease. Although viral titers determined from body fluids and tissues of the infected animals were generally moderate, oral and rectal shedding was observed in some, but not all, of the studies and could be a possible route of infection of humans [14–16]. However, MARV was not transmitted from experimentally infected to susceptible in-contact bats [17].

1.2. Virus Structure and Genome Oganization

1.2.1. Viral Proteins

This article only provides a brief overview of the MARV proteins and their functions. For a more detailed description see [3,18].

The filamentous MARV particles consist of a host cell-derived membrane, seven viral proteins and the nonsegmented negative-sense RNA genome. The single surface protein, glycoprotein (GP), is inserted into the viral membrane [19]. GP is required for attachment, receptor binding and fusion, and enhances budding (reviewed in [20,21]). After synthesis in the endoplasmic reticulum (ER) and during its transport to the cell membrane, GP is cleaved in the *trans*-Golgi network in two subunits, GP_1 and GP_2 which are covalently linked [22]. While GP_1 mediates attachment to the cell surface and receptor binding [20], membrane-bound GP_2 contains the fusion domain [23]. In addition to its function in viral entry and budding, GP plays an important role as target protein for the development of antiviral therapeutics, including therapeutic monoclonal antibodies, and vaccines [24–28].

The viral protein (VP) 40 is a typical viral matrix protein and mediates budding [29–31]. In contrast to EBOV VP40 (eVP40), MARV VP40 (mVP40) antagonizes the Janus kinase/Signal Transducer and Activator of Transcription (JAK/STAT) signaling pathway and plays a crucial role as a virulence factor in host adaptation [32–36].

The helical MARV ribonucleoprotein complex, or nucleocapsid, is composed of five viral proteins, nucleoprotein (NP), VP35, VP30, large protein (L) and VP24, and the viral RNA. VP24, a viral protein unique to filoviruses, is loosely attached to the nucleocapsid [37,38]. VP24 is involved in viral particle release, possibly in nucleocapsid maturation, and might also play a role in the regulation of viral genome replication [39,40]. Recently, it has been shown that MARV VP24 (mVP24) competes with the nuclear transcription factor erythroid-derived 2 (Nrf2) for binding with Kelch-like ECH-associated protein 1 (Keap1), a negative regulator of Nrf2. This leads to the constitutive activation of the anti-oxidative stress response, including the expression of cytoprotective genes, in MARV-infected cells [41,42]. Intriguingly, EBOV VP24 (eVP24) does not bind to Keap1.

Both the MARV genomic and antigenomic RNAs are encapsidated by NP. NP is the driving force for nucleocapsid formation. It self-assembles along the viral RNA and interacts with the other nucleocapsid proteins either directly or via protein linkers to form the nucleocapsids [37,38,43,44]. MARV NP (mNP) is also involved in the recruitment of cellular components of the endosomal sorting complexes required for transport (ESCRT) to support intracellular transport and budding of viral particles [45–47].

VP35 is a multifunctional protein involved in nucleocapsid formation, viral RNA synthesis, and the suppression of antiviral responses. It interacts with NP and seems to play an important role in chaperoning the formation of the NP-RNA complex. Recent crystal structure analyses revealed that EBOV VP35 (eVP35) contains an intrinsically disordered region that prevents EBOV NP (eNP) oligomerization and releases RNA from eNP-RNA complexes [48,49]. Although comparable structural analyses are not yet available for MARV VP35 (mVP35)-NP-RNA complexes, it is conceivable that similar encapsidation strategies may apply. Besides its function as a structural nucleocapsid component, VP35 is a polymerase cofactor and, together with L, forms the viral RNA-dependent RNA polymerase complex. VP35 binds to both NP and L and connects L to the NP-RNA complex [38,50]. L is the enzymatic subunit of the polymerase complex [3,18,51]. VP35 is not only required for viral replication and transcription, it also interferes with the host innate immune response by blocking retinoic acid-inducible gene 1 (RIG-I)-like receptor-mediated type I interferon (IFN) induction. mVP35 can bind to double-strand RNA (dsRNA), and the integrity of its dsRNA binding domain is essential for its antagonistic properties [52–56].

The function of the nucleoprotein VP30 in the MARV replication cycle is poorly understood. Although it is tightly associated with the nucleocapsid complex, it is dispensable for proper nucleocapsid formation [37,38,57]. In contrast to EBOV VP30 (eVP30), which is a transcription activator and significantly enhances transcriptional activity in an EBOV minigenome system [58,59], MARV VP30 (mVP30) is not essential for efficient transcription of MARV minigenomes and has only moderate effects on minigenome-driven reporter gene expression [40,60,61]. However, rescue of full-length MARV clones was only successful in the presence of VP30 indicating an important role of mVP30 during the MARV replication cycle [62]. This observation was supported by a small interfering RNA (siRNA) study, in which downregulation of VP30 in MARV-infected cells strongly reduced viral protein amounts [63]. Despite the obvious functional differences observed in the minigenome systems, mVP30 and eVP30 are structurally closely related. This close relationship is reflected by functional similarities. Thus, mVP30 was able to enhance transcription in an EBOV minigenome system, albeit with reduced efficiency [58]. Vice versa, rescue of MARV full-length clones was also successful when mVP30 was replaced by eVP30 [62]. In summary, mVP30 is an essential component of the MARV replication cycle, but its function is less well studied than that of its EBOV counterpart.

1.2.2. Genome Organization

The nonsegmented negative-sense RNA genome of MARV is about 19 kb in length and contains seven monocistronic genes encoding the seven viral proteins (Figure 1, top). Each gene is flanked by conserved gene start (GS) and gene end (GE) signals which are recognized by the viral polymerase as the sites of transcription initiation and termination. The genes are either separated by intergenic regions of variable length or they overlap [3,51,64,65]. Regulatory cis-acting elements containing the promoter regions are located at the 3′ and 5′ ends of the genome (Figure 1, top). These regions are the leader (3′ end of the negative-sense genome) and the trailer (5′ end of the genome). The leader of the MARV Musoke isolate is 48 nucleotides (nts) in length and contains the first promoter element of the bipartite replication promoter. The second promoter element is located within the 3′ untranslated region (negative-sense orientation) of the NP gene [66]. The transcription promoter is also located in the leader but is not mapped yet. The trailer region contains the complementary replication promoter which is used to produce genomes from the positive-sense antigenomic RNA template.

MARV transcription follows the stop-start model postulated for all nonsegmented negative-sense (NNS) RNA viruses [67]. The viral polymerase enters the genome at a single transcription promoter located within the leader and scans the genome until it reaches the GS signal of the first gene where it initiates transcription. The nascent mRNA is capped by the polymerase complex immediately after transcription initiation. The polymerase moves along the template until it recognizes a GE signal, leading to transcription termination. Using an unusual stuttering mechanism, the polymerase adds a poly-A tail to the nascent mRNA strand. It then scans for the next GS signal to initiate transcription of the following gene. If the polymerase complex falls off the template, it has to re-enter the genome at the transcription promoter located in the leader. This sequential transcription leads to an mRNA gradient with the 3′ proximal genes being more frequently transcribed than the 5′ proximal genes [65].

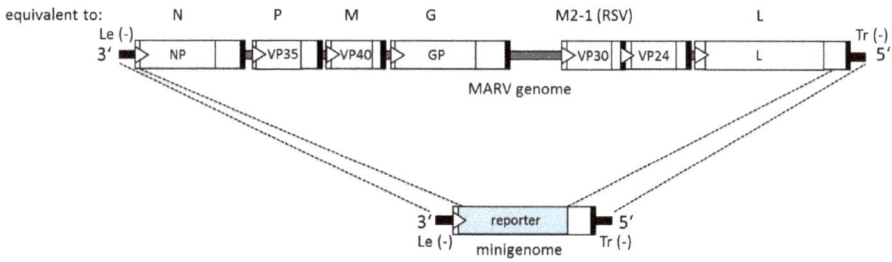

Figure 1. Scheme of the Marburg virus (MARV) genome (**top**) and the minigenome (**bottom**) in negative-sense orientation. Coding regions of the seven MARV genes are shown as grey boxes in the genome. The homologous proteins of other nonsegmented negative-sense (NNS) RNA viruses are indicated above the genome. In the minigenome, the viral genes are replaced by a single reporter gene, which is flanked by the MARV leader and trailer regions, the nontranslated region of the *nucleoprotein* (*NP*) gene containing the gene start (GS) signal, and the nontranslated region of the *large protein* (*L*) gene containing the gene end (GE) signal. Black lines, leader and trailer regions; gray lines, intergenic regions; white bars, nontranslated regions; white arrow heads, GS signals; black bars, GE signals; blue box, reporter gene.

During genome replication, the viral polymerase binds to the bipartite replication promoter and generates a full-length complementary copy of the genome, the antigenome. The GS and GE signals are ignored when the polymerase is in replication mode. The antigenome in turn serves as the template for the production of viral genomes (Figures 2 and 3). Both genome and antigenome are encapsidated by the nucleocapsid proteins. This is not the case for the viral mRNAs [3,18,51,60].

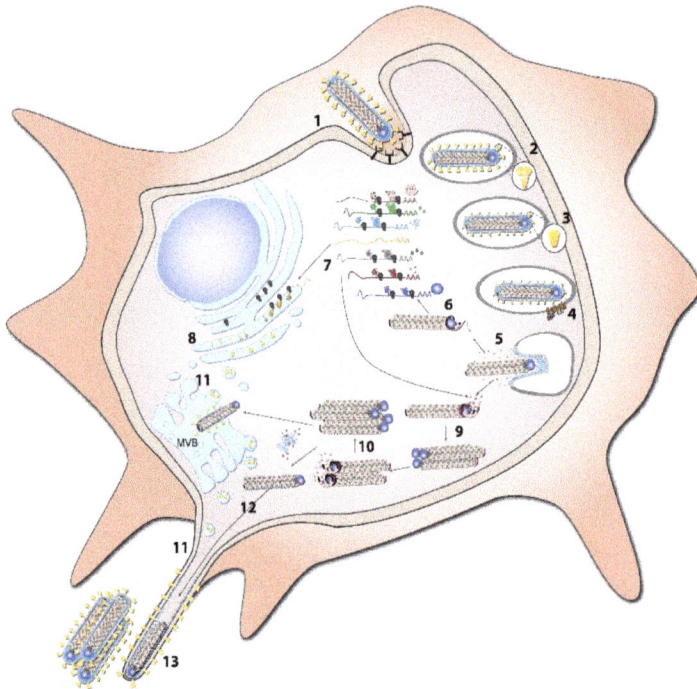

Figure 2. MARV replication cycle. MARV attaches to the surface of target cells by binding to attachment factors (**1**); following endocytosis (**2**); glycoprotein subunit GP$_1$ is cleaved by endosomal proteases (**3**) facilitating binding to Niemann-Pick C1 (NPC1); the entry receptor (**4**); fusion is mediated in a pH-dependent manner by glycoprotein subunit GP$_2$. Following release of viral nucleocapsid into the cytosol (**5**); transcription of the viral genome takes place (**6**); mRNA is subsequently translated by the host cell machinery (**7**); synthesis of GP takes place at the endoplasmic reticulum (ER) and undergoes multiple post-translational modifications on its way through the classical secretory pathway (**8**); positive sense antigenomes are synthesized from the incoming viral genomes (**9**); these intermediate products then serve as templates to replicate new negative-sense genomes (**10**); after cleavage in the Golgi, GP is transported to multivesicular bodies (MVB) and to the cell membrane where budding takes place preferentially from filopodia (**11**); nucleocapsids and viral protein (VP) 24 are also recruited to sites of viral budding (**12**); which is mainly driven by VP40 (**13**); figure and modified legend from [3].

154

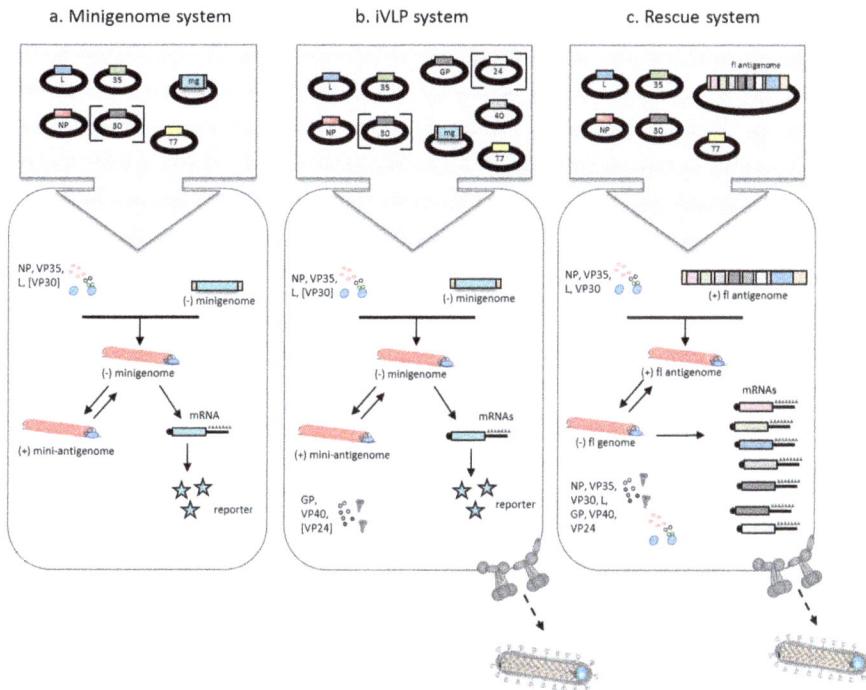

Figure 3. MARV reverse genetic systems. (**a**) Minigenome system; (**b**) infectious virus-like particles (iVLP) system; and (**c**) rescue system. In all systems, cells are transfected with the expression plasmids for the nucleocapsid proteins (NP, VP35, VP30, large protein (L)). For the iVLP system, the expression plasmids for GP, VP40 and VP24 are also included to facilitate budding. Viral proteins, which are not essential for the respective system, are bracketed. The T7 RNA polymerase (T7), which is required for the expression of the minigenome (mg), the full-length antigenome (fl antigenome) and, depending on the plasmid backbone, for the expression of the support proteins, can be provided by using T7-expressing cells, infection with T7-expressing viruses or by transfecting a T7 expression plasmid. (**a,b**) Minigenomic or (**c**) antigenomic cDNA is transcribed by T7 resulting in negative-sense minigenomic RNA or positive-sense fl antigenomic RNA. Encapsidation is mediated by the expressed viral nucleocapsid proteins. (**a,b**) The minigenome is replicated via a positive-sense mini-antigenome intermediate. Transcription of the minigenome leads to reporter gene expression; (**c**) The fl antigenome serves as the template for the production of viral genomic RNA which in turn is the template for viral mRNA as well as antigenome production. (**b,c**) In the iVLP and rescue systems, viral particles are released from the transfected cells. The iVLPs contain minigenome RNA, while the rescue system leads to the production of infectious MARV particles.

1.3. Replication Cycle

A detailed description of the MARV replication cycle is provided in [3]. In brief, uptake of MARV particles is mediated by GP (Figure 2). Following initial GP-mediated attachment to various cell surface proteins, the viral particles are endocytosed (reviewed in [20,68]). It is likely that macropinocytosis plays a crucial role in MARV entry as this mechanism was identified as a major uptake pathway for EBOV particles [69,70]. Endosomal cleavage of GP_1 by host cell proteases is required for binding to its receptor, the endosomal protein Niemann-Pick C1 (NPC1) [71–74] Fusion with the cellular membrane is preceded by a pH-dependent structural rearrangement of the GP_2 subunit which contains the fusion peptide [23,75]. After fusion of the viral and cellular membranes, the nucleocapsids are released into the cytoplasm where transcription and replication of the viral genomes take place (Figure 2). The viral monocistronic mRNAs are translated by cellular machinery to produce viral proteins. The newly synthesized viral genomes (and antigenomes) are encapsidated by the nucleocapsid proteins and assemble to large, highly ordered structures in the cytoplasm of the infected cells, the so-called inclusions [43,76,77]. These inclusions are believed to be the sites of viral genome replication and nucleocapsid maturation [37]. Mature nucleocapsids are transported along actin filaments from the inclusions to the sites of budding by exploiting components of the ESCRT complex, including tumor susceptibility gene 101 (Tsg101) [45–47]. Budding of viral particles is mainly mediated by VP40 and occurs internally at multivesicular bodies and at the plasma membrane preferentially from filopodia [29,77–80]. Viral particle release is enhanced by NP, GP and VP24 [39,45,46,81]. Studies on MARV budding in different cell models suggest that cell-type specific components determine whether the viral particles are released from the apical or the basolateral membrane [82–84].

2. MARV Reverse Genetics Systems

Per definition, reverse genetics is the functional analysis of genes by examining the phenotypic effects of targeted gene alterations. Reverse genetics systems have been successfully developed for numerous NNS RNA viruses of the order *Mononegavirales* [85,86]. This powerful technology has been used to address questions regarding all aspects of the viral infection, including viral genome replication, pathogenesis, and virus-host interactions. In addition, reverse genetics systems have been instrumental for the development of vaccines and antiviral screening assays (reviewed in [86]). In contrast to positive-sense RNA viruses, whose genome is used as an mRNA and is sufficient for virus particle formation when transfected into cells, the minimal infectious unit for NNS RNA viruses is the ribonucleoprotein complex, in which the viral RNA is encapsidated by the viral ribonucleoproteins before it can serve as a functional template to initiate viral transcription and genome replication (reviewed in [85–87]). Because neither the genome nor the antigenome of NNS RNA viruses can be used as an mRNA to generate viral proteins, the viral proteins required for viral transcription and genome replication have to be provided in *trans*.

Reverse genetics approaches for NNS RNA viruses range from minigenome systems to replicons, infectious virus-like particles (iVLPs) and full-length rescue systems. Since targeted genetic manipulation of viral RNA genomes is not possible yet, a tremendous advantage of all these systems is that the reverse transcribed cDNA copy of the RNA genome allows for the insertion of mutations and additional transcription units (ATUs; e.g., reporter genes). The first reverse genetics system established for filoviruses was the MARV minigenome system [60]. The different established MARV reverse genetics systems are summarized in Table 1 and described in the following overview.

Table 1. List of Marburg virus (MARV) reverse genetics systems.

System	Isolate	Used For	Publications
minigenome	Musoke	Defining the role of viral proteins in transcription and replication, genus-specificity of viral proteins in replication/transcription, amount and ratio of nucleocapsid proteins, functional studies of nucleocapsid proteins, analyzing *cis*-acting elements and their genus-specificity	[50,58,60,66,88–90]
minigenome	371Bat	Optimization of system with codon-optimized support plasmids, establishing high throughput antiviral screening assay, analyzing immune modulatory functions of viral proteins	[54,61,91]
iVLP	Musoke	Release and infectivity of iVLPs, titration of support plasmids for iVLP assay, genus-specificity of viral proteins in budding, testing of neutralizing antibodies, defining the role of viral proteins in the viral replication cycle, host adaptation	[30,36,40,92–94]
rescue system	Musoke	Role of VP30 for virus replication, analyzing *cis*-acting elements, live-cell imaging, transport and release of nucleocapsids, assembly of viral envelope, host adaptation	[36,46,62,66,77,80, 95,96]
rescue system	371Bat	Host response and immune modulatory activity, establishing high throughput antiviral screening assay	[54,61,91]
rescue system	Ebola virus Mayinga	Rescue of Ebola virus with MARV Musoke support plasmids	[97]

2.1. Minigenome Systems

Minigenomes are truncated versions of the viral genome as described below. The first cDNA-based minigenome system for NNS RNA viruses was generated for vesicular stomatitis virus (VSV) in the Ball and Wertz laboratories in 1992 [98]. The development of an exclusively cDNA-based minigenome system for NNS RNA viruses was hampered for a long time by the requirement of precise minigenome ends to initiate replication and transcription by the viral polymerase complex. To overcome this issue, Pattnaik and colleagues came up with an elegantly designed plasmid containing a VSV minigenome. In this plasmid, the minigenome was inserted immediately downstream of the T7 RNA polymerase promoter leading to the generation of discrete 5' ends (negative-sense orientation) when the minigenome was transcribed by the T7 RNA polymerase in transfected cells. To generate precise 3' ends, the hepatitis delta virus (HDV) ribozyme sequence was inserted downstream of the minigenome sequence. Following T7 RNA polymerase-mediated transcription of the minigenome-ribozyme hybrid RNA, the ribozyme sequence was removed through its autocatalytic cleavage activity thereby generating exact 3' minigenome ends [98–100]. This elegant approach led to a tremendous boost in generating reverse genetics systems for NNS RNA viruses including MARV [85].

The first MARV minigenome system was generated in the Mühlberger laboratory based on A. Ball's vector p 2.0 in which the MARV minigenome sequence was inserted between the T7 RNA polymerase promoter and the HDV ribozyme [60,98]. To our knowledge, all subsequently generated MARV minigenomes follow a similar design [61]. The minimal requirement for replicating minigenomes is the possession of the replication promoters at both minigenome ends [60]. Transcribing minigenomes also need the transcription promoter located at the 3' end of the minigenome (negative-sense orientation) and virus-specific GS and GE signals flanking each gene [60,66]. The viral genes are removed and replaced by one (monocistronic minigenomes) or more (polycistronic minigenomes) non-viral genes, usually reporter genes. Reporter gene expression is an easy readout for successful minigenome transcription and replication.

In the first generation of monocistronic MARV minigenomes, the reporter gene was flanked by 106 nts of the 3' end of the MARV genome and 439 nts of the 5' end of the genome. The 3' end of the minigenome preceding the reporter gene spans the leader, the NP GS signal and the 3' non-translated region of the NP gene (negative-sense orientation). The bipartite replication promoter of MARV is contained within this sequence [66]. The 5' end of the minigenome downstream of the reporter gene consists of the 5' non-translated region of the L gene, the L GE signal and the trailer region of the viral genome. Reporter gene transcription is driven by the GS signal of NP for initiation and the GE signal of L for termination [60] (Figure 1).

Variations of this minigenome design include replacing the leader region with a copy-back trailer region, inserting the minigenome in positive-sense orientation [60], generating chimeric MARV/EBOV minigenomes [66] and extending the monocistronic minigenome to bicistronic MARV minigenomes by inserting two reporter genes separated by a MARV-specific intergenic region [90]. In addition, minigenomes with different reporter genes have been generated, including chloramphenicol acetyltransferase (CAT), firefly luciferase (fLuc; Mühlberger, unpublished), Renilla luciferase (rLuc), Gaussia luciferase (gLuc), green fluorescent protein (GFP) and Cypridina luciferase (cLuc) [40,60,61,90,91]. The rLuc and gLuc reporter minigenome systems have the potential to be used as high-throughput antiviral screening platforms [40,91].

All recombinant MARV replication and transcription systems follow the same basic approach. The MARV minigenome plasmid is transfected along with the plasmids encoding the nucleocapsid proteins NP, VP35, and L which are the minimal requirement for MARV minigenome replication and transcription [60] (Figure 3a). In contrast to the Ebola virus minigenome assay [58], the nucleocapsid protein VP30 is not required for efficient transcription in the MARV minigenome system [60], although addition of VP30 moderately enhanced minigenome activity [40] and led to an increase of GFP-positive cells in a GFP-based MARV minigenome system [61]. However, VP30 is essential for the rescue of MARV from full-length cDNA clones (see Section 2.3). Initial transcription of the minigenome by the T7 RNA polymerase takes place in the cytoplasm. There are different approaches to provide the T7 RNA polymerase in *trans*, including (i) expression from a plasmid encoding the T7 RNA polymerase under the control of a eukaryotic promoter (e.g., pCAGGS-T7 [101]); (ii) use of a cell line constitutively expressing the T7 RNA polymerase (e.g., the baby hamster kidney cell line BSR-T5/7 [102]); or (iii) infection with a recombinant vaccinia virus encoding the T7 RNA polymerase (MVA-T7) [103].

The T7 RNA polymerase-derived minigenome RNA is then used as a template for transcription and replication by the newly synthesized nucleocapsid proteins (Figure 3a). There are different ways to express the support proteins: the nucleocapsid proteins genes are either transcribed by the T7 RNA polymerase in the cytoplasm [60] or by the cellular RNA polymerase II in the nucleus of the transfected cells [40, 61]. It is conceivable that the MARV mRNAs contain cryptic splice sites which can be targeted by the spliceosome when transcription takes place in the nucleus, leading to unwanted splicing events and consequently, reduced protein expression. This issue can be overcome using codon-optimized constructs in which putative cryptic splice sites are eliminated [104]. Using codon-optimized support plasmids significantly increased reporter gene expression in a MARV minigenome system and was instrumental for the successful recovery of full-length MARV clones [61,91].

Minigenome systems are particularly well suited to dissect *cis*-acting genetic elements, including replication and transcription promoters, transcription start and stop signals, intergenic regions and RNA editing signals. The MARV minigenome has been used to analyze *cis*-acting signals involved in viral transcription and replication [58,60], to map the replication promoter and to characterize the genus-specificity of *cis*-acting signals in chimeric MARV/EBOV minigenomes [66]. Further, chimeric minigenomes were used to compare the impact of the intergenic regions of the MARV isolates Musoke and Angola on transcriptional activity [90].

In addition to *cis*-acting functions, MARV minigenome systems have been used to analyze *trans*-acting factors relevant for minigenome transcription and replication.

This includes determining the protein requirement for replication and transcription and optimizing the amount and ratio of the support plasmids [40,60,61]. The optimal protein amounts may vary and have to be adapted to each system, especially the ratio of NP to VP35, which has been found to be very critical [60]. Functional studies of the nucleocapsid proteins have been instrumental to map NP phosphorylation sites critical for minigenome activity and protein binding domains on VP35 involved in the formation of a functional nucleocapsid complex [50,88,89,92].

2.2. Infectious VLP Systems

To overcome the limitations that minigenome systems are restricted to replication and transcription, other systems to model multiple steps of the viral life cycle have been developed. Expression of VP40 in the absence of other viral proteins leads to the formation and release of VLPs that are structurally similar to MARV particles and can incorporate GP [105,106]. The iVLP system combines the VLP approach with the minigenome system.

The first iVLP system for MARV was published by the Becker laboratory in 2010 [40]. In this system, multiple steps of the MARV replication cycle including replication/transcription, morphogenesis, budding, and infection of target cells can be examined (Figure 3b). In contrast to VLPs, the iVLPs contain nucleocapsids that are able to mediate replication and transcription (Figure 3b). To generate iVLPs, cells are transfected with the minigenome plasmid along with expression plasmids for all viral proteins and the T7 RNA polymerase. Expression of the viral proteins leads to the formation of iVLPs containing nucleocapsids. The iVLPs are released from the cell by budding and can be used to infect naive target cells. A second replication cycle can only be mediated by this system if the target cells express the viral nucleocapsid proteins [40,92,94]. Similar to the minigenome system, NP, VP35 and L, but not VP30, were essential for iVLP formation. iVLP formation was further strictly dependent on the expression of VP40 and was increased in the presence of GP, whereas VP24 did not affect iVLP formation [40]. Titration experiments to optimize the system revealed that higher amounts of VP40 and VP24 had a negative impact on viral replication and/or transcription. While the release mechanism of iVLPs appeared to be similar to that of MARV particles from infected cells, the morphology and length of these iVLPs and their nucleocapsid cores varied, which influenced their infectivity [40]. Chimeric iVLPs composed of different combinations of EBOV and MARV components revealed a strict genus-specific interaction of VP40 with nucleocapsids, while VP40 tolerated GP from a different filovirus genus [94]. Further functional studies using the iVLP system include the characterization of viral proteins, protein domains and post-translational modifications crucial for various steps of the viral replication cycle [30,92,93]. This system has also been used to analyze the role of

VP40 in host adaptation [36] and to determine neutralization titers of MARV-specific antisera [40].

2.3. Rescue Systems

The first rescue of NNS RNA viruses entirely from cDNA was achieved for rabies virus by the Conzelmann laboratory in 1994 [107]. Their approach was based on generating full-length positive-sense antigenomes from plasmid DNA instead of negative-sense genome transcripts. Rescue of recombinant NNS RNA viruses from cDNA requires de novo formation of replication-competent nucleocapsids using the T7 RNA polymerase derived-genomic (or antigenomic) RNA and the nucleocapsid proteins. In unsuccessful approaches that used the negative-sense RNA genome, the mRNAs of the support proteins (N, P, and L) might hybridize to the primary transcript of the T7 polymerase-derived genomic RNA, preventing the assembly of the genome into nucleocapsid complexes. The new approach, starting with the full-length positive-strand transcript, overcame this issue because the antigenome and the nucleocapsid protein mRNAs are in the same orientation and cannot hybridize. Despite ongoing efforts to develop rescue systems based on negative-sense full-length viral RNA, only few laboratories were successful using this approach. In addition, the established negative-sense RNA rescue systems were much less efficient compared to the respective positive-sense RNA approach [108,109]. As of date, rescue systems have been established for multiple members of the *Rhabdo-, Paramyxo-, Borna-* and *Filoviridae* families, and the list is still growing (reviewed in [85,86]).

The first MARV rescue system was published by the Mühlberger laboratory in 2006 using the positive-sense approach [62]. The cDNA encoding the MARV Musoke antigenomic RNA was inserted into an expression vector under the control of the T7 RNA polymerase promoter followed by the HDV ribozyme and a T7 RNA polymerase termination motif. Concomitantly with synthesis by the T7 RNA polymerase, the MARV antigenome is encapsidated by the viral support proteins, forming the nucleocapsid which is used as a template for the synthesis of the negative-sense genomic RNA (Figure 3c). Nucleocapsid complexes containing the genomic RNA are the templates for synthesis of the viral mRNAs and the antigenomic RNA. Since all viral proteins are expressed from the viral genome, all steps of the viral replication cycle, including particle formation and budding, are mediated. Successful rescue was achieved by transfecting T7 RNA polymerase-expressing cells with plasmids encoding NP, VP35, L, and VP30 along with the full-length antigenome plasmid (Figure 3c). In contrast to the minigenome and iVLP systems, VP30 was essential for virus rescue [62]. To our knowledge, recombinant MARV (rMARV) systems have been established in three laboratories [61,62,95]. In contrast to the minigenome and iVLP systems, the rescue system can only be used in a BSL-4 setting because infectious MARV is generated. Rescue of rMARV can be challenging and

161

was significantly improved by using codon-optimized support plasmids for the expression of the viral support proteins [61,91]. Attempts to rescue MARV from negative-sense genomic RNA were not successful [61].

Recombinant NNS RNA viruses, including MARV, are useful tools to study multiple aspects of the viral replication cycle, host-virus interactions, immune responses, and host adaptation mechanisms. Furthermore, they have been used in translational approaches for vaccine development and oncolytic therapy [110]. There are two main approaches of how recombinant NNS RNA viruses can be modified: (i) targeted mutagenesis of viral genes for functional analyses or to generate attenuated vaccine candidates; (ii) insertion of ATUs to image infection or to generate vectors for the expression of foreign genes (e.g., vaccine candidates, oncolytic viruses) [110,111]. rMARV clones have been generated by both approaches.

2.3.1. Targeted Mutagenesis of rMARV

The MARV full-length system was used to verify the function of *cis*-acting elements located within the viral replication promoter, which had originally been identified using the minigenome system [66]. It was also used to analyze the function of viral proteins involved in replication and transcription, including VP30 and L [61,62]. Functional studies on viral entry and particle release revealed that truncations of the cytoplasmic domain of GP led to growth defects and impaired entry [96]. Mutational analysis of a late domain motif within MARV NP, which is involved in the recruitment of the ESCRT protein Tsg101 [45], has provided insights into the role of NP in the transport of nucleocapsids to the sites of budding [46]. The filovirus rescue systems have also been used to study genus-specificity of the support plasmids. MARV was rescued using eVP30 as a support plasmid [62], and EBOV was successfully rescued using all four MARV support proteins (NP, VP35, VP30, L) [97]. Intriguingly, EBOV minigenomes were not replicated and transcribed using MARV support proteins [58], illustrating the sensitivity of the full-length rescue system that leads to autonomous viral replication and transcription after the jump-start transfection, using the viral proteins produced from the viral genomes. Targeted mutagenesis of viral proteins was further used to study the impact of adaptive mutations in the VP40 gene of guinea pig-adapted MARV and revealed increased fitness and higher infectivity in guinea pig cells infected with rMARV containing the adaptive mutations in the VP40 gene [36]. A bat-derived rMARV carrying mutations in the IFN inhibitory domain of VP35 was used to confirm the immunosuppressive functions of VP35 in the context of viral infection [54].

2.3.2. rMARV Containing ATUs

Insertion of a gene encoding a fluorescent protein into the MARV genome has been used as a strategy to visualize viral infection. The first MARV construct

(Musoke isolate) expressing a foreign gene contained the GFP gene as an independent transcription unit inserted between the VP35 and VP40 genes. To do this, the intergenic region between the VP35 and VP40 genes was mutated to an *Avr*II restriction site which was used to insert the GFP gene flanked by MARV-specific GS and GE sequences. The rescued virus showed slightly reduced replication kinetics in Vero cells compared to the MARV Musoke wild type. This might be attributed to a decrease in downstream gene expression due to the insertion of an ATU [77]. rMARV-GFP was used to detect virus-infected cells using live-cell imaging. It was observed that viral spread occurred mainly through cell-to-cell contact and from cell protrusions, and was promoted by viral replication in actively dividing cells. Intriguingly, GFP accumulated in MARV-derived inclusions in the infected cells [77]. The same strategy was used to insert a second copy of the MARV VP40 gene fused to the red fluorescent protein (RFP) gene into the *Avr*II restriction site, leading to the expression of a fluoresecentVP40 fusion protein in addition to unlabeled VP40 [80]. This virus was used for high resolution live-cell imaging studies to analyze the transport of MARV nucleocapsids to the sites of budding. Imaging of single nucleocapsid particles was achieved by providing VP30 fused to GFP in *trans*. This dual-color live-cell imaging approach allowed for simultaneous visualization of the nucleocapsids and VP40 in infected cells. It was shown that the nucleocapsids are transported along actin filaments to the plasma membrane where they associate with VP40 to be released from filopodia [80].

ATUs have also been inserted in rMARV clones based on the 371Bat isolate. 371Bat rMARV, in which the GFP gene was inserted between the NP and VP35 genes, grew to similar titers as the wild type virus in Vero cells, but was attenuated in primary human macrophages [61]. Similar to the GFP-containing MARV Musoke clone [77], the expression levels of viral genes located downstream of the ATU were decreased in cells infected with 371Bat rMARV-GFP [61]. Compared to wild type virus, infection with 371Bat rMARV-GFP elicited an increased inflammatory response in primary human macrophages. As a possible explanation for this observation, Albarino and colleagues suggested that the ability of the GFP-containing virus to counteract antiviral responses might be perturbed by the decreased expression levels of viral immune suppressors, such as VP35 and VP40 [61]. Bat-derived rMARV expressing gLuc from an ATU inserted between the NP and VP35 genes was generated to establish a high-throughput screening platform to test antivirals [91].

3. Conclusions

The various MARV reverse genetics systems have considerably contributed to our current understanding of the MARV replication cycle, MARV-host interactions, host specificity, pathogenesis, and antiviral treatment options. While the rescue systems result in the production of genetically engineered virus which can be used for all kinds of infection studies, including live-cell imaging approaches, they are restricted to a BSL-4 setting. Model systems that can be used to study various aspects of the MARV replication cycle under BSL-2 conditions because they do not lead to the production of infectious MARV include minigenome and iVLP systems. Although these model systems have certain restrictions compared to the rescue system, they are highly beneficial due to the lower biosafety level requirements. Despite the threat that filovirus disease poses to global public health, MARV, and to an even larger extent RAVV, are severely under-investigated, and there are many gaps in our understanding of MARV and RAVV infections. It is, therefore, desirable that the availability of MARV reverse genetics systems will stimulate more research activity on these important emerging pathogens.

Acknowledgments: This work was supported by the National Institute of Allergy and Infectious Diseases (NIAID) of the National Institutes of Health (NIH) under award numbers R03-AI114293 and UC6AI058618, and by the Defense Threat Reduction Agency (DTRA) under grant number HDTRA1-14-1-0016.

Conflicts of Interest: The authors declare no conflict of interest.

References

1. Bukreyev, A.A.; Chandran, K.; Dolnik, O.; Dye, J.M.; Ebihara, H.; Leroy, E.M.; Muhlberger, E.; Netesov, S.V.; Patterson, J.L.; Paweska, J.T.; *et al.* Discussions and decisions of the 2012–2014 international committee on taxonomy of viruses (ICTV) *filoviridae* study group, January 2012–June 2013. *Arch. Virol.* **2014**, *159*, 821–830.
2. Cross, R.W.; Fenton, K.A.; Geisbert, J.B.; Ebihara, H.; Mire, C.E.; Geisbert, T.W. Comparison of the pathogenesis of the Angola and RAVN strains of Marburg virus in the outbred guinea pig model. *J. Infect. Dis.* **2015**, *212* (Suppl. S2), S258–S270.
3. Brauburger, K.; Hume, A.J.; Mühlberger, E.; Olejnik, J. Forty-five years of Marburg virus research. *Viruses* **2012**, *4*, 1878–1927.
4. Towner, J.S.; Khristova, M.L.; Sealy, T.K.; Vincent, M.J.; Erickson, B.R.; Bawiec, D.A.; Hartman, A.L.; Comer, J.A.; Zaki, S.R.; Ströher, U.; *et al. Marburgvirus* genomics and association with a large hemorrhagic fever outbreak in Angola. *J. Virol.* **2006**, *80*, 6497–6516.
5. Ligon, B.L. Outbreak of Marburg hemorrhagic fever in Angola: A review of the history of the disease and its biological aspects. *Semin. Pediatr. Infect. Dis.* **2005**, *16*, 219–224.
6. Uyeki, T.M.; Mehta, A.K.; Davey, R.T., Jr.; Liddell, A.M.; Wolf, T.; Vetter, P.; Schmiedel, S.; Grunewald, T.; Jacobs, M.; Arribas, J.R.; *et al.* Clinical management of Ebola virus disease in the United States and Europe. *N. Engl. J. Med.* **2016**, *374*, 636–646.
7. Towner, J.S.; Pourrut, X.; Albarino, C.G.; Nkogue, C.N.; Bird, B.H.; Grard, G.; Ksiazek, T.G.; Gonzalez, J.P.; Nichol, S.T.; Leroy, E.M. Marburg virus infection detected in a common African bat. *PLoS ONE* **2007**, *2*, e764.

8. Swanepoel, R.; Smit, S.B.; Rollin, P.E.; Formenty, P.; Leman, P.A.; Kemp, A.; Burt, F.J.; Grobbelaar, A.A.; Croft, J.; Bausch, D.G.; *et al.* Studies of reservoir hosts for Marburg virus. *Emerg. Infect. Dis.* **2007**, *13*, 1847–1851.

9. Pourrut, X.; Souris, M.; Towner, J.S.; Rollin, P.E.; Nichol, S.T.; Gonzalez, J.P.; Leroy, E. Large serological survey showing cocirculation of Ebola and Marburg viruses in Gabonese bat populations, and a high seroprevalence of both viruses in *Rousettus aegyptiacus*. *BMC Infect. Dis.* **2009**.

10. Towner, J.S.; Amman, B.R.; Sealy, T.K.; Carroll, S.A.; Comer, J.A.; Kemp, A.; Swanepoel, R.; Paddock, C.D.; Balinandi, S.; Khristova, M.L.; *et al.* Isolation of genetically diverse Marburg viruses from Egyptian fruit bats. *PLoS Pathog.* **2009**, *5*, e1000536.

11. Adjemian, J.; Farnon, E.C.; Tschioko, F.; Wamala, J.F.; Byaruhanga, E.; Bwire, G.S.; Kansiime, E.; Kagirita, A.; Ahimbisibwe, S.; Katunguka, F.; *et al.* Outbreak of Marburg hemorrhagic fever among miners in Kamwenge and Ibanda Districts, Uganda, 2007. *J. Infect. Dis.* **2011**, *204* (Suppl. S3), S796–S799.

12. Amman, B.R.; Carroll, S.A.; Reed, Z.D.; Sealy, T.K.; Balinandi, S.; Swanepoel, R.; Kemp, A.; Erickson, B.R.; Comer, J.A.; Campbell, S.; *et al.* Seasonal pulses of Marburg virus circulation in juvenile *Rousettus aegyptiacus* bats coincide with periods of increased risk of human infection. *PLoS Pathog.* **2012**, *8*, e1002877.

13. Amman, B.R.; Nyakarahuka, L.; McElroy, A.K.; Dodd, K.A.; Sealy, T.K.; Schuh, A.J.; Shoemaker, T.R.; Balinandi, S.; Atimnedi, P.; Kaboyo, W.; *et al. Marburgvirus* resurgence in Kitaka Mine bat population after extermination attempts, Uganda. *Emerg. Infect. Dis.* **2014**, *20*, 1761–1764.

14. Paweska, J.T.; Jansen van Vuren, P.; Masumu, J.; Leman, P.A.; Grobbelaar, A.A.; Birkhead, M.; Clift, S.; Swanepoel, R.; Kemp, A. Virological and serological findings in *Rousettus aegyptiacus* experimentally inoculated with Vero cells-adapted hogan strain of Marburg virus. *PLoS ONE* **2012**, *7*, e45479.

15. Amman, B.R.; Jones, M.E.; Sealy, T.K.; Uebelhoer, L.S.; Schuh, A.J.; Bird, B.H.; Coleman-McCray, J.D.; Martin, B.E.; Nichol, S.T.; Towner, J.S. Oral shedding of Marburg virus in experimentally infected Egyptian fruit bats (*Rousettus aegyptiacus*). *J. Wildl. Dis.* **2015**, *51*, 113–124.

16. Jones, M.E.; Schuh, A.J.; Amman, B.R.; Sealy, T.K.; Zaki, S.R.; Nichol, S.T.; Towner, J.S. Experimental inoculation of Egyptian Rousette bats (*Rousettus aegyptiacus*) with viruses of the *Ebolavirus* and *Marburgvirus* genera. *Viruses* **2015**, *7*, 3420–3442.

17. Paweska, J.T.; Jansen van Vuren, P.; Fenton, K.A.; Graves, K.; Grobbelaar, A.A.; Moolla, N.; Leman, P.; Weyer, J.; Storm, N.; McCulloch, S.D.; *et al.* Lack of Marburg virus transmission from experimentally infected to susceptible in-contact Egyptian fruit bats. *J. Infect. Dis.* **2015**, *212* (Suppl. S2), S109–S118.

18. Brauburger, K.; Deflubé, L.R.; Mühlberger, E. Filovirus transcription and replication. In *Biology and Pathogenesis of Rhabdo- and Filoviruses*; Pattnaik, A.K., Whitt, M.A., Eds.; World Scientific Publishing Co. Pte. Ltd.: Singapore, 2015; pp. 515–555.

19. Feldmann, H.; Will, C.; Schikore, M.; Slenczka, W.; Klenk, H.D. Glycosylation and oligomerization of the spike protein of Marburg virus. *Virology* **1991**, *182*, 353–356.

20.	Hunt, C.L.; Lennemann, N.J.; Maury, W. Filovirus entry: A novelty in the viral fusion world. *Viruses* **2012**, *4*, 258–275.

21.	Miller, E.H.; Chandran, K. Filovirus entry into cells—New insights. *Curr. Opin. Virol.* **2012**, *2*, 206–214.

22.	Volchkov, V.E.; Volchkova, V.A.; Stroher, U.; Becker, S.; Dolnik, O.; Cieplik, M.; Garten, W.; Klenk, H.D.; Feldmann, H. Proteolytic processing of Marburg virus glycoprotein. *Virology* **2000**, *268*, 1–6.

23.	Koellhoffer, J.F.; Malashkevich, V.N.; Harrison, J.S.; Toro, R.; Bhosle, R.C.; Chandran, K.; Almo, S.C.; Lai, J.R. Crystal structure of the Marburg virus GP2 core domain in its postfusion conformation. *Biochemistry* **2012**, *51*, 7665–7675.

24.	Fusco, M.L.; Hashiguchi, T.; Cassan, R.; Biggins, J.E.; Murin, C.D.; Warfield, K.L.; Li, S.; Holtsberg, F.W.; Shulenin, S.; Vu, H.; *et al.* Protective mAbs and cross-reactive mAbs raised by immunization with engineered Marburg virus GPs. *PLoS Pathog.* **2015**, *11*, e1005016.

25.	Hashiguchi, T.; Fusco, M.L.; Bornholdt, Z.A.; Lee, J.E.; Flyak, A.I.; Matsuoka, R.; Kohda, D.; Yanagi, Y.; Hammel, M.; Crowe, J.E., Jr.; *et al.* Structural basis for Marburg virus neutralization by a cross-reactive human antibody. *Cell* **2015**, *160*, 904–912.

26.	Flyak, A.I.; Ilinykh, P.A.; Murin, C.D.; Garron, T.; Shen, X.; Fusco, M.L.; Hashiguchi, T.; Bornholdt, Z.A.; Slaughter, J.C.; Sapparapu, G.; *et al.* Mechanism of human antibody-mediated neutralization of Marburg virus. *Cell* **2015**, *160*, 893–903.

27.	Kibuuka, H.; Berkowitz, N.M.; Millard, M.; Enama, M.E.; Tindikahwa, A.; Sekiziyivu, A.B.; Costner, P.; Sitar, S.; Glover, D.; Hu, Z.; *et al.* Safety and immunogenicity of Ebola virus and Marburg virus glycoprotein DNA vaccines assessed separately and concomitantly in healthy Ugandan adults: A phase 1b, randomised, double-blind, placebo-controlled clinical trial. *Lancet* **2015**, *385*, 1545–1554.

28.	Daddario-DiCaprio, K.M.; Geisbert, T.W.; Stroher, U.; Geisbert, J.B.; Grolla, A.; Fritz, E.A.; Fernando, L.; Kagan, E.; Jahrling, P.B.; Hensley, L.E.; *et al.* Postexposure protection against Marburg hemorrhagic fever with recombinant vesicular stomatitis virus vectors in non-human primates: An efficacy assessment. *Lancet* **2006**, *367*, 1399–1404.

29.	Kolesnikova, L.; Bohil, A.B.; Cheney, R.E.; Becker, S. Budding of *Marburgvirus* is associated with filopodia. *Cell. Microbiol.* **2007**, *9*, 939–951.

30.	Kolesnikova, L.; Mittler, E.; Schudt, G.; Shams-Eldin, H.; Becker, S. Phosphorylation of Marburg virus matrix protein VP40 triggers assembly of nucleocapsids with the viral envelope at the plasma membrane. *Cell. Microbiol.* **2012**, *14*, 182–197.

31.	Wijesinghe, K.J.; Stahelin, R.V. Investigation of the lipid binding properties of the Marburg virus matrix protein VP40. *J. Virol.* **2015**, *90*, 3074–3085.

32.	Valmas, C.; Basler, C.F. Marburg virus VP40 antagonizes interferon signaling in a species-specific manner. *J. Virol.* **2011**, *85*, 4309–4317.

33.	Valmas, C.; Grosch, M.N.; Schümann, M.; Olejnik, J.; Martinez, O.; Best, S.M.; Krähling, V.; Basler, C.F.; Mühlberger, E. Marburg virus evades interferon responses by a mechanism distinct from Ebola virus. *PLoS Pathog.* **2010**, *6*, e1000721.

34. Feagins, A.R.; Basler, C.F. Amino acid residue at position 79 of Marburg virus VP40 confers interferon antagonism in mouse cells. *J. Infect. Dis.* **2015**, *212* (Suppl. S2), S219–S225.

35. Feagins, A.R.; Basler, C.F. The VP40 protein of Marburg virus exhibits impaired budding and increased sensitivity to human tetherin following mouse adaptation. *J. Virol.* **2014**, *88*, 14440–14450.

36. Koehler, A.; Kolesnikova, L.; Welzel, U.; Schudt, G.; Herwig, A.; Becker, S. A single amino acid change in the Marburg virus matrix protein VP40 provides a replicative advantage in a species-specific manner. *J. Virol.* **2015**, *90*, 1444–1454.

37. Bharat, T.A.; Riches, J.D.; Kolesnikova, L.; Welsch, S.; Krahling, V.; Davey, N.; Parsy, M.L.; Becker, S.; Briggs, J.A. Cryo-electron tomography of Marburg virus particles and their morphogenesis within infected cells. *PLoS Biol.* **2011**, *9*, e1001196.

38. Becker, S.; Rinne, C.; Hofsäss, U.; Klenk, H.-D.; Mühlberger, E. Interactions of Marburg virus nucleocapsid proteins. *Virology* **1998**, *249*, 406–417.

39. Bamberg, S.; Kolesnikova, L.; Möller, P.; Klenk, H.D.; Becker, S. VP24 of Marburg virus influences formation of infectious particles. *J. Virol.* **2005**, *79*, 13421–13433.

40. Wenigenrath, J.; Kolesnikova, L.; Hoenen, T.; Mittler, E.; Becker, S. Establishment and application of an infectious virus-like particle system for Marburg virus. *J. Gen. Virol.* **2010**, *91 Pt 5*, 1325–1334.

41. Page, A.; Volchkova, V.A.; Reid, S.P.; Mateo, M.; Bagnaud-Baule, A.; Nemirov, K.; Shurtleff, A.C.; Lawrence, P.; Reynard, O.; Ottmann, M.; *et al.* *Marburgvirus* hijacks Nrf2-dependent pathway by targeting Nrf2-negative regulator Keap1. *Cell Rep.* **2014**, *6*, 1026–1036.

42. Edwards, M.R.; Johnson, B.; Mire, C.E.; Xu, W.; Shabman, R.S.; Speller, L.N.; Leung, D.W.; Geisbert, T.W.; Amarasinghe, G.K.; Basler, C.F. The Marburg virus VP24 protein interacts with Keap1 to activate the cytoprotective antioxidant response pathway. *Cell Rep.* **2014**, *6*, 1017–1025.

43. Kolesnikova, L.; Mühlberger, E.; Ryabchikova, E.; Becker, S. Ultrastructural organization of recombinant Marburg virus nucleoprotein: Comparison with Marburg virus inclusions. *J. Virol.* **2000**, *74*, 3899–3904.

44. Mavrakis, M.; Kolesnikova, L.; Schoehn, G.; Becker, S.; Ruigrok, R.W. Morphology of Marburg virus NP-RNA. *Virology* **2002**, *296*, 300–307.

45. Dolnik, O.; Kolesnikova, L.; Stevermann, L.; Becker, S. Tsg101 is recruited by a late domain of the nucleocapsid protein to support budding of Marburg virus-like particles. *J. Virol.* **2010**, *84*, 7847–7856.

46. Dolnik, O.; Kolesnikova, L.; Welsch, S.; Strecker, T.; Schudt, G.; Becker, S. Interaction with Tsg101 is necessary for the efficient transport and release of nucleocapsids in marburg virus-infected cells. *PLoS Pathog.* **2014**, *10*, e1004463.

47. Dolnik, O.; Stevermann, L.; Kolesnikova, L.; Becker, S. Marburg virus inclusions: A virus-induced microcompartment and interface to multivesicular bodies and the late endosomal compartment. *Eur. J. Cell Biol.* **2015**, *94*, 323–331.

48. Leung, D.W.; Borek, D.; Luthra, P.; Binning, J.M.; Anantpadma, M.; Liu, G.; Harvey, I.B.; Su, Z.; Endlich-Frazier, A.; Pan, J.; *et al.* An intrinsically disordered peptide from Ebola virus VP35 controls viral RNA synthesis by modulating nucleoprotein-RNA interactions. *Cell Rep.* **2015**, *11*, 376–389.

49. Kirchdoerfer, R.N.; Abelson, D.M.; Li, S.; Wood, M.R.; Saphire, E.O. Assembly of the Ebola virus nucleoprotein from a chaperoned VP35 complex. *Cell Rep.* **2015**, *12*, 140–149.

50. Möller, P.; Pariente, N.; Klenk, H.D.; Becker, S. Homo-oligomerization of *Marburgvirus* VP35 is essential for its function in replication and transcription. *J. Virol.* **2005**, *79*, 14876–14886.

51. Mühlberger, E. Filovirus replication and transcription. *Future Virol.* **2007**, *2*, 205–215.

52. Bale, S.; Julien, J.P.; Bornholdt, Z.A.; Kimberlin, C.R.; Halfmann, P.; Zandonatti, M.A.; Kunert, J.; Kroon, G.J.; Kawaoka, Y.; MacRae, I.J.; *et al.* Marburg virus VP35 can both fully coat the backbone and cap the ends of dsRNA for interferon antagonism. *PLoS Pathog.* **2012**, *8*, e1002916.

53. Ramanan, P.; Edwards, M.R.; Shabman, R.S.; Leung, D.W.; Endlich-Frazier, A.C.; Borek, D.M.; Otwinowski, Z.; Liu, G.; Huh, J.; Basler, C.F.; *et al.* Structural basis for Marburg virus VP35-mediated immune evasion mechanisms. *Proc. Natl. Acad. Sci. USA* **2012**, *109*, 20661–20666.

54. Albarino, C.G.; Wiggleton Guerrero, L.; Spengler, J.R.; Uebelhoer, L.S.; Chakrabarti, A.K.; Nichol, S.T.; Towner, J.S. Recombinant Marburg viruses containing mutations in the IID region of VP35 prevent inhibition of host immune responses. *Virology* **2015**, *476*, 85–91.

55. Edwards, M.R.; Liu, G.; Mire, C.E.; Sureshchandra, S.; Luthra, P.; Yen, B.; Shabman, R.S.; Leung, D.W.; Messaoudi, I.; Geisbert, T.W.; *et al.* Differential regulation of interferon responses by Ebola and Marburg virus VP35 proteins. *Cell Rep.* **2016**, *14*, 1632–1640.

56. Yen, B.C.; Basler, C.F. Effects of filovirus IFN antagonists on responses of human monocyte-derived dendritic cells to RNA virus infection. *J. Virol.* **2016**, *90*, 5108–5118.

57. Modrof, J.; Moritz, C.; Kolesnikova, L.; Konakova, T.; Hartlieb, B.; Randolf, A.; Muhlberger, E.; Becker, S. Phosphorylation of Marburg virus VP30 at serines 40 and 42 is critical for its interaction with NP inclusions. *Virology* **2001**, *287*, 171–182.

58. Mühlberger, E.; Weik, M.; Volchkov, V.E.; Klenk, H.-D.; Becker, S. Comparison of the transcription and replication strategies of marburg virus and Ebola virus by using artificial replication systems. *J. Virol.* **1999**, *73*, 2333–2342.

59. Weik, M.; Modrof, J.; Klenk, H.-D.; Becker, S.; Mühlberger, E. Ebola virus VP30-mediated transcription is regulated by RNA secondary structure formation. *J. Virol.* **2002**, *76*, 8532–8539.

60. Mühlberger, E.; Lötfering, B.; Klenk, H.-D.; Becker, S. Three of the four nucleocapsid proteins of Marburg virus, NP, VP35, and L, are sufficient to mediate replication and transcription of Marburg virus-specific monocistronic minigenomes. *J. Virol.* **1998**, *72*, 8756–8764.

61. Albarino, C.G.; Uebelhoer, L.S.; Vincent, J.P.; Khristova, M.L.; Chakrabarti, A.K.; McElroy, A.; Nichol, S.T.; Towner, J.S. Development of a reverse genetics system to generate recombinant Marburg virus derived from a bat isolate. *Virology* **2013**, *446*, 230–237.

62. Enterlein, S.; Volchkov, V.; Weik, M.; Kolesnikova, L.; Volchkova, V.; Klenk, H.D.; Mühlberger, E. Rescue of recombinant Marburg virus from cDNA is dependent on nucleocapsid protein VP30. *J. Virol.* **2006**, *80*, 1038–1043.

63. Fowler, T.; Bamberg, S.; Möller, P.; Klenk, H.D.; Meyer, T.F.; Becker, S.; Rudel, T. Inhibition of Marburg virus protein expression and viral release by RNA interference. *J. Gen. Virol.* **2005**, *86 Pt 4*, 1181–1188.

64. Feldmann, H.; Mühlberger, E.; Randolf, A.; Will, C.; Kiley, M.P.; Sanchez, A.; Klenk, H.D. Marburg virus, a filovirus: Messenger RNAs, gene order, and regulatory elements of the replication cycle. *Virus Res.* **1992**, *24*, 1–19.

65. Mühlberger, E.; Trommer, S.; Funke, C.; Volchkov, V.; Klenk, H.-D.; Becker, S. Termini of all mRNA species of Marburg virus: Sequence and secondary structure. *Virology* **1996**, *223*, 376–380.

66. Enterlein, S.; Schmidt, K.M.; Schümann, M.; Conrad, D.; Krahling, V.; Olejnik, J.; Mühlberger, E. The Marburg virus 3' non-coding region structurally and functionally differs from that of Ebola virus. *J. Virol.* **2009**, *83*, 4508–4519.

67. Whelan, S.P.; Barr, J.N.; Wertz, G.W. Transcription and replication of nonsegmented negative-strand RNA viruses. *Curr. Top. Microbiol. Immunol.* **2004**, *283*, 61–119.

68. Hofmann-Winkler, H.; Kaup, F.; Pohlmann, S. Host cell factors in filovirus entry: Novel players, new insights. *Viruses* **2012**, *4*, 3336–3362.

69. Saeed, M.F.; Kolokoltsov, A.A.; Albrecht, T.; Davey, R.A. Cellular entry of Ebola virus involves uptake by a macropinocytosis-like mechanism and subsequent trafficking through early and late endosomes. *PLoS Pathog.* **2010**, *6*, e1001110.

70. Nanbo, A.; Imai, M.; Watanabe, S.; Noda, T.; Takahashi, K.; Neumann, G.; Halfmann, P.; Kawaoka, Y. *Ebolavirus* is internalized into host cells via macropinocytosis in a viral glycoprotein-dependent manner. *PLoS Pathog.* **2010**, *6*, e1001121.

71. Misasi, J.; Chandran, K.; Yang, J.Y.; Considine, B.; Filone, C.M.; Cote, M.; Sullivan, N.; Fabozzi, G.; Hensley, L.; Cunningham, J. Filoviruses require endosomal cysteine proteases for entry but exhibit distinct protease preferences. *J. Virol.* **2012**, *86*, 3284–3292.

72. Gnirss, K.; Kuhl, A.; Karsten, C.; Glowacka, I.; Bertram, S.; Kaup, F.; Hofmann, H.; Pohlmann, S. Cathepsins B and L activate Ebola but not Marburg virus glycoproteins for efficient entry into cell lines and macrophages independent of TMPRSS2 expression. *Virology* **2012**, *424*, 3–10.

73. Carette, J.E.; Raaben, M.; Wong, A.C.; Herbert, A.S.; Obernosterer, G.; Mulherkar, N.; Kuehne, A.I.; Kranzusch, P.J.; Griffin, A.M.; Ruthel, G.; *et al.* Ebola virus entry requires the cholesterol transporter niemann-pick C1. *Nature* **2011**, *477*, 340–343.

74. Cote, M.; Misasi, J.; Ren, T.; Bruchez, A.; Lee, K.; Filone, C.M.; Hensley, L.; Li, Q.; Ory, D.; Chandran, K.; *et al.* Small molecule inhibitors reveal Niemann-Pick C1 is essential for Ebola virus infection. *Nature* **2011**, *477*, 344–348.

75. Liu, N.; Tao, Y.; Brenowitz, M.D.; Girvin, M.E.; Lai, J.R. Structural and functional studies on the Marburg virus GP2 fusion loop. *J. Infect. Dis.* **2015**, *212* (Suppl. S2), S146–S153.

76. Ryabchikova, E.; Price, B.B.S. *Ebola and Marburg Viruses: A View of Infection Using Electron Microscopy*; Battelle Press: Columbus, Ohio, USA, 2004.

77. Schmidt, K.M.; Schümann, M.; Olejnik, J.; Krähling, V.; Mühlberger, E. Recombinant Marburg virus expressing EGFP allows rapid screening of virus growth and real-time visualization of virus spread. *J. Infect. Dis.* **2011**, *204* (Suppl. 3), S861–S870.

78. Kolesnikova, L.; Berghofer, B.; Bamberg, S.; Becker, S. Multivesicular bodies as a platform for formation of the Marburg virus envelope. *J. Virol.* **2004**, *78*, 12277–12287.

79. Welsch, S.; Kolesnikova, L.; Krahling, V.; Riches, J.D.; Becker, S.; Briggs, J.A. Electron tomography reveals the steps in filovirus budding. *PLoS Pathog.* **2010**, *6*, e1000875.

80. Schudt, G.; Kolesnikova, L.; Dolnik, O.; Sodeik, B.; Becker, S. Live-cell imaging of Marburg virus-infected cells uncovers actin-dependent transport of nucleocapsids over long distances. *Proc. Natl. Acad. Sci. USA* **2013**, *110*, 14402–14407.

81. Mittler, E.; Kolesnikova, L.; Strecker, T.; Garten, W.; Becker, S. Role of the transmembrane domain of marburg virus surface protein GP in assembly of the viral envelope. *J. Virol.* **2007**, *81*, 3942–3948.

82. Sänger, C.; Mühlberger, E.; Ryabchikova, E.; Kolesnikova, L.; Klenk, H.D.; Becker, S. Sorting of Marburg virus surface protein and virus release take place at opposite surfaces of infected polarized epithelial cells. *J. Virol.* **2001**, *75*, 1274–1283.

83. Kolesnikova, L.; Ryabchikova, E.; Shestopalov, A.; Becker, S. Basolateral budding of Marburg virus: VP40 retargets viral glycoprotein GP to the basolateral surface. *J. Infect. Dis.* **2007**, *196* (Suppl. S2), S232–S236.

84. Schnittler, H.J.; Mahner, F.; Drenckhahn, D.; Klenk, H.D.; Feldmann, H. Replication of Marburg virus in human endothelial cells. A possible mechanism for the development of viral hemorrhagic disease. *J. Clin. Investig.* **1993**, *91*, 1301–1309.

85. Conzelmann, K.K. Reverse genetics of mononegavirales. *Curr. Top. Microbiol. Immunol.* **2004**, *283*, 1–41.

86. Hoenen, T.; Groseth, A.; de Kok-Mercado, F.; Kuhn, J.H.; Wahl-Jensen, V. Minigenomes, transcription and replication competent virus-like particles and beyond: Reverse genetics systems for filoviruses and other negative stranded hemorrhagic fever viruses. *Antivir. Res.* **2011**, *91*, 195–208.

87. Neumann, G.; Kawaoka, Y. Reverse genetics systems for the generation of segmented negative-sense RNA viruses entirely from cloned cDNA. *Curr. Top. Microbiol. Immunol.* **2004**, *283*, 43–60.

88. Lötfering, B.; Mühlberger, E.; Tamura, T.; Klenk, H.D.; Becker, S. The nucleoprotein of Marburg virus is target for multiple cellular kinases. *Virology* **1999**, *255*, 50–62.

89. DiCarlo, A.; Moller, P.; Lander, A.; Kolesnikova, L.; Becker, S. Nucleocapsid formation and RNA synthesis of Marburg virus is dependent on two coiled coil motifs in the nucleoprotein. *Virol. J.* **2007**, *4*, 105.

90. Alonso, J.A.; Patterson, J.L. Sequence variability in viral genome non-coding regions likely contribute to observed differences in viral replication amongst MARV strains. *Virology* **2013**, *440*, 51–63.

91. Uebelhoer, L.S.; Albarino, C.G.; McMullan, L.K.; Chakrabarti, A.K.; Vincent, J.P.; Nichol, S.T.; Towner, J.S. High-throughput, luciferase-based reverse genetics systems for identifying inhibitors of Marburg and Ebola viruses. *Antivir. Res.* **2014**, *106*, 86–94.

92. DiCarlo, A.; Biedenkopf, N.; Hartlieb, B.; Klussmeier, A.; Becker, S. Phosphorylation of Marburg virus NP region II modulates viral RNA synthesis. *J. Infect. Dis.* **2011**, *204* (Suppl. S3), S927–S933.

93. Mittler, E.; Kolesnikova, L.; Hartlieb, B.; Davey, R.; Becker, S. The cytoplasmic domain of Marburg virus GP modulates early steps of viral infection. *J. Virol.* **2011**, *85*, 8188–8196.

94. Spiegelberg, L.; Wahl-Jensen, V.; Kolesnikova, L.; Feldmann, H.; Becker, S.; Hoenen, T. Genus-specific recruitment of filovirus ribonucleoprotein complexes into budding particles. *J. Gen. Virol.* **2011**, *92 Pt 12*, 2900–2905.

95. Krähling, V.; Dolnik, O.; Kolesnikova, L.; Schmidt-Chanasit, J.; Jordan, I.; Sandig, V.; Gunther, S.; Becker, S. Establishment of fruit bat cells (*Rousettus aegyptiacus*) as a model system for the investigation of filoviral infection. *PLoS Negl. Trop. Dis.* **2010**, *4*, e802.

96. Mittler, E.; Kolesnikova, L.; Herwig, A.; Dolnik, O.; Becker, S. Assembly of the Marburg virus envelope. *Cell. Microbiol.* **2013**, *15*, 270–284.

97. Theriault, S.; Groseth, A.; Neumann, G.; Kawaoka, Y.; Feldmann, H. Rescue of Ebola virus from cDNA using heterologous support proteins. *Virus Res.* **2004**, *106*, 43–50.

98. Pattnaik, A.K.; Ball, L.A.; LeGrone, A.W.; Wertz, G.W. Infectious defective interfering particles of VSV from transcripts of a cDNA clone. *Cell* **1992**, *69*, 1011–1020.

99. Pattnaik, A.K.; Wertz, G.W. Replication and amplification of defective interfering particle RNAs of vesicular stomatitis virus in cells expressing viral proteins from vectors containing cloned cDNAs. *J. Virol.* **1990**, *64*, 2948–2957.

100. Ball, L.A. Cellular expression of a functional nodavirus RNA replicon from vaccinia virus vectors. *J. Virol.* **1992**, *66*, 2335–2345.

101. Neumann, G.; Feldmann, H.; Watanabe, S.; Lukashevich, I.; Kawaoka, Y. Reverse genetics demonstrates that proteolytic processing of the Ebola virus glycoprotein is not essential for replication in cell culture. *J. Virol.* **2002**, *76*, 406–410.

102. Buchholz, U.J.; Finke, S.; Conzelmann, K.K. Generation of bovine respiratory syncytial virus (BRSV) from cDNA: BRSV NS2 is not essential for virus replication in tissue culture, and the human RSV leader region acts as a functional BRSV genome promoter. *J. Virol.* **1999**, *73*, 251–259.

103. Sutter, G.; Ohlmann, M.; Erfle, V. Non-replicating vaccinia vector efficiently expresses bacteriophage T7 RNA polymerase. *FEBS Lett.* **1995**, *371*, 9–12.

104. Bali, V.; Bebok, Z. Decoding mechanisms by which silent codon changes influence protein biogenesis and function. *Int. J. Biochem. Cell Biol.* **2015**, *64*, 58–74.

105. Swenson, D.L.; Warfield, K.L.; Kuehl, K.; Larsen, T.; Hevey, M.C.; Schmaljohn, A.; Bavari, S.; Aman, M.J. Generation of Marburg virus-like particles by co-expression of glycoprotein and matrix protein. *FEMS Immunol. Med. Microbiol.* **2004**, *40*, 27–31.

106. Kolesnikova, L.; Bamberg, S.; Berghofer, B.; Becker, S. The matrix protein of Marburg virus is transported to the plasma membrane along cellular membranes: Exploiting the retrograde late endosomal pathway. *J. Virol.* **2004**, *78*, 2382–2393.

107. Schnell, M.J.; Mebatsion, T.; Conzelmann, K.K. Infectious rabies viruses from cloned cDNA. *EMBO J.* **1994**, *13*, 4195–4203.

108. Kato, A.; Sakai, Y.; Shioda, T.; Kondo, T.; Nakanishi, M.; Nagai, Y. Initiation of Sendai virus multiplication from transfected cDNA or RNA with negative or positive sense. *Genes Cells* **1996**, *1*, 569–579.

109. Durbin, A.P.; Hall, S.L.; Siew, J.W.; Whitehead, S.S.; Collins, P.L.; Murphy, B.R. Recovery of infectious human parainfluenza virus type 3 from cDNA. *Virology* **1997**, *235*, 323–332.

110. Pfaller, C.K.; Cattaneo, R.; Schnell, M.J. Reverse genetics of Mononegavirales: How they work, new vaccines, and new cancer therapeutics. *Virology* **2015**, *479–480*, 331–344.

111. Falzarano, D.; Groseth, A.; Hoenen, T. Development and application of reporter-expressing mononegaviruses: Current challenges and perspectives. *Antivir. Res.* **2014**, *103*, 78–87.

Replication-Competent Influenza A Viruses Expressing Reporter Genes

Michael Breen, Aitor Nogales, Steven F. Baker and Luis Martínez-Sobrido

Abstract: Influenza A viruses (IAV) cause annual seasonal human respiratory disease epidemics. In addition, IAV have been implicated in occasional pandemics with inordinate health and economic consequences. Studying IAV, in vitro or in vivo, requires the use of laborious secondary methodologies to identify virus-infected cells. To circumvent this requirement, replication-competent IAV expressing an easily traceable reporter protein can be used. Here we discuss the development and applications of recombinant replication-competent IAV harboring diverse fluorescent or bioluminescent reporter genes in different locations of the viral genome. These viruses have been employed for in vitro and in vivo studies, such as the screening of neutralizing antibodies or antiviral compounds, the identification of host factors involved in viral replication, cell tropism, the development of vaccines, or the assessment of viral infection dynamics. In summary, reporter-expressing, replicating-competent IAV represent a powerful tool for the study of IAV both in vitro and in vivo.

Reprinted from *Viruses*. Cite as: Breen, M.; Nogales, A.; Baker, S.F.; Martínez-Sobrido, L. Replication-Competent Influenza A Viruses Expressing Reporter Genes. *Viruses* **2016**, *8*, 179.

1. Introduction

1.1. Influenza A Virus

Influenza A viruses (IAV) are enveloped viruses within the family *Orthomyxoviridae* [1]. The genome of IAV contains eight single-stranded, negative-sense viral RNA (vRNA) segments [1] (Figure 1A). The vRNAs contain a long central coding region that is flanked at both termini by non-coding regions (NCRs), which serve as promoters to initiate replication and transcription by the viral heterotrimeric polymerase complex [1–3]. vRNAs reside within the virion as viral ribonucleoprotein (vRNP) complexes bound to a viral polymerase and many copies of nucleoprotein (NP) (Figure 1B). IAV are important pathogens that exert a dramatic impact on public health and the global economy [4] and cause annually recurrent epidemics, which result in approximately three to five million cases of severe illness and 250,000 to 500,000 deaths worldwide [5]. IAV are classified on the basis of the antigenic properties of the enveloped glycoproteins hemagglutinin (HA) and neuraminidase (NA), into 18 HA (H1–H18) and 11 NA (N1–N11) subtypes [6,7]. The

173

HA protein is critical for binding to cellular receptors and fusion of the viral and endosomal membranes [8,9]. Additionally, infection with IAV results in protective immunity mediated, at least in part, by antibodies against the viral HA, which is the key immunogen in natural immunity and vaccine approaches. The NA protein cleaves sialic acid moieties from sialyloligosaccharides and facilitates the release of nascent virions [10,11]. Importantly, NA is a major target for antiviral drugs, such as oseltamivir, that block the aforementioned cleavage and prevent viral dissemination to prevent further infection [12,13].

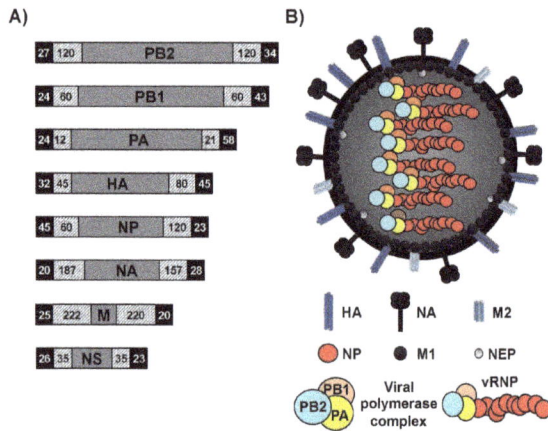

Figure 1. Influenza A virus (IAV) genome organization and virion structure. (**A**) Genome organization: The eight single-stranded, negative-sense, viral (v)RNA segments PB2, PB1, PA, HA, NP, NA, M and NS of IAV are indicated. Black boxes at the end of each of the vRNAs indicate the 3′ and 5′ non-coding regions (NCR). Hatched boxes indicate the packaging signals present at the 3′ and 5′ ends of each of the vRNAs that are responsible for efficient encapsidation into nascent virions. Numbers represent nucleotide lengths for each of the NCR and packaging signals; (**B**) Virion structure: IAV is surrounded by a lipid bilayer containing the two viral glycoproteins hemagglutinin (HA), responsible for binding to sialic acid-containing receptors; and neuraminidase (NA), responsible for viral release from infected cells. Also in the virion membrane is the ion channel matrix 2 (M2) protein. Under the viral lipid bilayer is a protein layer composed of the inner surface envelop matrix 1 (M1) protein, which plays a role in virion assembly and budding; and the nuclear export protein (NEP) involved in the nuclear export of the viral ribonucleoprotein (vRNP) complexes. Underneath is the core of the virus made of the eight vRNA segments that are encapsidated by the viral nucleoprotein (NP). Associated with each vRNP a complex is the viral RNA-dependent RNA polymerase (RdRp) complex made of the three polymerase subunits PB2, PB1 and PA that, together with the viral NP are the minimal components required for viral replication and transcription.

The replication and transcription process of influenza vRNAs are carried out by NP and the three polymerase subunits, an acidic (PA) and two basic (PB1 and PB2) proteins, which are encoded by the three largest vRNA segments [1]. Unlike many RNA viruses, influenza viral genome replication and transcription occurs in the nucleus of infected cells [14]. Newly synthesized vRNP complexes are then exported from the nucleus to the cytoplasm by the nuclear export protein (NEP) and the matrix protein 1 (M1), and are assembled into virions at the plasma membrane [1]. The small IAV genome is able to transcribe multiple viral genes from single segments through multiple mechanisms. These mechanisms include alternative splicing of viral mRNAs (M and NS segments), non-canonical translation, non-AUG initiation, or ribosomal frameshifting [1,15–19]. Moreover, to extend the coding capability of the viral genome, IAV encode proteins containing more than one function during virus infection. A well-studied multifunctional IAV protein is the non-structural protein 1 (NS1), which is expressed at very high levels in infected cells and is a determinant of virulence that functions in several ways to defeat cellular innate antiviral mechanisms [20]. NS1 is encoded on a collinear mRNA derived from vRNA segment eight (NS), which upon splicing results in the synthesis of NEP [21].

Although the natural reservoirs of IAV are wild waterfowl and shorebirds, IAV expand their host range to many avian and mammalian species through undefined adaptive processes involving mutation and genome reassortment [22,23], and this cross-species jumping characteristic allows the generation of potentially pandemic strains. In addition, antigenic drift occurs when the virus accumulates mutations that preclude binding by pre-existing antibodies, producing variant viruses that can escape immunity. IAV of three HA subtypes (H1, H2 and H3) thus gained the ability to be transmitted efficiently among humans [24]. In addition, IAV of the H5, H7 and H9 subtypes are also thought to represent pandemic threats because they have crossed the species barrier and infected humans [25–28]. Given the persistent threat posed by IAV infections, accelerating the development of novel countermeasures against IAV infections and increasing the biological understanding associated with viral infections are imperative.

Current available options to counter IAV include both vaccines and antivirals [12,29]. Only two classes of antivirals are approved for IAV that target either the ion channel function of the matrix 2 (M2) protein or the neuraminidase function of the NA protein [12]. However, these antiviral compounds have problems in terms of safety and the emergence of viral resistance [12,30,31]. Vaccines, due to the induction of sterilizing immunity, are the primary means to prevent IAV infections. However, currently available vaccines have moderate efficacy that changes seasonally [32]. Moreover, to generate vaccines against highly pathogenic IAV, as in the case of a pandemic outbreak, requires time. Therefore, developing new antiviral strategies to combat IAV infections are urgently needed. Current and traditional technologies to identify antivirals against IAV have been extensively reviewed [33]. This review will focus on the application and limitations of replication-competent IAV harboring fluorescent and/or luminescent reporter genes. Through better knowledge of the influenza virus genome, most importantly the identification of vRNA packaging signals [34–37], it has become possible to engineer replication-competent IAV encoding exogenous genes [38–41]. Stable incorporation of foreign genes in replication-competent, reporter-expressing IAV allows for effective tracking of viral infection in vitro and in vivo enabling a robust quantitative readout. This readout can be used with high throughput screenings (HTS) and to assess viral infection in tissue culture cells and animals models without the use of secondary approaches to identify the presence of the virus.

1.2. Comparison of Fluorescent and Luciferase Reporter Genes

The major advantage of using recombinant, replication-competent IAV is their flexibility to support the presence of reporter genes, such as fluorescence or luciferase proteins, in different viral segments. These reporter genes provide a good readout of viral replication and are compatible with HTS settings [33,39,42,43]. Moreover, reporter genes have a noteworthy role in multiple applications, both in vitro and in vivo. An ideal reporter gene encodes a protein whose activity can be detected with high sensitivity above any endogenous background and is amenable to assays that are sensitive, quantitative, and reproducible. In addition, reporter proteins can be detected directly by its inherent characteristics, such as fluorescence or enzymatic activity, as well as indirectly with antibody-based assays like Western blot. Although there are multiple reporter genes that can be used, this review will focus on replication-competent recombinant IAV harboring fluorescence or luciferases reporters, two categories of proteins that glow. Both types of systems (fluorescence and bioluminescence) create photons through energy transitions from excited states to their corresponding ground states. However, they differ in how the excited states are generated. The glow mechanism for fluorescent proteins is generated by first absorbing energy of one color light (excitation), and then emitting energy as a different wavelength [42]. On the other hand, the bioluminescence glow results from exothermic chemical reactions [44].

The first bioluminescent reporter identified was named aequorin, a calcium-activated photoprotein from the *Aequorea victoria* jellyfish [45]. However, the discovery of green fluorescent protein (GFP) in the early 1960s, with the gene first being cloned in 1992, ultimately heralded a new era in molecular biology [42]. GFP was identified as a protein that lacked the bioluminescent properties of aequorin, but was able to generate fluorescence when illuminated with UV light [42]. More recently, fluorescent and bioluminescence proteins from other species have been identified, and mutant variants have been developed with different glow properties, resulting in rapid expansion of the color spectrum [43]. Importantly, newly developed technologies to excite or detect emission has helped expand a range of applications [46].

Cloning of the *luc* gene from the firefly *Photinus pyralis* provided the first luciferase reporter system with widespread utility in mammalian cells [47]. The luciferase family of enzymes generates bioluminescent signals through mono-oxygenation of luciferin (substrate); utilizing ATP and O_2 as co-substrates with luciferin, luciferase catalysis produces light [44,46]. Luciferase enzymes isolated from different animal species have different variability in light emission, sensitivities, and emission duration times that accommodate different experimental designs [44,46]. Multiple luciferase enzymes can further be combined for multiplex analyses, including in vivo imaging [46]. Moreover, new secreted versions of luciferases [48] and shorter versions of luciferase genes [49] have been described to facilitate detection of reporter gene expression upon viral infection.

Properties of reporter genes must be considered on a per experiment basis because different genes will serve different purposes and choosing the best reporter gene assay depends on the type of study (Table 1) [39]. For example to observe localization, fluorescent genes like GFP are most convenient [38,42]. However, for quantitative purposes, luciferases are more useful [50,51]. Whole-body imaging is increasingly used in mice or other small laboratory animals [46]. For this purpose, luciferase reporters are preferred over fluorescent proteins because fluorescence requires excitation light to travel to the location of the fluorescent probe, while luciferase substrates can be administered systemically [46]. For fluorescent targets in vivo, excitation light is scattered from tissue above the plane of the target, which reduces the intensity. Moreover, the sensitivity and specificity of fluorescence imaging are frequently disturbed by tissue autofluorescence, resulting in substantial background [42,43]. Problems arising from tissue penetrance and autofluorescence are reduced when fluorescent proteins are visualized ex vivo. Bioluminescence reporters also have limitations, since the production of light requires the presence of all components involved in the oxidation reaction. Although some reagents like ATP or O_2 exist in tissues, the concentrations can vary by anatomical location and the physiological condition of the animal [51]. Moreover the substrate (luciferin) must be injected into the animal to generate the bioluminescent signal [51,52]. It should be noted that advances in instruments used for molecular imaging are improving in sensitivity and resolution, helping to minimize some of the aforementioned limitations with both types of reporter genes.

Table 1. Fluorescence versus bioluminescence, features and applications.

Properties	Fluorescence	Bioluminescence
Enzymatic amplification of signal	NO	YES
Substrate required for assay	NO	YES
High Reproducibility	YES	YES
FACS-compatible	YES	NO
In vitro applications	YES	YES
Ex vivo applications	YES	YES
In vivo applications	NO	YES
Flexible readout	YES	NO
Analysis of individual cells	YES	NO
HTS	YES	YES
Analysis of intermolecular interactions	YES	NO
Detection	Fluorescence	Luminescence

FACS: fluorescence-activated cell sorting; HTS: High-Throughput Screening.

2. Generation of Replication-Competent IAV Harboring Reporter Genes

IAV segmented genome allows the opportunity to tag various gene segments with fluorescent or luminescent reporters, thus allowing for visual and/or quantitative observation of reporter gene expression. Several caveats must be considered when designing IAV encoding reporter genes. First, the virus segments are small (~0.9–2.4 kb in length) and do not tolerate large insertions. Second, adding a reporter gene in the 3' or 5' end of the viral segment disrupts packaging signals located at the end of each viral RNA that are required for efficient virion assembly. Third, it is important to evaluate the stability of the inserted reporter gene since some replicating-competent IAV has been described to easily lose the inserted reporter gene. The causes associated to the reporter instability are not totally understood, however can be related with the size or the nature of the inserted foreign sequence. To overcome these hurdles, multiple strategies have been employed to rescue recombinant IAV harboring reporter genes, leading to an abundance of reporter-containing IAV (Tables 2–7).

2.1. Reporter-Expressing IAV Containing the Foreign Gene in the PB2 Segment

IAV segment 1 (Figure 2A) encodes for the PB2 protein that plays an important role in viral genome transcription initiation by generating 5'-capped RNA fragments from cellular pre-mRNA molecules, which serve as primers for viral transcription [53,54]. PB2 also modulates vRNP assembly and is thought to contribute to viral replicase or transcriptase activity [14]. Finally, IAV PB2 is a major host range and virulence determinant [55–57].

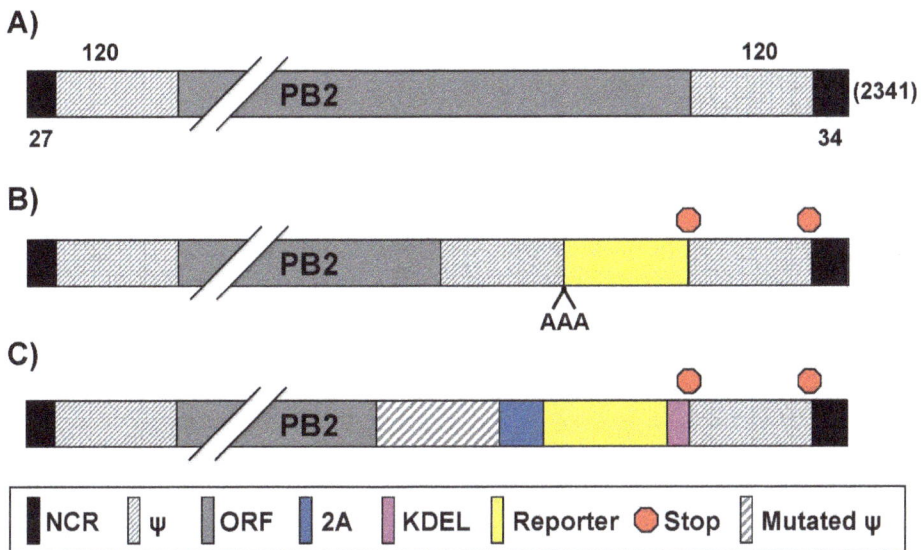

Figure 2. PB2 reporter influenza A viruses: Schematic representation of the PB2 segment from wild type (WT) (**A**) and reporter (**B,C**) IAV. Gene segments all contain non-coding regions (NCR), packaging signals (ψ) and open reading frames (ORF) for gene replication/transcription, virion incorporation, and protein expression, respectively. Nucleotide lengths for the NCR, ψ, and PB2 segment are indicated; (**B**) PB2 fusion proteins: Reporter genes were fused to native PB2 ORF with a triple alanine (AAA) spacer; (**C**) Bicistronic transcription of PB2 and reporter gene: Insertion of the 2A autocleavage site separates PB2 from reporter gene. Packaging signals encoding the 3′ terminus of PB2 were mutated to minimize interference with native ψ, which are duplicated at the 3′ NCR-proximal region. KDEL (lysine-aspartic acid-glutamic acid-leucine) signal sequence was inserted for endoplasmic reticulum-retention of the reporter gene. Packaging signals were duplicated after protein stop transcription signal and before the 3′ NCR terminus.

Studies to examine intracellular vRNP trafficking were previously limited due to the difficulty of visualizing their movement in living cells. To solve this, Avilov et al. [58] utilized "split-GFP" [59], where the 16 C-terminal amino acids (aa) of GFP were fused to PB2 (Figure 2B), and GFP reconstitution occurs in *trans*-complementing transiently transfected cells [58] (Table 2). The GFP-tagged virus in the backbone of influenza A/WSN/1933 H1N1 (WSN) was deemed "WSN-PB2-GFP11" [58]. This virus had a similar, albeit slightly diminished, growth kinetic and plaque phenotype compared to wild-type (WT) WSN virus in both parental and *trans*-complementing Madin-Darby canine kidney (MDCK) cells [58]. While it does propagate efficiently in vitro, a distinct disadvantage is that transfected cells are required to observe fluorescence, which is simplified by generating stable cell lines expressing the GFP *trans*-complementing domain [58]. In addition, the generation of mice expressing the same GFP domain, either in target tissues or constitutively, could be used for in or ex vivo analysis of viral infection. Avilov et al. used WSN-PB2-GFP11 to monitor trafficking of vRNPs during infection using dynamic light microscopy [58]. Live imaging of cells infected with a split-GFP-based virus demonstrated that over the course of infection, vRNPs accumulated pericentriolarly, followed by a wide distribution throughout the cytoplasm and an accumulation at the plasma membrane. Occasional quick movements of vRNPs were also detected in the cytoplasm and reported to be actin- and microtubule-dependent. These results were in agreement with previous observations of fixed cells [59,60]. Furthermore, Avilov et al. observed vRNP association with Rab11 [61] (Table 2), a host protein involved in cellular vesicle trafficking [62]. Their findings reiterate previous reports that vRNPs accumulate in Rab11 containing particles [63–65]. The authors used fluorescence resonance energy transfer microscopy to suggest a direct vRNP:Rab11 interaction and proposed that vRNPs traffic through the cytoplasm with recycling endosomes via Rab11 interactions [61].

Table 2. Influenza A viruses with reporter genes in the PB2 viral segment.

Gene	Virus Backbone [1]	Transgene [2]	Insertion Mechanism [3]	Application	Ref.
PB2	WSN	Split GFP	Fusion	Virus Biology	[58,61]
PB2	PR8, WSN	Gluc	2A site	Neutralizing antibodies Antivirals	[66,67]

[1] WSN: A/WSN/1933 (H1N1); PR8: A/Puerto Rico/8/1934 (H1N1); [2] GFP: Green fluorescent protein; Gluc: Gaussia luciferase; [3] Fusion: The C-terminal region of the reporter was fused to PB2; 2A site: The reporter gene was separated from the viral ORF via a 2A peptide sequence.

Influenza polymerase function is tightly regulated by protein-protein interactions, and the subunits do not tolerate large foreign protein additions. The insertion of viral 2A peptides has been extensively used to generate replication-competent IAV containing foreign sequences [40,41,48,52,66,68–73]. Viral 2A sequences mediate co-translational "ribosome skipping" or protein cleavage to separate two distinct polypeptides [74]. Equimolar amounts of collinear transcripts are therefore expressed from a single mRNA separated by 2A under the control of a single viral promoter. Heaton et al. [66] cloned the Gaussia luciferase (Gluc) gene into the C-terminal end of PB2 and separated the viral open reading frame (ORF) from the reporter via a foot-and-mouth disease virus (FMDV) 2A peptide sequence (Figure 2C, Table 2). Importantly, the addition of a foreign sequence at the end of the vRNA disrupts the packaging signals needed to assemble progeny virions [1]. To overcome packaging restrictions, the complete 5′ packaging signal of PB2 was duplicated after Gluc, upstream of the 5′ NCR. Moreover, silent mutations were introduced into the original 5′ packaging signals in the PB2 ORF, to eliminate the original packaging signals. Additionally, an endoplasmic reticulum (ER) retention sequence (KDEL) was added to the C-terminus of Gluc to prevent secretion, and the virus was rescued in the backbone of influenza A/Puerto Rico/8/1934 (PR8) [66]. In embryonated chicken eggs, this PR8-Gluc had 1 log lower replication levels compared to WT PR8, and the recombinant virus was stable for at least four serial passages in eggs [66]. A major interest in the IAV field is the characterization of antibodies that bind to the conserved "stalk" region of the HA glycoprotein. This domain is much less variable than the "head" region, and therapeutics targeting this domain potentially have the ability to cross-protect against multiple variants and subtypes of IAV [75,76]. PR8-Gluc was used in vivo to characterize the therapeutic potential of two stalk-reactive monoclonal antibodies (MAbs) GG3 and KB2 [77]. These stalk-reactive MAbs bind to many H1 and H5 IAV and exhibit neutralization activities in plaque reduction assays [78]. The ability of the broadly neutralizing GG3 and KB2 to impede PR8-Gluc infection in the lungs was examined by passive transfer experiments [79]. To this end, mice were given the GG3 and KB2 antibodies 2 h before being infected with a $5\times$ mouse lethal dose-50 (MLD_{50}). The studies showed no morbidity or mortality of mice receiving the antibody therapies [66]. These MAbs also protected against lethal challenge with influenza A/Netherlands/602/2009 (H1N1) and influenza A/Vietnam/1203/2004 (H5N1) in mice [66]. PR8-Gluc had a MLD_{50} of ~5000 plaque forming units (PFU), approximately 50–100 times less lethal than WT virus (MLD_{50} of ~50) [66], which suggests PR8-Gluc recapitulates a PR8 WT-like virus life cycle in vivo.

Table 3. Recombinant influenza A viruses expressing reporter gene in the viral PB1 segment.

Gene	Virus Backbone [1]	Transgene [2]	Insertion Mechanism [3]	Application	Ref.
PB2 PB1 PA	WSN	Split Gluc	Fusion	Virus-host interaction	[80]

[1] WSN: A/WSN/1933 (H1N1); [2] Gluc: Gaussia luciferase; [3] Fusion: A fragment of a split Gluc (Gluc1 or Gluc2) was fused to the C-terminus of PB1, PB2, or PA. To reconstitute the Gluc activity, both fragments needs to be in the same cell.

Yan et al. established a HTS protocol for the simultaneous identification of pathogen- and host-targeted hit candidates against either respiratory syncytial virus (RSV) or IAV [67]. To this end, the authors generated a recombinant WSN-Gluc (Figure 2C, Table 2), which was used with a recombinant RSV-firefly. The dual-pathogen protocol using replication-competent recombinant viruses shows superior cost and resource effectiveness. Moreover, the screening agents used in this new approach, IAV and RSV, are clinically important human respiratory pathogens.

2.2. IAV Containing a Viral Polymerase Subunit Fused to the Reporter Gene

A precise mapping of pathogen–host interactions is essential for a comprehensive understanding of the processes of infection and pathogenesis. Interactome studies have used multiple approaches to identify and characterize protein interactions. For instance the use of yeast two-hybrid screens, which for animal viruses, does not reflect the pathogen's microenvironment. Tandem affinity purification and mass spectrometry are also used, but this approach cannot distinguish direct from indirect interactions. New technologies are thus needed to improve the mapping of pathogen–host interactions, including IAV. Munier et al. generated a set of recombinant IAV that contain a fragment (Gluc1 or Gluc2) of a split Gluc fused to the C-terminus of PB1, PB2, or PA (vP-Gluc1 or vP-Gluc2) in the WSN backbone [80] (Figure 3, Table 3). To reconstitute Gluc activity, a cell must be co-infected by two viruses that, in combination, produce Gluc1 and Gluc2 [80]. Despite moderate attenuation in vitro relative to the WT virus, the viruses expressing a viral fusion protein (vP-Gluc1 or vP-Gluc2) were replication-competent, with the vP-Gluc1 viruses showing higher titers than their vP-Gluc2 counterparts upon multi-cycle amplification on MDCK cells [80]. The authors then used the split Gluc viruses to demonstrate a dose-dependent reduction of luciferase reconstitution in the presence of ribavirin or nucleozin, but not in the presence of amantadine, consistent with previously published data for WSN [81,82]. It is important to note that luciferase activity can be reduced by inhibiting viral protein-protein interactions or by reducing viral proteins abundance. The split Gluc viruses could thus be used with compound libraries, or knockdown or overexpression assays to identify host factors that affect viral replication, RNP assembly, or inhibitors of virus replication [80]. Munier et al. modified the experimental parameters to detect binary interactions between IAV polymerase and host proteins, whereby a single Gluc1-tagged virus was used to infect cells transfected with Gluc2-fused host proteins [80]. The assay detected viral–host protein–protein interactions within their exploratory set [80]. Among the host factors identified were those involved in the nuclear import pathway, components of the nuclear pore complex such as nucleoporin 62 (NUP62) and mRNA export factors such as nuclear RNA export factor 1 (NXF1), RNA binding motif protein 15B (RMB15B), and DDX19B [80].

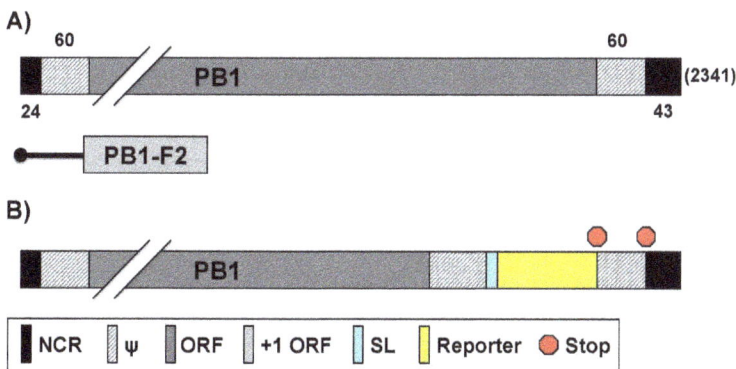

Figure 3. PB1 reporter influenza A viruses: Schematic representation of the PB1 segment from WT (A) and reporter (B) viruses as described in Figure 1. Influenza A WT PB1 viral segment encodes for both PB1 and PB1-F2 in the +1 ORF via an alternative start codon. Nucleotide lengths for the NCR, ψ, and PB1 segment are indicated; (B) PB1 fusion protein: Reporter genes were fused to native PB1 ORF with a short linker (SL). Packaging signals were duplicated after protein stop transcription signal and before the 3′ NCR terminus.

2.3. Reporter IAV Containing a Recombinant PA Segment Harboring the Foreign Gene

NanoLuc (NLuc) is a small molecular weight (19 kDa) luciferase, which has a light output 150 times greater than other popular luciferases like Renilla and Firefly [49]. Because of these advantages, Tran et al. described the generation of a recombinant WSN IAV that expresses NLuc from the PA segment, using a 2A autocleavage sequence [52] (Table 4). IAV segment 3 (Figure 4A) encodes the polymerase subunit PA that has previously been shown to tolerate fusions to its C terminus without disrupting polymerase function [83]. Moreover, as compared to the other polymerase subunits, IAV PA has minimal packaging sequences at the 3′ and the 5′ end [1,84]. PA is structurally required for polymerase activity, possesses endonuclease activity to cleave host capped pre-mRNAs, and is required for nuclear accumulation of PB1 [14]. The PA segment was altered such that PA-2A-NLuc was followed by a 50 nucleotide (nt) repeat of the 3′ packaging signal; and the segment was referred to as PA-2A-NLuc50, or PATN [85] (Figure 4B). Another virus was generated (PA-SWAP-2A-NLuc50 or PASTN) that differed from PATN by introducing 18 silent mutations within the 47 terminal nt of the PA coding sequence, possibly relieving competition of multiple packaging signals to achieve stable reporter gene maintenance over repeated passaging [52] (Figure 4C). The authors showed that these NLuc WSN viruses (PATN and PASTN) replicate with WT properties in culture and in vivo, and possesses remarkably similar pathogenicity and lethality in mice [52]. The WSN PATN virus was then used to investigate the dissemination of IAV in mouse lungs using an in vivo imaging system. WSN PATN viruses encoding either human-signature PB2 K627 or the avian-signature PB2 E627 were used in vivo to assess host range determinants of viral dissemination in living mice [86]. As expected, mice infected with WT (PB2 K627) WSN PATN lost weight and displayed robust bioluminescence that increased over time [86]. In contrast, the WSN PATN virus encoding avian-signature PB2 E627 was severely restricted. Infected mice showed little weight loss and bioluminescence was near background levels throughout the experiment [86]. Thus, the reporter virus faithfully recapitulated the known polymerase-mediated host range restriction and could therefore also be used with newly isolated IAV strains to determine their host range and the role of species-specific adaptive mutations. PATN and PASTN virus infection is therefore a viable model for WSN infection in vitro and in vivo. Similarly, Tran et al. used this system with multi-modal bioluminescence and positron emission tomography-computed tomography (PET/CT) imaging to evaluate the effect of oseltamivir treatment in viral load, dissemination and inflammation in mice [50] (Table 4).

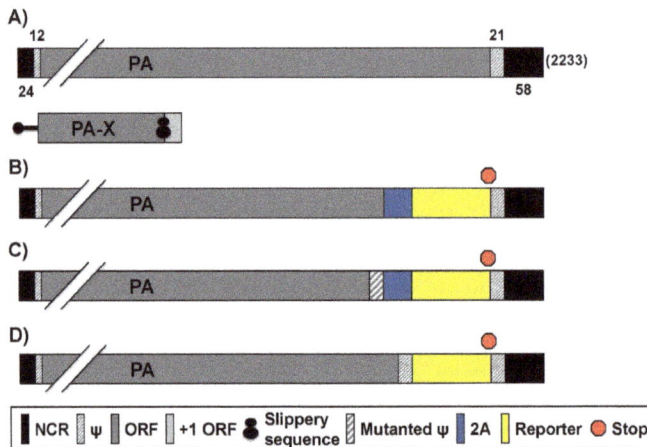

Figure 4. **PA reporter influenza A viruses:** Schematic representation of the PA segment from WT (**A**) and reporter (**B–D**) influenza A viruses as described in Figure 1. Influenza A WT PA gene segment encodes for both PA and PA-X, which shares the N-terminal amino acids with PA but the C-terminus is in the +1 ORF via ribosomal frame shift. Nucleotide lengths for the NCR, ψ, and PA segment are indicated; (**B,C**) Bicistronic transcription of PA and reporter gene: Insertion of the 2A autocleavage site separates PA from reporter gene. Packaging signals encoding the 3' terminus of PA were WT (**B**) or mutated (**C**) to minimize interference with native ψ, which are duplicated at the 3' NCR-proximal region; (**D**) PA fusion protein: Reporter genes were fused to native PA ORF. Packaging signals were duplicated after the protein stop transcription signal, before the 3' NCR terminus.

Table 4. Reporter-expressing recombinant influenza viruses in the viral PA.

Gene	Virus Backbone [1]	Transgene [2]	Insertion Mechanism [3]	Application	Ref.
PA	WSN	Nluc	2A site	Virus biology and transmission	[50, 52]
PA	pH1N1	Nluc	2A site	Virus biology and transmission	[69]
PA	PR8, Neth602, Ind5, Anh1	eGFP, fRFP, iRFP, Gluc, FFluc	2A site	Virus biology	[73]
PA	WSN	GFP	Fusion	Virus biology	[87]

[1] WSN: A/WSN/1933 (H1N1); pH1N1: A/California/04/2009 (pH1N1); PR8: A/Puerto Rico/8/1934 (H1N1); Neth602: A/Netherlands/602/2009 (H1N1); Ind5: A/Indonesia/5/2005 (H5N1); Anh1: A/Anhui/1/2013 (H7N9); [2] Nluc: Nanoluciferase; eGFP: Enhanced GFP; Gluc: Gaussia luciferase; fRFP: far-red fluorescent protein; iRFP: near-infrared fluorescent protein; FFluc: Firefly luciferase; GFP: Green fluorescent protein; [3] 2A site: The reporter gene was separated from the viral ORF via a 2A peptide sequence; Fusion: Fusing the entire GFP protein to the C-terminus of PA.

A similar approach was also used by Karlsson et al. to generate a PA-2A-NLuc IAV in the influenza A/California/04/2009 pandemic H1N1 (pH1N1) backbone [69] (Table 4). This recombinant virus (pH1N1-PA-NLuc) was used in a transmission study in ferrets. pH1N1-PA-NLuc had WT-like kinetics both in vitro and in vivo, and was transmissible by direct and respiratory contact [69]. Titers of pH1N1-PA-NLuc were determined by both bioluminescence and tissue culture infective dose 50 ($TCID_{50}$) and were found to be nearly identical [69]. A benefit of using pH1N1-PA-NLuc is that measuring bioluminescence reduced the turnaround time of titer determination by 54 h [69].

Spronken and Short et al. also generated several recombinant replication-competent IAV harboring reporter genes in the PA segment [73] (Table 4). To optimize the strategy to make an IAV expressing a reporter gene, different constructs were cloned using the near-infrared fluorescent protein (iRFP) [73]. Firstly, a construct consisting of the PA 5′ untranslated region (UTR), the PA ORF without stop codon, a short linker, the 2A sequence, iRFP and the 3′ UTR was produced [73]. This construct was then further modified by inserting a duplication of the packaging region (dPR) [73]. Finally, two or three mutations in the promoter region (2UP and 3UP) were also introduced [73]. The duplication of the packaging region was essential to rescue iRFP reporter virus efficiently, whereas introduction of the 3UP mutation did not result in virus production [73]. On the other hand, the 2UP_PA_iRFP_dPR construct was the only one that resulted in recombinant virus expressing iRFP in vitro, although the virus titers were lower than that of WT [73]. A reduction in virus titer was also observed when the 2UP mutation was introduced into the WT PA gene segment [73]. The 2UP_PA_iRFP_dPR cloning strategy was later used to insert different reporters into PR8, including enhanced GFP (eGFP), far-red fluorescent protein (fRFP), Gluc and FFluc (Table 4). The levels of virus replication, reporter expression and stability of the reporter were evaluated. This strategy was then used to generate eGFP-expressing viruses in the backbone of influenza A/Netherlands/602/2009 H1N1, or the highly pathogenic avian influenza (HPAI) A/Indonesia/5/2005 H5N1, and A/Anhui/1/2013 H7N9 [73,88]. 2UP_PA-Gluc_dPR had the greatest reporter stability tested followed by 2UP_PA-eGFP_dPR. 2UP_PA-fRFP_dPR, _iRFP, and _FFLuc lost considerable reporter activity after four or five passages in vitro [73]. Further optimization, which consisted of shortening the duplicated packaging signal from 149 nt to 50 nt (2UP_PA-eGFP_sPR), had limited impact on viral replication but did enhance eGFP expression versus 2UP_PA-eGFP_dPR. When evaluating infection using an in vivo imaging system (IVIS), 2UP_PA-Gluc_sPR, _eGFP, and _fRFP showed strong signals in mouse lungs whereas 2UP_PA-iRFP_dPR did not [73]. The authors hypothesized that this is due to 2UP_PA-iRFP_dPR's low reporter expression and not the loss of the reporter gene in vivo since MDCK cells infected with lung homogenates became fluorescent [73]. Due to strong signal output, reporter stability, and WT-like replication characteristics, 2UP_PA_eGFP_sPR was chosen as a model for further studying aspects of IAV infection, such as detecting morphological changes in infected cells by fluorescence and electron microscopy (EM) [73]. Green fluorescent (infected) MDCK cells were examined using EM and found to have microvillar projections with virus-like particles budding from these structures [73]. Finally, 2UP_PA_eGFP_sPR and _dPR pH1N1 and HPAI H5N1 and H7N9 viruses were rescued and characterized in vitro [73]. Each of the reporter viruses rescued grew ~2 logs less than WT virus in vitro [73]. Importantly, the H5N1 and H7N9 eGFP reporter viruses exhibited fluorescent stability for up to four serial passages [73]. Conversely, 2UP_PA_eGFP_sPR pH1N1 lacked stability while the dPR isolate showed reduced stability over the avian strains [73]. In vivo experiments were conducted with the 2UP_PA_eGFP_dPR H5N1 isolate exclusively. No signal came from live imaging, and only ex vivo lung images using IVIS showed a diffuse reporter signature, thus providing a model of avian influenza infection in mice [73].

To track vRNA movement in infected cells, a fluorescent tag can be fused to a viral protein involved in vRNA trafficking. Although some fluorescent IAV have been generated, most contain GFP fused with a viral protein not involved in vRNA transport or as a separate fluorescent polypeptide [71,89,90]. Lakdawala et al. overcame this limitation by fusing the entire GFP protein to the C-terminus of PA in the backbone of a WSN IAV (WSN-PA GFP) [87] (Figure 4D, Table 4). The segment constructed contained, after the GFP stop codon, a duplication of the PA 5' packaging signals containing approximately 150 nt of the coding region, upstream of the

5′ NCR [87]. The authors developed two novel imaging tools: a system to visualize four different vRNA segments within an infected cell and a fluorescent influenza virus (WSN-PA GFP) to track vRNA dynamics in live cells during a productive infection [87]. Using the WSN-PA GFP virus and live cell fluorescence microscopy, the investigators were able to show PA-GFP foci fuse in the cytoplasm and remain in this state as they traveled to the plasma membrane [87]. Additionally, using fluorescent in situ hybridization probes for specific vRNAs, it was shown that PA-GFP fuses with PB2 and HA viral segments in the cytoplasm of infected cells [87]. Overall, the data suggested that vRNA segments are not exported as individual segments since the majority of foci at the external nuclear periphery contain more than one vRNA segment [87]. Moreover, many foci with fewer than 4 vRNA segments were observed in the cytoplasm, implying that all 8 vRNA segments are not exported from the nucleus together [87]. Therefore, the authors concluded that vRNA assembly includes the formation of flexible subcomplexes that export from the nucleus and then undergo further assembly *en route* to the plasma membrane via dynamic co-localization events [87].

2.4. Generation of Reporter-Expressing IAV Containing a Modified NA Segment

Segment 6 from IAV (Figure 5A) encodes the NA protein that functions to promote viral release and is one of the major surface viral antigens [1]. Its principal biological role is the cleavage of the terminal sialic acid residues that are receptors for the HA glycoprotein [10,11]. The receptor-destroying activity in NA resides in the distal head domain that is linked to the viral membrane by an N-terminal hydrophobic transmembrane domain [91]. The ability to cleave sialic acid is also thought to help the virus penetrate mucus [92].

Table 5. NA recombinant reporter-expressing influenza A viruses.

Gene	Virus Backbone [1]	Transgene [2]	Insertion Mechanism [3]	Application	Ref.
NA	PR8	eGFP	2A site	Virus biology	[70]
NA	PR8	Gluc	2A site	Virus biology	[72]
NA	WSN	GFP	Viral promoter	Virus biology	[93,94]

[1] PR8: A/Puerto Rico/8/1934 (H1N1); WSN: A/WSN/1933 (H1N1); [2] eGFP: Enhanced GFP; Gluc: Gaussia luciferase; GFP: Green fluorescent protein; [3] 2A site: The reporter gene was separated from the viral ORF via a 2A peptide sequence; Viral promoter: The reporter was introduced under the control of a duplicated 3′ NCR (viral promoter).

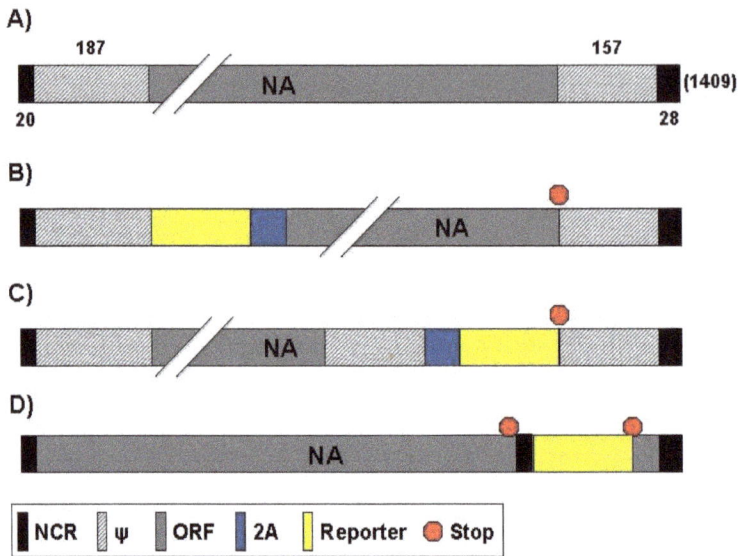

Figure 5. NA reporter influenza A viruses: Schematic representation of the NA segment from WT (**A**) and reporter (**B–D**) influenza A viruses as described in Figure 1. Nucleotide lengths for the NCR, ψ, and NA segment are indicated; (**B,C**) Bicistronic transcription of NA and reporter gene: Insertion of the 2A autocleavage site before (**B**) or after (**C**) the NA ORF separates the viral gene from the reporter gene. In (**C**), the packaging signals were duplicated before the 3' NCR; (**D**) Dicistronic recombinant NA segment: The NA coding sequence is followed by a duplicated 3' NCR, the reporter gene and the 5' NCR.

Feng et al. generated two eGFP-expressing PR8 viruses with the reporter linked to the NA segment [70] (Table 5). The NA vRNA packaging signals, including both the 3' NCR (19 nt) and the adjacent 183 nt of the coding region; and the 5' NCR (28 nt) and the adjacent 157 nt of the coding region, were maintained for the efficient packaging of the modified NA vRNA [70]. Each of the reporter PR8 viruses utilized the 2A autocleavage site to allow for collinear expression of both NA and eGFP [70]. In the "rPR8-eGFP+NA" virus, the eGFP gene precedes the 2A site, followed by NA (Figure 5B). In the "rPR8-NA+eGFP" the NA gene precedes eGFP and are separated by the 2A site (Figure 5C). What the authors found is that the order of viral gene and fluorescent reporter did have an effect on viral growth kinetics, plaque phenotype, NA activity, or eGFP localization. In embryonated eggs, MDCK, and human lung epithelial A549 cell lines, rPR8-NA-eGFP had a viral growth kinetic similar to WT virus, while rPR8-eGFP+NA replicated at lower levels (1 log or more) [70]. Likewise, and as expected, rPR8-NA+eGFP formed WT-like plaques that were large in comparison to rPR8-eGFP+NA [70]. Next, the amount of NA in rPR8-eGFP+NA and rPR8-NA+eGFP purified viruses were compared with those in WT virus. The rPR8-NA+eGFP virion possessed nearly as much NA activity (~90%) as the WT virus, while the rPR8-eGFP+NA virion possessed much less NA activity (~17%), which may be the result of eGFP interference and truncation of the N-terminal region of the NA protein [70]. In addition, NA vRNA packaging efficiency was tested. Compared to WT, nearly 80% of the NA+eGFP vRNA segments were packaged into the rPR8-NA+eGFP virions, while only 50% of eGFP+NA vRNA segments were packaged into the rPR8-eGFP+NA virions [70]. The localization of eGFP of the two recombinant viruses was also shown to be different. eGFP produced by rPR8-NA+eGFP was uniformly distributed throughout the infected cells, including the nucleus, while rPR8-eGFP + NA reporter was found only on the cell membrane [70]. The authors attribute the dissimilarities between viruses to the differences in the recombinant NA segments. rPR8-eGFP+NA has the reporter fused to the NA transmembrane anchoring region, thus localizing it to membrane. This could reduce the presence of NA on the membrane, limiting viral release. The authors suggest that this membrane bound eGFP would be useful for live infection monitoring in vivo but rPR8-eGFP+NA's attenuated viral kinetics make rPR8-NA+eGFP the more logical choice [70]. However, replication-competent IAV harboring fluorescent reporters have been ineffective for such studies and only used ex vivo [38,39,41]. Bioluminescent harboring models have been far more effective for IAV in vivo research [39,46]. The authors also used the rPR8-NA+eGFP recombinant virus for the identification of neutralizing antibodies (NAbs) by fluorescence-activated cell sorting (FACS). FACS provides a rapid, highly accurate measurement of fluorescence measured by mean fluorescent intensity (MFI). Serial dilutions of mouse anti-PR8 serum were pre-incubated with rPR8-NA+eGFP and then the serum-fluorescent virus mixture was used to infect MDCK cells [70]. The authors showed incrementally increasing MFI, indicative of ineffective serum neutralization, in the lower concentration dilutions of serum, demonstrating the use of rPR8-NA+eGFP to evaluate the presence IAV NAbs.

Studying viral dissemination in vivo using replication-competent, reporter-expressing IAV remains a challenge because it requires that the reporter signal to be detected through tissue and skin. Pan et al. published the generation of replication-competent IAV harboring Gluc in the NA segment using PR8 as the backbone (IAV-Gluc) [72] (Table 5). The viral NA segment was similar to the described above for rPR8-NA + eGFP. IAV-Gluc replicated at lower levels in MDCK cells (2–3 log) and eggs (1–2 log) as compared with WT virus [72]. The authors were also able to visualize viral dissemination in live mice with a model virus that causes pathology. However, IAV-Gluc virus was attenuated in vivo; it took 1,000x more IAV-Gluc to achieve PR8 WT-like pathology as characterized by body weight change, percent survival, and lung histopathology [72]. Live-mouse imaging and antiviral therapeutic studies using

IVIS was found to be optimal using a high dose (10^6 PFU) for dissemination [72]. Importantly, Gluc signal was specific to the site of viral infection after the injection of the luciferase substrate. Finally, the authors tested whether IAV-Gluc can be used for evaluating antiviral therapeutics in vitro. As a proof-of-principle, they used an antiviral serum collected from convalescent mice previously infected with PR8. IAV-Gluc was pre-incubated with serial dilutions of the antiviral serum, the serum-virus mixture was used to infect MDCK cells in 96-well plates, and the bioluminescence intensity was determined at 24 h post-infection. Results showed that IAV-Gluc could be completely inhibited by the antiviral sera, demonstrating the potential of using IAV-Gluc for developing viral neutralization assays to evaluate antiviral drugs in vitro.

Vieira Machado et al. attempted the generation of a replication-competent IAV in the backbone of WSN harboring a dicistronic NA segment containing NA and foreign sequences with different sizes, either a foreign GFP (239 aa), chloramphenicol acetyl transferase (CAT; 220 aa) or a fragment of the *Mengovirus* VP0 capsid (101 aa) under the control of a duplicated 3' promoter sequence (NA35-foreign gene) [93] (Table 5). To this end, the 3' NCR and a multiple cloning site were inserted between the stop codon and the 5' promoter sequence of the NA segment [93] (Figure 5D). Despite numerous attempts, a virus expressing GFP was not rescued. However, recombinant viruses expressing CAT or VP0 were successfully generated, suggesting that reporter genes other than GFP could be included and it can exist constraints on the size or the nature of the inserted foreign sequences. The authors demonstrated that the duplicated 3' promoter was used to drive foreign gene expression [93]. Northern blot analysis for vRNA, cRNA and mRNA showed that two NA-derived RNA species were detected, corresponding with the full-length and a shorter subgenomic molecule comprising the reporter gene sequences flanked by 5' and 3' noncoding sequences [93]. Despite slightly reduced NA expression, the recombinant viruses replicated efficiently and proved to be stable upon serial passage in MDCK cells or in the pulmonary tissue of infected mice [93]. Later, Vieira Machado et al. generated a novel recombinant IAV (vNA38) harboring a dicistronic NA segment with an extended 5' terminal sequence of 70 nts comprised of the last 42 nts of the NA ORF and the 5' NCR [94] (Figure 5D, Table 5). vNA38 viruses, containing the same foreign genes as vNA35 viruses [93], replicated stably and more efficiently than vNA35 viruses with a dicistronic NA segment comprised of the native 5' NCR only [94]. Moreover, whereas the NA35-GFP dicistronic vRNAs could not be rescued into infectious viruses, all three NA38-CAT, NA38-VP0 and NA38-GFP viruses were rescued [93]. In addition, vNA38 viruses expressed the foreign gene to higher levels than vNA35 viruses in cell culture and in the pulmonary tissue of infected mice [94]. The authors proposed this later recombinant IAV harboring the dicistronic NA segment for the development of live bivalent vaccines.

2.5. Generation of Reporter-Expressing IAV by Rearrangement of the PB1 and NS Viral Segment

Avian influenza virus subtypes H5N1 and H9N2 top the World Health Organization's list for the greatest pandemic potential [95,96]. Inactivated H5N1 vaccines induce limited immune responses and, in the case of live-attenuated influenza vaccines (LAIV), there are safety concerns regarding the possibility of reassortment between the H5N1 viral segments and circulating IAV strains. To overcome these drawbacks, Pena et al. introduced a novel method of generating a bivalent vaccine against both influenza A/Guinea fowl/Hong Kong/WF10/1999 (H9N2) and influenza A/Vietnam/120320/04 (H5N1) using viral genome rearrangement [97] (Table 6). This was achieved by first removing NEP from the H9N2 NS viral segment. Then NEP was replaced with the H5 HA ORF separated by the FMDV 2A autocleavage site, allowing for collinear expression of both proteins (Figure 6A). The transgene was inserted by cloning it downstream of either a full-length or a truncated (expressing the first N-terminal 99 aa) *NS1* gene [97]. To prevent the normal splicing activity in the NS segment, the donor site and branch point within the full-length NS1 were mutated, and a stop codon was inserted early in the residual open reading frame of NEP. NEP was then fused to the H9N2 PB1 segment and separated by FMDV 2A (Figure 6B). In addition, the corresponding packaging signals previously determined for RNA segments 2 (PB1) and 8 (NS) were maintained at the $5'$ end of each segment. As a proof of concept, the authors first used GFP or Gluc with the truncated NS1 ORF instead of the H5 HA viral protein and were able to rescue infectious, reporter expressing viruses (H9N2-GFP and -Gluc, respectively) [97]. These recombinant rearranged IAV reached titers on the order of six to seven $\log_{10}$ egg infectious dose (EID_{50})/mL, and transgene expression was maintained for up to ten passages [97]. A 10- to 100-fold reduction in virus titers of rearranged recombinant H9N2-GFP and H9N2-H5 HA (in the NS1-99aa backbone) was observed compared to parental recombinant virus (containing the same NS1 deletion) was observed [97]. In addition rearranged viruses were also attenuated in vivo. H9N2-GFP was not able to provide complete protection to mice after a single immunization. However, the H9N2-H5 HA virus provided complete protection against lethal challenge with influenza A/Vietnam/1203/2004 H5N1 in mice and ferrets, and also against a potentially pandemic H9:pH1N1 IAV reassortant virus [97,98]. Altogether, these studies, demonstrated that rearrangement of the IAV genome has great potential for the development of improved vaccines against multiple IAV, as well as other pathogens, and for the expression of reporter genes [97].

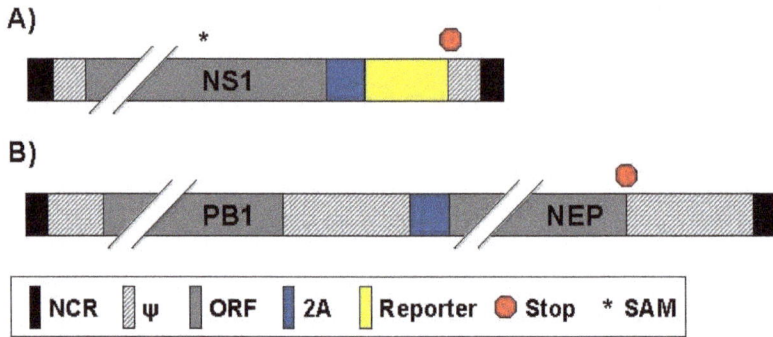

Figure 6. Reporter influenza A viruses with genome rearrangement: Schematic representation of mutant NS (**A**) and PB1 (**B**) viral segments as described in Figure 1. (**A**) Bicistronic transcription of NS1 and reporter gene: A splice acceptor mutation (SAM; *) inhibits alternative splicing. Reporter protein expression occurs after 2A cleavage; (**B**) Bicistronic transcription of PB1 and NEP: Expression of PB1 and NEP gene products occurs by insertion of the 2A autocleavage site sequence. PB1 3′ packaging signals were duplicated after the NEP ORF and before the 3′ NCR terminus.

Table 6. Reporter-expressing recombinant influenza A viruses with a rearranged genome.

Gene	Virus Backbone [1]	Transgene [2]	Insertion Mechanism [3]	Application	Ref.
NS	HK99 VN1203	GFP, Gluc	Genome rearrangement	Vaccine	[97]
NS	pH1N1	GFP, Gluc	Genome rearrangement	Antivirals	[99]

[1] HK99: A/Guinea Fowl/Hong Kong/WF10/1999 (H9N2); VN1203: A/Vietnam/1203/2004 (H5N1); pH1N1: A/California/04/2009 (H1N1); [2] GFP: Green fluorescent protein; Gluc: Gaussia luciferase; [3] Genome rearrangement: The NEP was removed from the NS viral segment, and replaced with the foreign gene separated by a 2A autocleavage site. Then, the NEP was fused to the PB1 segment also separated by a 2A site.

The genome rearrangement strategy was also extended to influenza A/California/04/2009 (pH1N1 or Ca04) by Sutton et al. where Gluc or GFP were expressed downstream of NS1 ORF (designated GlucCa04 and GFPCa04, respectively) [99] (Table 6). The researchers also rescued an amantadine-resistant GlucCa04 virus for anti-viral drug screening purposes and denoted (Res/GlucCa04) [99]. GlucCa04 and Res/GlucCa04 grew to significantly reduced titer levels compared to the recombinant Ca04 and Res/Ca04 WT counterparts in MDCK cells over multicycle growth [99]. Both the GlucCa04 and Res/GlucCa04 grew to similar titers, although the sensitive virus appears to grow to slightly higher titers than the resistant virus [99]. The genetic stability of three separate replicates of GlucCa04 was evaluated by serial passaging five times in MDCK cells. Of the three replicates, two had a ten-fold decrease in Gluc expression while one had a ten-fold increase in reporter expression after passaging, although the cause for these differences was not further explored. As a proof-of-concept, Sutton et al. used these viruses for an in vitro anti-viral screening microneutralization assay. Amantadine treatment significantly decreased Gluc expression of GlucCa04 compared to Res/GlucCa04, with half maximal inhibitory concentration (IC_{50}) results comparable to previously published literature. Microneutralization assay outcomes also showed similar results between pH1N1 WT and GlucCa04 [99]. Finally, the authors showed that GlucCa04 could be used as an in vivo screening tool for compounds with antiviral activity [99]. When taken all together, GlucCa04 provides another luciferase-based tool to use as a screening technique to identify novel antiviral drugs, shortening the time required for virus detection for in vitro and in vivo studies.

Table 7. Recombinant influenza A viruses expressing reporter genes from the NS viral segment.

Gene	Virus Backbone [1]	Transgene [2]	Insertion Mechanism [3]	Application	Ref.
NS	PR8	GFP	Caspase recognition site	Virus biology	[90]
NS	PR8	GFP	Stop/start	Vaccine	[100]
NS	PR8	maxGFP	2A site	Virus pathogenesis	[71,101–104]
NS	PR8	maxGFP, turboRFP, Gluc	2A site	Antiviral and virus-host interaction	[48]
NS	PR8 pH1N1	mCherry	2A site	Antivirals, neutralizing antibodies, virus pathogenesis	[41]
NS	pH1N1	Timer	2A site	Virus propagation	[68]
NS	PR8 VN1203	Venus, eGFP, eCFP, mCherry	2A site	Virus-host interaction and virus pathogenesis	[38]
NS	PR8 WSN	GFP	2 × 2A site	Virus pathogenesis	[105,106]

[1] PR8: A/Puerto Rico/8/1934 (H1N1); pH1N1: A/California/04/2009 (H1N1); VN1203: A/Vietnam/1203/2004 (H5N1); [2] GFP: Green fluorescent protein; maxGFP: advanced version of eGFP; Gluc: Gaussia luciferase; mCherry: monomeric Cherry fluorescent protein; Timer: modified Discosoma red fluorescent protein; Venus: advanced version of yellow fluorescent protein; eGFP: Enhanced GFP; eCFP: Enhanced cyan fluorescent protein; [3] Caspase recognition site: The reporter gene was fused to NS1 protein separated by a peptide sequence containing a caspase recognition site; Stop/Start: The stop-start pentanucleotide (UAAUG) from BM2 of influenza B virus was inserted between NS1 and the reporter gene; 2A site: The reporter gene was separated from the viral ORF via a 2A peptide sequence.

2.6. Generation of Reporter-Expressing, Replicating-Competent IAV Containing a Recombinant NS Segment Harboring the Foreign Gene

Research using a replication-competent IAV containing a modified NS segment (Figure 7A) started as early as 2004. Kittel et al. published a study where GFP was fused to a truncated (only expressing the first 125 aa) NS1 protein separated by a peptide sequence that contained a caspase recognition site (CRS) [90] (Figure 7B, Table 7). The truncated NS1 maintained the ability to antagonize type I interferon (IFN) and allowed for the insertion of a foreign gene, like GFP. The GFP-harboring IAV (NS1-GFP) was able to replicate in protein kinase R (PKR) knockout mice and reached similar viral lung titers with or without the GFP gene (5×10^4 PFU/g), but were attenuated in WT mice [90]. In vitro, NS1-GFP was capable of replicating to WT-like titers in IFN-deficient Vero cells [90]. In contrast to Vero, virus passaged in IFN-competent MDCK cells resulted in the selection of NS1-GFP deletion mutants [90]. Sequence data of these mutants confirmed that loss of GFP expression correlated with partial deletion of the GFP ORF. The deletions were very heterogeneous, ranging from several nucleotides up to the removal of approximately 80% of the C terminal end of the GFP protein. Interestingly, the deletion mutants that outcompeted NS1-GFP only contained GFP deficiencies and not NS1 mutations. While certainly not the most practical replication-competent, fluorescent-expressing IAV, Kittel et al. provided an early, efficacious genetic strategy for inserting foreign genes into a viral genome that yielded viable progeny [90].

Kittel et al. also incorporated genetic characteristics of influenza B virus (IBV), which uses a unique strategy to encode the matrix 2 (BM2) protein [107]. The initiation codon of the BM2 overlaps with the termination codon of the upstream gene for the M1 protein, forming a stop-start pentanucleotide (UAAUG) [107]. Utilizing the IBV strategy, Kittel et al. developed a bicistronic recombinant IAV in which the stop codon of the stop-start cassette terminates the translation of NS1 after 125 aa and the start codon reinitiates the translation of GFP (A/PR8/NS1-GFPStSt) (Figure 7C, Table 7) [100]. Although the expression level of GFP was significantly lower than that obtained previously with the NS1-GFP [90], the bicistronic IAV appeared to be replication competent in mice and showed higher genetic stability. In fact, all viral isolates derived from infected mouse lungs were still capable of expressing GFP in infected cells [100]. Utilizing this bicistronic virus, authors also expressed interleukin-2 (IL-2) instead of GFP [100]. Although the IL-2-expressing IAV showed high titers in mouse lungs, it did not display any mortality rate in infected animals. In addition, the IL-2-expressing virus showed an enhanced CD8+ response to viral antigens in mice after a single intranasal immunization. These results suggested that influenza viruses could be engineered for the expression of biologically active molecules such as cytokines for immune modulation purposes [100].

Manicassamy et al. described a replication-competent IAV that contained GFP fused to NS1 protein (NS1-GFP) of PR8 [71]. It had been previously shown that functionally active IAV NS1 and NEP could be expressed as a single polyprotein with a FMDV 2A autoproteolytic cleavage site [108]. Manicassamy et al. modified the NS segment to express NS1-GFP and NEP as a single polyprotein with a porcine teschovirus-1 (PTV-1) 2A autoproteolytic cleavage site between them, allowing NEP to be released from the upstream NS1-GFP protein during translation (Figure 7D, Table 7) [71]. In order to avoid the splicing of NS mRNA, two silent mutations in the splice acceptor site were introduced [71]. NS1-GFP replicated like WT in a single-cycle replication assay in MDCK cells but showed 100x lower replication levels in a multicycle assay [71]. A well-characterized function of IAV NS1 protein is to counteract the host type I IFN response [20]. It was shown that NS1-GFP was also able to suppress IFN activity. In vivo, NS1-GFP was 100x attenuated when evaluated by MLD_{50} and body weight loss compared to WT virus [71]. Using the NS1-GFP virus, the authors studied viral dissemination in lungs. Not only was GFP observable in the lungs of infected mice ex vivo using IVIS for whole-organ imaging, but could also be observed using flow cytometry to analyze the infection progression in antigen presenting cells [71]. Imaging of murine lungs showed that infection starts in the respiratory tract in areas close to large conducting airways and later spreads to deeper sections of the lungs [71]. The authors found that using a 10^6 PFU intranasal inoculation, 10% of dendritic cells (DCs) were observed to express GFP and 2%–3% of macrophages and neutrophils were also GFP+ at 48 h post-infection [71]. At 96 h, the percent of GFP-expressing DCs declined but percentages of fluorescing macrophages and neutrophils increased [71]. Dosing animals with oseltamivir limited GFP expression in all antigen-presenting cells while amantadine was only effective in specific cell types [71]. The authors do however acknowledge that antigen presenting cells could be GFP+ due to phagocytizing infected cells. In fact, Helft et al. used the NS1-GFP IAV to establish the kinetics of infection and transport to the draining lymph nodes [109] (Table 7). The authors provided evidence that lung DCs that transport viral antigens to the draining lymph nodes are protected from influenza virus infection in vivo and that induction of viral-specific CD8+ T cell immunity is mainly dependent on cross-presentation of virally infected cells by lung migratory non-infected CD103+ DCs [109]. They also reported that lung migratory CD103+ DCs express a natural anti-viral state that is further strengthened by type I IFN released during the first few hours following influenza virus infection.

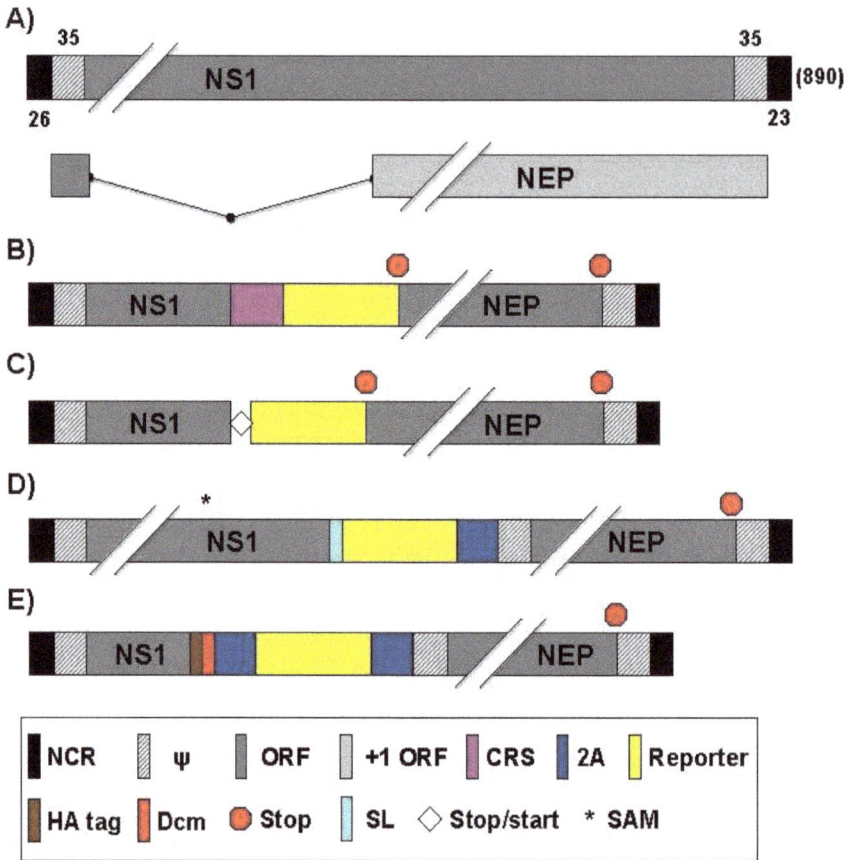

Figure 7. NS reporter influenza A viruses: Schematic representation of the NS segment from WT (**A**) and reporter (**B–E**) influenza A viruses as described in Figure 1. Influenza A WT NS gene segment encodes for NS1 and NEP via alternative splicing. Nucleotide lengths for the NCR, ψ, and NS segment are indicated; (**B**) Multicistronic transcription using a caspase recognition site: Multicistronic NS reporter influenza A viruses were generated by insertion of a caspase recognition site (CRS) after the NS1 ORF; (**C**) Multicistronic transcription using stop-start sequence: Multicistronic NS reporter influenza A viruses were generated by insertion of a stop/start transcription site after the NS1 ORF for independent translation of NS1, reporter gene, and NEP; (**D**) NS1 fusion protein: Reporter genes were fused to native NS1 ORF with a short linker (SL). A splice acceptor mutation (SAM; *) inhibits NEP alternative splicing. NEP expression occurs after 2A cleavage. The 5′ ψ are duplicated and contain NEP N-terminal amino acid codons; (**E**) Tricistronic transcription of the NS segment: Reporter gene and NEP expression occurs after two 2A cleavage sites. The 5′ ψ are duplicated and contain NEP N-terminal amino acid codons. An HA tag and a heterologous dimerization domain (Dcm) were added after the NS1 ORF.

196

This NS1-GFP has been also used in other studies. For instance, Hufford and Richardson et al. utilized NS1-GFP to demonstrate that lung-resident neutrophils are infected by IAV [110] (Table 7). The authors suggest that they can act not only to stimulate the innate immune response, but to also activate CD8+ T lymphocytes to begin viral clearance in the lung. While the innate immune system is critical for antiviral clearance, its signaling can promote IAV replication. A study by Pang et al. has shown that innate immune activation triggered by toll-like receptor 7 (TLR7) and retinoic acid inducible gene-1 (RIG-I) is required for efficient IAV replication in the reparatory tract [102] (Table 7). NS1-GFP was utilized here by infecting bone marrow-derived DCs (BMDCs) that are deficient in genes required for TLR7 and RIG-I signalling. Consistent with previous studies [111], WT BMDCs were less susceptible to infection than TLR7 and RIG-I deficient cells as determined by GFP fluorescence. Comparison of IAV-infected cells showed that inflammatory mediators elicited by TLR7 and RIG-I signaling recruit viral target cells to the airway, thereby increasing viral load within the respiratory tract. The authors suggested that IAV uses physiological levels of inflammatory responses to its replicative advantage, highlighting the complex interplay between viruses and the host innate-immune responses [102].

A study by Resa-Infante et al. showed that without importin-α7, NS1-GFP is unable to efficiently replicate in the alveolar epithelium [103] (Table 7). Direct observation of IAV infection provides a means to evaluate antiviral therapies. In a study by Kim et al., NS1-GFP infected cells were treated with an NA inhibitor (oseltamivir) and various natural compounds [112] (Table 7). Oseltamivir was also used in vivo by treating infected mice [112]. In both cases, reduced GFP expression was indicative of limited or neutralized infection.

Mice are commonly used to model IAV infection in vivo, but Gabor et al. have reported IAV infection in zebrafish as a new and inexpensive model to reproduce viral infection [101] (Table 7). Using IAV PR8 (H1N1), X31 (H3N2) and NS1-GFP, they showed that zebrafish can support infection and mount an immune response [101]. Interestingly, infected (GFP+) cells were localized within the cardiovascular system and the swim bladder [101]. The authors point out that the swim bladder can be likened to IAV infection in human lung endothelial tissue, further supporting the relevance of the model.

Due to technical limitations, it has been historically difficult to monitor the process of IAV infection. Modern fluorescent microscopy has given researchers the ability to use video to capture infection of fluorescent-expressing, replication-competent IAV. Using NS1-GFP, Roberts et al. demonstrated that IAV-infected cells can infect neighbouring cells by passing the virus through actin protrusions [104] (Table 7). Previous observations using NA-deficient viruses [36], and treating infected cells with NA inhibitors by Roberts et al. [104], showed that microplaques of IAV appear in cell culture [104]. However, the addition of actin inhibitors or microtubule stabilizers to oseltamivir treatment prohibited microplaque formation [104]. Live video microscopy shows compelling evidence for cell-to-cell IAV infection, NS1-GFP-infected cells show GFP movement from one cell to another through an intercellular connection [104]. This, coupled with immunofluorescence data showing actin filaments containing vRNPs connecting infected to uninfected cells, led to the hypothesis that cell-free virion infection is not the only method of IAV dissemination [104].

Using a similar NS1 construct to NS1-GFP, Eckert et al. generated IAV encoding maxGFP, turboRFP, or Gluc that replicated comparably to WT PR8 [48] (Table 7). The researchers moved forward only with NS1-Gluc because it showed the greatest reporter stability of the three viruses, while NS1-maxGFP and -turboRFP dropped below 5% fluorescent positive cells on the third passage in tissue culture [48]. Since Gluc is secreted, it is well suited as a reporter for high throughput antiviral compound screening. After establishing a correlation between the viral titer and luciferase activity, zanamivir (an IAV NA inhibitor) was used to benchmark NS1-Gluc in infected human epithelial colorectal adenocarcinoma CaCo-2 cells [48]. Numerous antiviral factors are expressed upon type I IFN induction. Members of the IFN-inducible transmembrane (IFITM) protein family, specifically IFITM1, 2 and 3, have been identified as inhibitors of IAV [113]. The authors showed that NS1-Gluc virus can be used to investigate cellular proteins that exhibit inhibitory functions against IAV infection [48]. Remarkably, IFITM2 and 3 proteins, when transduced into A549, MDCK, and 293T cells, inhibit the proliferation of IAV [48]. In contrast, the inhibitory effects observed upon expression of IFITM1 were minor (A549 and human embryonic kidney 293T cells) or absent (MDCK cells), again in agreement with published data [113,114]. Therefore, the data indicated that NS1-Gluc virus could be used to evaluate the antiviral activity of host cell proteins [48]. In fact, previous studies have suggested that the IFITM proteins act by increasing endosomal cholesterol [115]. The authors found that U18666A, a compound that increases endosomal cholesterol [116], displayed a dose dependent inhibitory effect against IAV infection in vitro [48].

Likewise, Nogales et al. used a similar approach to NS1-GFP to generate an IAV expressing the monomeric (m)Cherry fluorescent protein in the backbone of PR8 or pH1N1 (PR8- and pH1N1-mCherry, respectively) [41] (Table 7). PR8-mCherry replicated at lower levels than WT in MDCK cells [41]. However, the PR8-mCherry virus was able to abrogate type I IFN induction similarly to PR8 WT [41]. Importantly, both mCherry-expressing viruses were inhibited with antivirals or type-specific NAbs to levels comparable to WT viruses, representing an excellent option for the rapid identification of antivirals or NAbs using HTS methodologies [41]. The authors also presented the potential use for PR8-mCherry for in vivo studies [41]. Even though PR8-mCherry was attenuated in mice, inoculation with 10^4 PFU resulted in infection-specific fluorescence, and replication could be directly visualized and quantified from whole excised lung using IVIS [41]. These results offer a promising option to directly study the biology of IAV and to evaluate experimental outcomes from treating IAV infections in vitro and ex vivo [41].

The reporter-expressing IAV discussed thus far all encode static reporters, which are not optimal for determining the origin or chronology of infection. A dynamic fluorescent protein Timer was engineered that changes its emission spectra from green to red over time and could allow for tracking IAV in more detail [117]. Timer is derived from the red fluorescent protein of *Discosoma* (DsRed) and contains two point mutations that confer a strong quantum yield and the spectral shift phenotype [117]. Breen et al. described the generation of replication-competent viruses expressing Timer fused to the viral protein NS1 in the backbone of pH1N1 (IAV-Timer) and influenza B/Brisbane/60/2008 (IBV-Timer) viruses [68] (Table 7). The recombinant IAV-Timer, in vitro, showed similar growth kinetics compared to the WT virus [68]. Using multiple approaches, including fluorescent microscopy and plaque assays, the authors were able to differentiate primary from secondarily infected cells [68]. Timer expression and spectral shift were quantified in infected cells using a fluorescence plate reader and flow cytometry. Importantly, IAV-Timer was useful to evaluate the dynamics of viral infections in mouse lungs using IVIS [68]. These studies constitute proof-of-principle of the usefulness for recombinant IAV expressing the dynamic Timer protein to study viral infection dynamics both, in vitro and in vivo [68].

A drawback of many fluorescent or luminescent IAV is that they are commonly attenuated in vivo [38,39,41]. Fukuyama et al. generated a series of Color-flu viruses in the backbone of PR8, expressing fluorescent proteins of different colors: Venus, eCFP, mCherry and eGFP [38] (Table 7). Fukuyama et al. then mouse-adapted (MA) each of the PR8 reporter-expressing IAV by serial passaging in mice [38]. The MA virus variants (MA-PR8-eGFP, -eCFP, -Venus, and -mCherry) showed higher pathogenicity compare to parental viruses, although they were still less pathogenic than WT PR8 as determined by MLD_{50} [38]. The ability to use the four unique viruses for multiplex was tested in infected mice. Each reporter could be observed in clusters of whole-lung explants using stereomicroscopy. Additionally, co-infection could be observed in lung tissue slices from viruses containing two or more fluorescent reporters [38]. This suggests that Color-flu can be used to study reassortment, which is implicated in the generation of pandemic strains of IAV along with the generation of human adapted strains [22,24,118,119]. The researchers also applied MA-PR8-Venus to analyze the macrophage response to IAV infection [38]. Macrophages were observed to infiltrate the bronchial epithelium around MA-PR8-Venus positive cells. Transcriptome analysis of infected

(Venus-positive) macrophages showed elevated message levels of genes involved in innate immune response [38]. In addition, a HPAI virus was engineered to contain Venus (MA-HPAI-Venus). Mice infected with MA-HPAI-Venus resulted in a greater Venus-positive bronchial epithelium than MA-PR8-Venus as observed by two-photon microscopy [38]. There were also increased numbers of Venus-positive macrophages in MA-HPAI-Venus infections, supporting findings that H5N1 HPAI viruses induce more severe inflammatory responses in the lung of infected mice than PR8 [38]. Therefore, these studies demonstrated the utility of Color-flu for comparative studies of IAV pathogenesis [38].

Most of the recombinant IAV discussed use a single 2A autocleavage site to collinearly express an NS1-reporter fusion protein and NEP separately, without the need for alternate splicing [38,39,41,48,52,66,68–72]. Recently, De Baets et al. have reported the generation of an IAV expressing GFP from a tri-cistronic NS segment in the backbone of PR8 (Figure 7E, Table 7) [105]. To reduce the size of this engineered gene segment, they used a truncated NS1 protein of 73 aa combined with a heterologous dimerization domain of the *Drosophila melanogaster* nonclaret disjunctional (Ncd) protein [120], to increase protein stability. GFP and NEP sequences were in frame after the truncated NS1, and were each separated by 2A self-processing sites (from FMDV and PTV-1, respectively) [105]. An HA-tag was also fused to NS1 to facilitate protein detection. The resulting PR8-NS1(1–73)GFP virus replicated as efficiently as PR8 WT in vitro and retained reporter expression in 100% of plaques from 5 passages in MDCK-V cells, which have their IFN response blocked by stable expression of parainfluenza virus type 5 V protein [105]. However, when passaged in parental MDCK cells, only 23% of plaques were GFP-positive after five passages [105]. The recombinant virus was slightly attenuated in vivo but maintained 96.4% GFP-positive plaques when recovered five days post-infection from mouse lungs [105]. The cellular tropism of PR8-NS1(1–73)GFP was also evaluated during treatment with either oseltamivir or a MAbs targeted against the IAV matrix protein 2 ectodomain (M2e) [105]. The latter treatment is of interest based on data that show efficacy of M2e IAV vaccines [121]. Both treatments protected mice from weight loss post PR8-NS1(1–73)GFP infection [105]. Finally, the authors demonstrated the usefulness of this virus to study the viral cell tropism ex vivo. The prophylactic treatment of mice with anti-M2e MAb or oseltamivir resulted in a decrease in the percentage of GFP-expressing cells [105]. Future studies must address if resistance to this MAb through antigenic mutation of the M2 protein could nullify this treatment.

Reuther et al. also have reported the generation of IAVs expressing GFP from a tri-cistronic NS segment in the backbone of A/SC35M (H7N7) (Figure 7E, Table 7) [106]. The resulting viruses encoding luciferases or fluorescent proteins maintained high genetic stability in vitro up to 4 rounds of passaging in human cells, and were characterized in vivo. Therefore, the recombinant viruses generated could be readily employed for antiviral compound screenings in addition to visualization of infected cells or cells that survived acute infection.

3. Conclusions and Future Directions

The purpose of this review is to discuss the biology and applications of replication-competent IAV expressing the most commonly used fluorescent or luciferase reporter genes. Plasmid-based reverse genetics techniques allow for the simultaneous expression of the IAV RNA-dependent RNA polymerase (RdRp) and negative-stranded genome viral segments in transiently transfected mammalian cells, which together generate de novo, or rescue, recombinant IAV [122,123]. Moreover, these techniques allowed the use of recombinant DNA technology to modify the genome of IAV and to engineer viruses expressing foreign genes. Recombinant reporter-expressing, replicating-competent IAV are applicable to translational research and have been demonstrated in screening platforms to identify specific or broadly reactive NAbs, antiviral compounds, or host proteins involved in viral replication. In fact, the generation of recombinant, reporter-expressing IAV lead the design of cell-based assays that capture all stages of the virus life-cycle. With these replicating-competent reporter viruses, there is greater flexibility in the choice of cells to perform these assays, providing increased potential for identifying inhibitors of both viral and cellular functions that are critical for optimal virus replication. Moreover, these fluorescent- or luciferase-expressing IAV have been shown to be a valuable asset for in vivo studies when used in conjunction with new imaging technologies. The combination of fluorescent or luminescent genes plus the generation of reporter IAV using different backbones, including HPAI or pandemic strains, have increased the spectrum of tools that can be used to facilitate the study of these IAV. Other future areas of replication-competent IAV application include, but are not limited to, virus tropism and the analysis of viral reassortments during replication or vaccine development. In conclusion, there are potential advantages and disadvantages associated with replication-competent IAV; however, the number of applications as well as the abundant economic, quantitative, and biological advantages of these IAV, highlight their promising applications in basic and translational influenza research in the imminent future.

Acknowledgments: We want to thank all influenza virologists whose work has contributed to the generation of recombinant replicating-competent, reporter-expressing IAV. We would want to apologize if we inadvertently omitted any manuscript(s) describing the generation of recombinant replicating-competent IAV expressing reporter genes. Influenza virus research in Luis Martínez-Sobrido laboratory was partially funded by the NIAID Centers of Excellence for Influenza Research and Surveillance (CEIRS HHSN266200700008C).

Author Contributions: All these authors wrote the manuscript.

Conflicts of Interest: The authors declare not conflict of interest.

Abbreviations

WSN	A/WSN/1933 H1N1
PR8	A/Puerto Rico/8/1934
pH1N1	A/California/04/2009 H1N1
Neth602	A/Netherlands/602/2009 H1N1
Ind5	A/Indonesia/5/2005 H5N1
Anh1	A/Anhui/1/2013 H7N9
HK99	A/Guinea Fowl/Hong Kong/WF10/1999 H9N2
VN1203	A/Vietnam/1203/2004 H5N1
maxGFP	advanced version of eGFP
Venus	advanced version of yellow fluorescent protein
aa	amino acid
2A	autocleavage 2A site
BMDCs	bone marrow-derived DCs
CRS	caspase recognition site
CAT	chloramphenicol acetyl transferase
DCs	dendritic cells
dPR	duplicated packaging region
EM	electron microscopy
ER	endoplasmic reticulum
eCFP	enhanced cyan fluorescent protein
eGFP	enhanced green fluorescent protein
fRFP	far-red fluorescent protein
FFluc	firefly luciferase
FACS	fluorescence-activated cell sorting
FMDV	foot-and-mouth disease virus
Gluc	Gaussia luciferase
GFP	green fluorescent protein
HA	hemagglutinin
HTS	high-throughput screening
IAV	influenza A virus
BM2	influenza B virus M2 protein
IL-2	interleukin-2
IVIS	in vivo imaging system
MDKC	Madin-Darby canine kidney epithelial cells
M1	matrix 1 protein
M2	matrix 2 protein
M2e	matrix 2 protein ectodomain
MFI	mean fluorescent intensity

Timer	modified DsRed fluorescent protein Timer
MAbs	monoclonal antibodies
mCherry	monomeric cherry fluorescent protein
MLD_{50}	mouse lethal dose-50
NLuc	NanoLuc
iRFP	near-infrared fluorescent protein
NA	neuraminidase
NAb	neutralizing antibody
Ncd protein	nonclaret disjunctional protein
NCRs	non-coding regions
NS1	non-structural protein 1
NEP	nuclear export protein
NP	nucleoprotein
nt	nucleotide
ORF	open reading frame
ψ	packaging signals
PFU	plaque forming units
PTV-1	polymerase acidic (PA) and basic (PB1 and PB2) subunits, porcine teschovirus-1
PET/CT	positron emission tomography-computed tomography
PKR	protein kinase R
DsRed	red fluorescent protein of Discosoma
RIG-I	retinoic acid inducible gene-1
RdRp	RNA-dependent RNA polymerase
SL	short linker
SAM	splice acceptor mutation
$TCID_{50}$	tissue culture infectious dose 50
TLR7	toll-like receptor 7
IFN	type I interferon
IFITM proteins	type I IFN-induced transmembrane proteins
UTR	untranslated region
vRNA	viral RNA
vRNP	viral ribonucleoprotein
WT	wild-type

References

1. Palese, P.; Shaw, M.L. Orthomyxoviridae: The viruses and their replication. In *Fields Virology*, 5th ed.; Knipe, D.M., Howley, P.M., Griffin, D.E., Lamb, R.A., Martin, M.A., Eds.; Lippincott Williams and Wilkins: Philadelphia, PA, USA, 2007.

2. Pritlove, D.C.; Fodor, E.; Seong, B.L.; Brownlee, G.G. In vitro transcription and polymerase binding studies of the termini of influenza A virus cRNA: Evidence for a cRNA panhandle. *J. Gen. Virol.* **1995**, *76*, 2205–2213.

3. Flick, R.; Neumann, G.; Hoffmann, E.; Neumeier, E.; Hobom, G. Promoter elements in the influenza vRNAterminal structure. *RNA* **1996**, *2*, 1046–1057.

4. Medina, R.A.; Garcia-Sastre, A. Influenza A viruses: New research developments. *Nat. Rev. Microbiol.* **2011**, *9*, 590–603.

5. WHO. *fluenza (Seasonal) Fact Sheet No. 211*; WHO: Geneva, Switzerland, 2009.

6. Tong, S.; Zhu, X.; Li, Y.; Shi, M.; Zhang, J.; Bourgeois, M.; Yang, H.; Chen, X.; Recuenco, S.; Gomez, J.; et al. New world bats harbor diverse influenza A viruses. *PLoS Pathog.* **2013**, *9*, e1003657.

7. Tong, S.; Li, Y.; Rivailler, P.; Conrardy, C.; Castillo, D.A.; Chen, L.M.; Recuenco, S.; Ellison, J.A.; Davis, C.T.; York, I.A.; et al. A distinct lineage of influenza A virus from bats. *Proc. Natl. Acad. Sci. USA* **2012**, *109*, 4269–4274.

8. Skehel, J.J.; Wiley, D.C. Receptor binding and membrane fusion in virus entry: The influenza hemagglutinin. *Annu. Rev. Biochem.* **2000**, *69*, 531–569.

9. Taubenberger, J.K. Influenza hemagglutinin attachment to target cells: 'Birds do it, we do it...'. *Future Virol.* **2006**, *1*, 415–418.

10. Varghese, J.N.; Colman, P.M.; van Donkelaar, A.; Blick, T.J.; Sahasrabudhe, A.; McKimm-Breschkin, J.L. Structural evidence for a second sialic acid binding site in avian influenza virus neuraminidases. *Proc. Natl. Acad. Sci. USA* **1997**, *94*, 11808–11812.

11. Varghese, J.N.; McKimm-Breschkin, J.L.; Caldwell, J.B.; Kortt, A.A.; Colman, P.M. The structure of the complex between influenza virus neuraminidase and sialic acid, the viral receptor. *Proteins* **1992**, *14*, 327–332.

12. Jackson, R.J.; Cooper, K.L.; Tappenden, P.; Rees, A.; Simpson, E.L.; Read, R.C.; Nicholson, K.G. Oseltamivir, zanamivir and amantadine in the prevention of influenza: A systematic review. *J. Infect.* **2011**, *62*, 14–25.

13. Oxford, J.S.; Mann, A.; Lambkin, R. A designer drug against influenza: The NA inhibitor oseltamivir (Tamiflu). *Expert Rev. Anti Infect. Ther.* **2003**, *1*, 337–342.

14. Resa-Infante, P.; Jorba, N.; Coloma, R.; Ortin, J. The influenza virus rna synthesis machine: Advances in its structure and function. *RNA Biol.* **2011**, *8*, 207–215.

15. Wise, H.M.; Hutchinson, E.C.; Jagger, B.W.; Stuart, A.D.; Kang, Z.H.; Robb, N.; Schwartzman, L.M.; Kash, J.C.; Fodor, E.; Firth, A.E.; et al. Identification of a novel splice variant form of the influenza A virus M2 ion channel with an antigenically distinct ectodomain. *PLoS Pathog.* **2012**, *8*, e1002998.

16. Paterson, D.; Fodor, E. Emerging roles for the influenza A virus nuclear export protein (NEP). *PLoS Pathog.* **2012**, *8*, e1003019.

17. Hai, R.; Schmolke, M.; Varga, Z.T.; Manicassamy, B.; Wang, T.T.; Belser, J.A.; Pearce, M.B.; Garcia-Sastre, A.; Tumpey, T.M.; Palese, P. Pb1-F2 expression by the 2009 pandemic H1N1 influenza virus has minimal impact on virulence in animal models. *J. Virol.* **2010**, *84*, 4442–4450.

18. Jagger, B.W.; Wise, H.M.; Kash, J.C.; Walters, K.A.; Wills, N.M.; Xiao, Y.L.; Dunfee, R.L.; Schwartzman, L.M.; Ozinsky, A.; Bell, G.L.; et al. An overlapping protein-coding region in influenza A virus segment 3 modulates the host response. *Science* **2012**, *337*, 199–204.

19. Wise, H.M.; Foeglein, A.; Sun, J.; Dalton, R.M.; Patel, S.; Howard, W.; Anderson, E.C.; Barclay, W.S.; Digard, P. A complicated message: Identification of a novel pb1-related protein translated from influenza A virus segment 2 mRNA. *J. Virol.* **2009**, *83*, 8021–8031.

20. Hale, B.G.; Randall, R.E.; Ortin, J.; Jackson, D. The multifunctional ns1 protein of influenza A viruses. *J. Gen. Virol.* **2008**, *89*, 2359–2376.

21. Lamb, R.A.; Lai, C.J. Sequence of interrupted and uninterrupted mRNAs and cloned DNA coding for the two overlapping nonstructural proteins of influenza virus. *Cell* **1980**, *21*, 475–485.

22. Smith, G.J.D.; Bahl, J.; Vijaykrishna, D.; Zhang, J.X.; Poon, L.L.M.; Chen, H.L.; Webster, R.G.; Peiris, J.S.M.; Guan, Y. Dating the emergence of pandemic influenza viruses. *Proc. Natl. Acad. Sci. USA* **2009**, *106*, 11709–11712.

23. Jadhao, S.J.; Nguyen, D.C.; Uyeki, T.M.; Shaw, M.; Maines, T.; Rowe, T.; Smith, C.; Huynh, L.P.; Nghiem, H.K.; Nguyen, D.H.; et al. Genetic analysis of avian influenza A viruses isolated from domestic waterfowl in live-bird markets of Hanoi, Vietnam, preceding fatal H5N1 human infections in 2004. *Arch. Virol.* **2009**, *154*, 1249–1261.

24. Kilbourne, E.D. Influenza pandemics of the 20th century. *Emerg. Infect. Dis.* **2006**, *12*, 9–14.

25. Herfst, S.; Schrauwen, E.J.; Linster, M.; Chutinimitkul, S.; de Wit, E.; Munster, V.J.; Sorrell, E.M.; Bestebroer, T.M.; Burke, D.F.; Smith, D.J.; et al. Airborne transmission of influenza A/h5n1 virus between ferrets. *Science* **2012**, *336*, 1534–1541.

26. Imai, M.; Watanabe, T.; Hatta, M.; Das, S.C.; Ozawa, M.; Shinya, K.; Zhong, G.; Hanson, A.; Katsura, H.; Watanabe, S.; et al. Experimental adaptation of an influenza H5 HA confers respiratory droplet transmission to a reassortant H5 HA/H1N1 virus in ferrets. *Nature* **2012**, *486*, 420–428.

27. Uyeki, T.M.; Cox, N.J. Global concerns regarding novel influenza A (H7N9) virus infections. *N. Engl. J. Med.* **2013**, *368*, 1862–1864.

28. Aamir, U.B.; Naeem, K.; Ahmed, Z.; Obert, C.A.; Franks, J.; Krauss, S.; Seiler, P.; Webster, R.G. Zoonotic potential of highly pathogenic avian H7N3 influenza viruses from pakistan. *Virology* **2009**, *390*, 212–220.

29. Osterholm, M.T.; Kelley, N.S.; Sommer, A.; Belongia, E.A. Efficacy and effectiveness of influenza vaccines: A systematic review and meta-analysis. *Lancet Infect. Dis.* **2012**, *12*, 36–44.

30. Beigel, J.; Bray, M. Current and future antiviral therapy of severe seasonal and avian influenza. *Antivir. Res.* **2008**, *78*, 91–102.

31. Garcia-Sastre, A. Antiviral response in pandemic influenza viruses. *Emerg. Infect. Dis.* **2006**, *12*, 44–47.

32. Carrat, F.; Flahault, A. Influenza vaccine: The challenge of antigenic drift. *Vaccine* **2007**, *25*, 6852–6862.

33. Beyleveld, G.; White, K.M.; Ayllon, J.; Shaw, M.L. New-generation screening assays for the detection of anti-influenza compounds targeting viral and host functions. *Antivir. Res.* **2013**, *100*, 120–132.

34. Fujii, K.; Ozawa, M.; Iwatsuki-Horimoto, K.; Horimoto, T.; Kawaoka, Y. Incorporation of influenza A virus genome segments does not absolutely require wild-type sequences. *J. Gen. Virol.* **2009**, *90*, 1734–1740.

35. Watanabe, T.; Watanabe, S.; Noda, T.; Fujii, Y.; Kawaoka, Y. Exploitation of nucleic acid packaging signals to generate a novel influenza virus-based vector stably expressing two foreign genes. *J. Virol.* **2003**, *77*, 10575–10583.

36. Fujii, Y.; Goto, H.; Watanabe, T.; Yoshida, T.; Kawaoka, Y. Selective incorporation of influenza virus RNA segments into virions. *Proc. Natl. Acad. Sci. USA* **2003**, *100*, 2002–2007.

37. Hutchinson, E.C.; von Kirchbach, J.C.; Gog, J.R.; Digard, P. Genome packaging in influenza A virus. *J. Gen. Virol.* **2010**, *91*, 313–328.

38. Fukuyama, S.; Katsura, H.; Zhao, D.; Ozawa, M.; Ando, T.; Shoemaker, J.E.; Ishikawa, I.; Yamada, S.; Neumann, G.; Watanabe, S.; et al. Multi-spectral fluorescent reporter influenza viruses (Color-flu) as powerful tools for in vivo studies. *Nat. Commun.* **2015**, *6*, 6600.

39. Fiege, J.K.; Langlois, R.A. Investigating influenza A virus infection: Tools to track infection and limit tropism. *J. Virol.* **2015**, *89*, 6167–6170.

40. Nogales, A.; Rodriguez-Sanchez, I.; Monte, K.; Lenschow, D.J.; Perez, D.R.; Martinez-Sobrido, L. Replication-competent fluorescent-expressing influenza b virus. *Virus Res.* **2015**, *213*, 69–81.

41. Nogales, A.; Baker, S.F.; Martinez-Sobrido, L. Replication-competent influenza A viruses expressing a red fluorescent protein. *Virology* **2014**, *476*, 206–216.

42. Shaner, N.C.; Patterson, G.H.; Davidson, M.W. Advances in fluorescent protein technology. *J. Cell Sci.* **2007**, *120*, 4247–4260.

43. Shaner, N.C.; Steinbach, P.A.; Tsien, R.Y. A guide to choosing fluorescent proteins. *Nat. Methods* **2005**, *2*, 905–909.

44. Welsh, D.K.; Noguchi, T. Cellular bioluminescence imaging. *Cold Spring Harb. Protoc.* **2012**, *2012*.

45. Shimomura, O.; Johnson, F.H.; Saiga, Y. Extraction, purification and properties of aequorin, a bioluminescent protein from the Luminous Hydromedusan, *Aequorea. J. Cell. Comp. Physiol.* **1962**, *59*, 223–239.

46. Kelkar, M.; De, A. Bioluminescence based in vivo screening technologies. *Curr. Opin. Pharmacol.* **2012**, *12*, 592–600.

47. De Wet, J.R.; Wood, K.V.; DeLuca, M.; Helinski, D.R.; Subramani, S. Firefly luciferase gene: Structure and expression in mammalian cells. *Mol. Cell. Biol.* **1987**, *7*, 725–737.

48. Eckert, N.; Wrensch, F.; Gartner, S.; Palanisamy, N.; Goedecke, U.; Jager, N.; Pohlmann, S.; Winkler, M. Influenza A virus encoding secreted gaussia luciferase as useful tool to analyze viral replication and its inhibition by antiviral compounds and cellular proteins. *PLoS ONE* **2014**, *9*, e97695.

49. Hall, M.P.; Unch, J.; Binkowski, B.F.; Valley, M.P.; Butler, B.L.; Wood, M.G.; Otto, P.; Zimmerman, K.; Vidugiris, G.; Machleidt, T.; et al. Engineered luciferase reporter from a deep sea shrimp utilizing a novel imidazopyrazinone substrate. *ACS Chem. Biol.* **2012**, *7*, 1848–1857.

50. Tran, V.; Poole, D.S.; Jeffery, J.J.; Sheahan, T.P.; Creech, D.; Yevtodiyenko, A.; Peat, A.J.; Francis, K.P.; You, S.; Mehle, A. Multi-modal imaging with a toolbox of influenza A reporter viruses. *Viruses* **2015**, *7*, 5319–5327.

51. Zhao, H.; Doyle, T.C.; Coquoz, O.; Kalish, F.; Rice, B.W.; Contag, C.H. Emission spectra of bioluminescent reporters and interaction with mammalian tissue determine the sensitivity of detection in vivo. *J. Biomed. Opt.* **2005**, *10*, 41210.

52. Tran, V.; Moser, L.A.; Poole, D.S.; Mehle, A. Highly sensitive real-time in vivo imaging of an influenza reporter virus reveals dynamics of replication and spread. *J. Virol.* **2013**, *87*, 13321–13329.

53. Guilligay, D.; Tarendeau, F.; Resa-Infante, P.; Coloma, R.; Crepin, T.; Sehr, P.; Lewis, J.; Ruigrok, R.W.; Ortin, J.; Hart, D.J.; et al. The structural basis for cap binding by influenza virus polymerase subunit PB2. *Nat. Struct. Mol. Biol.* **2008**, *15*, 500–506.

54. Engelhardt, O.G.; Fodor, E. Functional association between viral and cellular transcription during influenza virus infection. *Rev. Med. Virol.* **2006**, *16*, 329–345.

55. Hutchinson, E.C.; Charles, P.D.; Hester, S.S.; Thomas, B.; Trudgian, D.; Martinez-Alonso, M.; Fodor, E. Conserved and host-specific features of influenza virion architecture. *Nat. Commun.* **2014**, *5*, 4816.

56. Taubenberger, J.K.; Kash, J.C. Influenza virus evolution, host adaptation, and pandemic formation. *Cell Host Microbe* **2010**, *7*, 440–451.

57. Konig, R.; Stertz, S.; Zhou, Y.; Inoue, A.; Hoffmann, H.H.; Bhattacharyya, S.; Alamares, J.G.; Tscherne, D.M.; Ortigoza, M.B.; Liang, Y.; et al. Human host factors required for influenza virus replication. *Nature* **2010**, *463*, 813–817.

58. Avilov, S.V.; Moisy, D.; Munier, S.; Schraidt, O.; Naffakh, N.; Cusack, S. Replication-competent influenza A virus that encodes a split-green fluorescent protein-tagged PB2 polymerase subunit allows live-cell imaging of the virus life cycle. *J. Virol.* **2012**, *86*, 1433–1448.

59. Cabantous, S.; Terwilliger, T.C.; Waldo, G.S. Protein tagging and detection with engineered self-assembling fragments of green fluorescent protein. *Nat. Biotechnol.* **2005**, *23*, 102–107.

60. Crescenzo-Chaigne, B.; Naffakh, N.; van der Werf, S. Comparative analysis of the ability of the polymerase complexes of influenza viruses type A, B and C to assemble into functional rnps that allow expression and replication of heterotypic model RNA templates in vivo. *Virology* **1999**, *265*, 342–353.

61. Avilov, S.V.; Moisy, D.; Naffakh, N.; Cusack, S. Influenza A virus progeny vrnp trafficking in live infected cells studied with the virus-encoded fluorescently tagged PB2 protein. *Vaccine* **2012**, *30*, 7411–7417.

62. Jing, J.; Prekeris, R. Polarized endocytic transport: The roles of rab11 and rab11-fips in regulating cell polarity. *Histol. Histopathol.* **2009**, *24*, 1171–1180.

63. Eisfeld, A.J.; Kawakami, E.; Watanabe, T.; Neumann, G.; Kawaoka, Y. Rab11a is essential for transport of the influenza virus genome to the plasma membrane. *J. Virol.* **2011**, *85*, 6117–6126.

64. Jo, S.; Kawaguchi, A.; Takizawa, N.; Morikawa, Y.; Momose, F.; Nagata, K. Involvement of vesicular trafficking system in membrane targeting of the progeny influenza virus genome. *Microbes Infect.* **2010**, *12*, 1079–1084.

65. Bruce, E.A.; Digard, P.; Stuart, A.D. The rab11 pathway is required for influenza A virus budding and filament formation. *J. Virol.* **2010**, *84*, 5848–5859.

66. Heaton, N.S.; Leyva-Grado, V.H.; Tan, G.S.; Eggink, D.; Hai, R.; Palese, P. in vivo bioluminescent imaging of influenza A virus infection and characterization of novel cross-protective monoclonal antibodies. *J. Virol.* **2013**, *87*, 8272–8281.

67. Yan, D.; Weisshaar, M.; Lamb, K.; Chung, H.K.; Lin, M.Z.; Plemper, R.K. Replication-competent influenza virus and respiratory syncytial virus luciferase reporter strains engineered for co-infections identify antiviral compounds in combination screens. *Biochemistry* **2015**, *54*, 5589–5604.

68. Breen, M.; Nogales, A.; Baker, S.F.; Perez, D.R.; Martinez-Sobrido, L. Replication-competent influenza A and b viruses expressing a fluorescent dynamic timer protein for in vitro and in vivo studies. *PLoS ONE* **2016**, *11*, e0147723.

69. Karlsson, E.A.; Meliopoulos, V.A.; Savage, C.; Livingston, B.; Mehle, A.; Schultz-Cherry, S. Visualizing real-time influenza virus infection, transmission and protection in ferrets. *Nat. Commun.* **2015**, *6*, 6378.

70. Li, F.; Feng, L.; Pan, W.; Dong, Z.; Li, C.; Sun, C.; Chen, L. Generation of replication-competent recombinant influenza A viruses carrying a reporter gene harbored in the neuraminidase segment. *J. Virol.* **2010**, *84*, 12075–12081.

71. Manicassamy, B.; Manicassamy, S.; Belicha-Villanueva, A.; Pisanelli, G.; Pulendran, B.; Garcia-Sastre, A. Analysis of in vivo dynamics of influenza virus infection in mice using a GFP reporter virus. *Proc. Natl. Acad. Sci. USA* **2010**, *107*, 11531–11536.

72. Pan, W.; Dong, Z.; Li, F.; Meng, W.; Feng, L.; Niu, X.; Li, C.; Luo, Q.; Li, Z.; Sun, C.; et al. Visualizing influenza virus infection in living mice. *Nat. Commun.* **2013**, *4*, 2369.

73. Spronken, M.I.; Short, K.R.; Herfst, S.; Bestebroer, T.M.; Vaes, V.P.; van der Hoeven, B.; Koster, A.J.; Kremers, G.J.; Scott, D.P.; Gultyaev, A.P.; et al. Optimisations and challenges involved in the creation of various bioluminescent and fluorescent influenza A virus strains for in vitro and in vivo applications. *PLoS ONE* **2015**, *10*, e0133888.

74. Sharma, P.; Yan, F.; Doronina, V.A.; Escuin-Ordinas, H.; Ryan, M.D.; Brown, J.D. 2A peptides provide distinct solutions to driving stop-carry on translational recoding. *Nucleic Acids Res.* **2012**, *40*, 3143–3151.

75. Sui, J.; Hwang, W.C.; Perez, S.; Wei, G.; Aird, D.; Chen, L.M.; Santelli, E.; Stec, B.; Cadwell, G.; Ali, M.; et al. Structural and functional bases for broad-spectrum neutralization of avian and human influenza A viruses. *Nat. Struct. Mol. Biol.* **2009**, *16*, 265–273.

76. He, W.; Mullarkey, C.E.; Miller, M.S. Measuring the neutralization potency of influenza A virus hemagglutinin stalk/stem-binding antibodies in polyclonal preparations by microneutralization assay. *Methods* **2015**, *90*, 95–100.

77. Hai, R.; Krammer, F.; Tan, G.S.; Pica, N.; Eggink, D.; Maamary, J.; Margine, I.; Albrecht, R.A.; Palese, P. Influenza viruses expressing chimeric hemagglutinins: Globular head and stalk domains derived from different subtypes. *J. Virol.* **2012**, *86*, 5774–5781.

78. Hoffmann, H.H.; Kunz, A.; Simon, V.A.; Palese, P.; Shaw, M.L. Broad-spectrum antiviral that interferes with de novo pyrimidine biosynthesis. *Proc. Natl. Acad. Sci. USA* **2011**, *108*, 5777–5782.

79. Baker, S.F.; Guo, H.; Albrecht, R.A.; Garcia-Sastre, A.; Topham, D.J.; Martinez-Sobrido, L. Protection against lethal influenza with a viral mimic. *J. Virol.* **2013**, *87*, 8591–8605.

80. Munier, S.; Rolland, T.; Diot, C.; Jacob, Y.; Naffakh, N. Exploration of binary virus-host interactions using an infectious protein complementation assay. *Mol. Cell. Proteom.* **2013**, *12*, 2845–2855.

81. Sidwell, R.W.; Bailey, K.W.; Wong, M.H.; Barnard, D.L.; Smee, D.F. In vitro and in vivo influenza virus-inhibitory effects of viramidine. *Antivir. Res.* **2005**, *68*, 10–17.

82. Takeda, M.; Pekosz, A.; Shuck, K.; Pinto, L.H.; Lamb, R.A. Influenza A virus M2 ion channel activity is essential for efficient replication in tissue culture. *J. Virol.* **2002**, *76*, 1391–1399.

83. Fodor, E.; Smith, M. The pa subunit is required for efficient nuclear accumulation of the PB1 subunit of the influenza A virus RNA polymerase complex. *J. Virol.* **2004**, *78*, 9144–9153.

84. Nogales, A.; Baker, S.F.; Domm, W.; Martinez-Sobrido, L. Development and applications of single-cycle infectious influenza A virus (sciIAV). *Virus Res* **2016**, *216*, 26–40.

85. Liang, Y.; Hong, Y.; Parslow, T.G. Cis-acting packaging signals in the influenza virus PB1, PB2, and PA genomic RNA segments. *J. Virol.* **2005**, *79*, 10348–10355.

86. Steel, J.; Lowen, A.C.; Mubareka, S.; Palese, P. Transmission of influenza virus in a mammalian host is increased by PB2 amino acids 627K or 627E/701N. *PLoS Pathog.* **2009**, *5*, e1000252.

87. Lakdawala, S.S.; Wu, Y.; Wawrzusin, P.; Kabat, J.; Broadbent, A.J.; Lamirande, E.W.; Fodor, E.; Altan-Bonnet, N.; Shroff, H.; Subbarao, K. Influenza A virus assembly intermediates fuse in the cytoplasm. *PLoS Pathog.* **2014**, *10*, e1003971.

88. Fouchier, R.A.; Garcia-Sastre, A.; Kawaoka, Y.; Barclay, W.S.; Bouvier, N.M.; Brown, I.H.; Capua, I.; Chen, H.; Compans, R.W.; Couch, R.B.; et al. Transmission studies resume for avian flu. *Science* **2013**, *339*, 520–521.

89. Perez, J.T.; Garcia-Sastre, A.; Manicassamy, B. Insertion of a GFP reporter gene in influenza virus. *Curr. Protoc. Microbiol.* **2013**.

90. Kittel, C.; Sereinig, S.; Ferko, B.; Stasakova, J.; Romanova, J.; Wolkerstorfer, A.; Katinger, H.; Egorov, A. Rescue of influenza virus expressing gfp from the NS1 reading frame. *Virology* **2004**, *324*, 67–73.

209

91. Da Silva, D.V.; Nordholm, J.; Dou, D.; Wang, H.; Rossman, J.S.; Daniels, R. The influenza virus neuraminidase protein transmembrane and head domains have coevolved. *J. Virol.* **2015**, *89*, 1094–1104.

92. Yang, X.; Steukers, L.; Forier, K.; Xiong, R.; Braeckmans, K.; Van Reeth, K.; Nauwynck, H. A beneficiary role for neuraminidase in influenza virus penetration through the respiratory mucus. *PLoS ONE* **2014**, *9*, e110026.

93. Machado, A.V.; Naffakh, N.; van der Werf, S.; Escriou, N. Expression of a foreign gene by stable recombinant influenza viruses harboring a dicistronic genomic segment with an internal promoter. *Virology* **2003**, *313*, 235–249.

94. Machado, A.V.; Naffakh, N.; Gerbaud, S.; van der Werf, S.; Escriou, N. Recombinant influenza A viruses harboring optimized dicistronic na segment with an extended native 5′ terminal sequence: Induction of heterospecific B and T cell responses in mice. *Virology* **2006**, *345*, 73–87.

95. Perez, D.R.; Garcia-Sastre, A. H5N1, a wealth of knowledge to improve pandemic preparedness. *Virus Res.* **2013**, *178*, 1–2.

96. Perez, D.R.; Lim, W.; Seiler, J.P.; Yi, G.; Peiris, M.; Shortridge, K.F.; Webster, R.G. Role of quail in the interspecies transmission of H9 influenza A viruses: Molecular changes on ha that correspond to adaptation from ducks to chickens. *J. Virol.* **2003**, *77*, 3148–3156.

97. Pena, L.; Sutton, T.; Chockalingam, A.; Kumar, S.; Angel, M.; Shao, H.; Chen, H.; Li, W.; Perez, D.R. Influenza viruses with rearranged genomes as live-attenuated vaccines. *J. Virol.* **2013**, *87*, 5118–5127.

98. Kimble, J.B.; Sorrell, E.; Shao, H.; Martin, P.L.; Perez, D.R. Compatibility of H9N2 avian influenza surface genes and 2009 pandemic H1N1 internal genes for transmission in the ferret model. *Proc. Natl. Acad. Sci. USA* **2011**, *108*, 12084–12088.

99. Sutton, T.C.; Obadan, A.; Lavigne, J.; Chen, H.; Li, W.; Perez, D.R. Genome rearrangement of influenza virus for anti-viral drug screening. *Virus Res* **2014**, *189*, 14–23.

100. Kittel, C.; Ferko, B.; Kurz, M.; Voglauer, R.; Sereinig, S.; Romanova, J.; Stiegler, G.; Katinger, H.; Egorov, A. Generation of an influenza A virus vector expressing biologically active human interleukin-2 from the ns gene segment. *J. Virol.* **2005**, *79*, 10672–10677.

101. Gabor, K.A.; Goody, M.F.; Mowel, W.K.; Breitbach, M.E.; Gratacap, R.L.; Witten, P.E.; Kim, C.H. Influenza A virus infection in zebrafish recapitulates mammalian infection and sensitivity to anti-influenza drug treatment. *Dis. Models Mech.* **2014**, *7*, 1227–1237.

102. Pang, I.K.; Pillai, P.S.; Iwasaki, A. Efficient influenza A virus replication in the respiratory tract requires signals from TLR7 and RIG-I. *Proc. Natl. Acad. Sci. USA* **2013**, *110*, 13910–13915.

103. Resa-Infante, P.; Thieme, R.; Ernst, T.; Arck, P.C.; Ittrich, H.; Reimer, R.; Gabriel, G. Importin-alpha7 is required for enhanced influenza A virus replication in the alveolar epithelium and severe lung damage in mice. *J. Virol.* **2014**, *88*, 8166–8179.

104. Roberts, K.L.; Manicassamy, B.; Lamb, R.A. Influenza A virus uses intercellular connections to spread to neighboring cells. *J. Virol.* **2015**, *89*, 1537–1549.

105. De Baets, S.; Verhelst, J.; Van den Hoecke, S.; Smet, A.; Schotsaert, M.; Job, E.R.; Roose, K.; Schepens, B.; Fiers, W.; Saelens, X. A GFP expressing influenza A virus to report in vivo tropism and protection by a matrix protein 2 ectodomain-specific monoclonal antibody. *PLoS ONE* **2015**, *10*, e0121491.

106. Reuther, P.; Gopfert, K.; Dudek, A.H.; Heiner, M.; Herold, S.; Schwemmle, M. Generation of a variety of stable influenza A reporter viruses by genetic engineering of the ns gene segment. *Sci. Rep.* **2015**, *5*, 11346.

107. Horvath, C.M.; Williams, M.A.; Lamb, R.A. Eukaryotic coupled translation of tandem cistrons: Identification of the influenza B virus BM2 polypeptide. *EMBO J.* **1990**, *9*, 2639–2647.

108. Basler, C.F.; Reid, A.H.; Dybing, J.K.; Janczewski, T.A.; Fanning, T.G.; Zheng, H.; Salvatore, M.; Perdue, M.L.; Swayne, D.E.; Garcia-Sastre, A.; et al. Sequence of the 1918 pandemic influenza virus nonstructural gene (NS) segment and characterization of recombinant viruses bearing the 1918 NS genes. *Proc. Natl. Acad. Sci. USA* **2001**, *98*, 2746–2751.

109. Helft, J.; Manicassamy, B.; Guermonprez, P.; Hashimoto, D.; Silvin, A.; Agudo, J.; Brown, B.D.; Schmolke, M.; Miller, J.C.; Leboeuf, M.; et al. Cross-presenting CD103+ dendritic cells are protected from influenza virus infection. *J. Clin. Investig.* **2012**, *122*, 4037–4047.

110. Hufford, M.M.; Richardson, G.; Zhou, H.; Manicassamy, B.; Garcia-Sastre, A.; Enelow, R.I.; Braciale, T.J. Influenza-infected neutrophils within the infected lungs act as antigen presenting cells for anti-viral CD8(+) T cells. *PLoS ONE* **2012**, *7*, e46581.

111. Kato, H.; Sato, S.; Yoneyama, M.; Yamamoto, M.; Uematsu, S.; Matsui, K.; Tsujimura, T.; Takeda, K.; Fujita, T.; Takeuchi, O.; et al. Cell type-specific involvement of RIG-I in antiviral response. *Immunity* **2005**, *23*, 19–28.

112. Kim, J.I.; Park, S.; Lee, I.; Lee, S.; Shin, S.; Won, Y.; Hwang, M.W.; Bae, J.Y.; Heo, J.; Hyun, H.E.; et al. GFP-expressing influenza A virus for evaluation of the efficacy of antiviral agents. *J. Microbiol.* **2012**, *50*, 359–362.

113. Brass, A.L.; Huang, I.C.; Benita, Y.; John, S.P.; Krishnan, M.N.; Feeley, E.M.; Ryan, B.J.; Weyer, J.L.; van der Weyden, L.; Fikrig, E.; et al. The ifitm proteins mediate cellular resistance to influenza A H1N1 virus, west nile virus, and dengue virus. *Cell* **2009**, *139*, 1243–1254.

114. Huang, I.C.; Bailey, C.C.; Weyer, J.L.; Radoshitzky, S.R.; Becker, M.M.; Chiang, J.J.; Brass, A.L.; Ahmed, A.A.; Chi, X.; Dong, L.; et al. Distinct patterns of ifitm-mediated restriction of filoviruses, sars coronavirus, and influenza A virus. *PLoS Pathog.* **2011**, *7*, e1001258.

115. Amini-Bavil-Olyaee, S.; Choi, Y.J.; Lee, J.H.; Shi, M.; Huang, I.C.; Farzan, M.; Jung, J.U. The antiviral effector IFITM3 disrupts intracellular cholesterol homeostasis to block viral entry. *Cell Host Microbe* **2013**, *13*, 452–464.

116. Cenedella, R.J. Cholesterol synthesis inhibitor U18666A and the role of sterol metabolism and trafficking in numerous pathophysiological processes. *Lipids* **2009**, *44*, 477–487.

117. Terskikh, A.; Fradkov, A.; Ermakova, G.; Zaraisky, A.; Tan, P.; Kajava, A.V.; Zhao, X.; Lukyanov, S.; Matz, M.; Kim, S.; et al. "Fluorescent timer": Protein that changes color with time. *Science* **2000**, *290*, 1585–1588.

118. Yen, H.L.; Webster, R.G. Pandemic influenza As a current threat. *Curr. Top. Microbiol. Immunol.* **2009**, *333*, 3–24.

119. Taubenberger, J.K.; Morens, D.M. 1918 influenza: The mother of all pandemics. *Emerg. Infect. Dis.* **2006**, *12*, 15–22.

120. Wang, X.; Basler, C.F.; Williams, B.R.; Silverman, R.H.; Palese, P.; Garcia-Sastre, A. Functional replacement of the carboxy-terminal two-thirds of the influenza A virus NS1 protein with short heterologous dimerization domains. *J. Virol.* **2002**, *76*, 12951–12962.

121. Neirynck, S.; Deroo, T.; Saelens, X.; Vanlandschoot, P.; Jou, W.M.; Fiers, W. A universal influenza A vaccine based on the extracellular domain of the M2 protein. *Nat. Med.* **1999**, *5*, 1157–1163.

122. Martinez-Sobrido, L.; Garcia-Sastre, A. Generation of recombinant influenza virus from plasmid DNA. *J. Vis. Exp.* **2010**, *42*, e2057.

123. Hoffmann, E.; Neumann, G.; Kawaoka, Y.; Hobom, G.; Webster, R.G. A DNA transfection system for generation of influenza A virus from eight plasmids. *Proc. Natl. Acad. Sci. USA* **2000**, *97*, 6108–6113.

Development of Neutralization Assay Using an eGFP Chikungunya Virus

Cheng-Lin Deng, Si-Qing Liu, Dong-Gen Zhou, Lin-Lin Xu, Xiao-Dan Li,
Pan-Tao Zhang, Peng-Hui Li, Han-Qing Ye, Hong-Ping Wei, Zhi-Ming Yuan,
Cheng-Feng Qin and Bo Zhang

Abstract: Chikungunya virus (CHIKV), a member of the *Alphavirus* genus, is an important human emerging/re-emerging pathogen. Currently, there are no effective antiviral drugs or vaccines against CHIKV infection. Herein, we construct an infectious clone of CHIKV and an eGFP reporter CHIKV (eGFP-CHIKV) with an isolated strain (assigned to Asian lineage) from CHIKV-infected patients. The eGFP-CHIKV reporter virus allows for direct visualization of viral replication through the levels of eGFP expression. Using a known CHIKV inhibitor, ribavirin, we confirmed that the eGFP-CHIKV reporter virus could be used to identify inhibitors against CHIKV. Importantly, we developed a novel and reliable eGFP-CHIKV reporter virus-based neutralization assay that could be used for rapid screening neutralizing antibodies against CHIKV.

Reprinted from *Viruses*. Cite as: Deng, C.-L.; Liu, S.-Q.; Zhou, D.-G.; Xu, L.-L.; Li, X.-D.; Zhang, P.-T.; Li, P.-H.; Ye, H.-Q.; Wei, H.-P.; Yuan, Z.-M.; Qin, C.-F.; Zhang, B. Development of Neutralization Assay Using an eGFP Chikungunya Virus. *Viruses* **2016**, *8*, 181.

1. Introduction

Chikungunya virus (CHIKV) is an important mosquito-borne virus that belongs to the genus *Alphavirus*, family *Togaviridae*. The complete viral genome consists of a linear, positive-sense, single-stranded RNA around 12 kilobases (Kb) [1,2]. The 5′ two-thirds of the genome encodes the nonstructural proteins (nsP1-nsP2-nsP3-nsP4) that were required for replication and transcription of the viral genome in the cytoplasm of infected cells. The structural proteins (C-E3-E2-6K-E1) that are responsible for viral particles formation are translated from a subgenomic (sg) messenger RNA. The sg RNA is transcribed from the remaining 3′ one-third of the genome with a 26S promoter that is present on the full-length negative-stranded RNA replication intermediate.

CHIKV mainly causes abrupt high fever, headache, rashes, myalgia, and arthralgia [3,4], and occasionally the arthritis may last for months or years. Although it is historically regarded as a none life-threatening pathogen [5], the large outbreak of CHIKV on the Island of La Réunion in 2005 indicated that the virus may have become more virulent for humans [6]. Furthermore, lethal cases of encephalitis have also been reported for CHIKV infection [6]. CHIKV is primarily transmitted by *Aedes aegypti* mosquitoes. Recently, it was shown that A226V mutation in the structural protein E1 significantly increased fitness of the virus for *Ae. albopictus* mosquitoes and likely contributed to the epidemic [7]. So far, CHIKV has resulted in numerous outbreaks and been considered as a re-emerging pathogen [1,2,8]. The importation of CHIKV to mainland China was firstly reported in 2008 [9], and the first documented outbreak of CHIKV in China occurred in 2010 in Guangdong province [10,11].

Currently, there are no effective antiviral treatments and vaccines against CHIKV infection. Although mechanisms of protective immunity against CHIKV are poorly understood, different vaccine strategies [12–17] and monoclonal antibodies (mAbs) [18,19] that neutralized CHIKV were developed. A reliable neutralization (NT) assay is an important method to validate the efficiency of potential vaccines during vaccine development, and is also essential for screening of neutralizing mAbs. Furthermore, NT assay is also necessary to determine the immune status of a patient.

Traditionally, plaque reduction neutralization tests (PRNTs) and inhibition of the cytopathogenic effects (CPEs) [20] were used for NT assay. Recently, some new methods based on the non-infectious virus replicon particles (VRPs) and the pseudotyped lentiviral vector have been developed for the CHIKV NT assay [21,22]. CHIKV VRPs were produced by co-transfecting BHK-21 cells with a CHIKV replicon expressing Gaussia luciferase (Gluc) and two helper RNAs expressing the CHIKV capsid and other structural proteins, respectively [21]. The CHIKV-pseudotyped lentiviral vector was prepared by co-transfection of plasmids encoding the CHIKV glycoproteins E3, E2, 6K, and E1, packaging elements, and a luciferase reporter [22]. The luciferase activities were used as readout to determine CHIKV neutralization sera/antibodies for both assays [21,22].

In this study, we developed another convenient alternative NT assay for CHIKV. We firstly constructed an infectious clone of CHIKV and a stable eGFP reporter CHIKV (eGFP-CHIKV) with a newly isolated strain of Asian lineage. The eGFP-CHIKV replicated efficiently that was comparable with wild type CHIKV (WT CHIKV) and genetically stable at least after five rounds of viral passages. Using the known inhibitor of CHIKV, we confirmed that the expression levels of eGFP could be used to quantify the replication of CHIKV. Based on eGFP-CHIKV reporter virus, a new neutralization assay for detection of CHIKV neutralizing serum/antibody was developed. Our method is a rapid and quantitative assay for studying patient or animal serum samples and screening neutralizing antibodies against CHIKV. Furthermore, the eGFP-CHIKV also has the potential to be used in large-scale and high-content assays.

2. Materials and Methods

2.1. Cell Lines, Viruses, Antibodies

BHK-21 cells were cultured in Dulbecco's modified Eagle's medium (DMEM; Invitrogen, Darmstadt, Germany) with 10% Fetal Bovine Serum (FBS), 100 U/mL of penicillin and 100 μg/mL of streptomycin at 37 °C with 5% CO_2. The mosquito cells of C6/36 *Aedes albopictus* clone were cultured in RPMI-1640 medium with 10% FBS at 28 °C. The CHIKV strain (GenBank accession No. KC488650) was isolated from a clinically CHIKV-positive patient in China through seven rounds of serial passages in C6/36 cells (Figure 1A). The CHIKV stock was stored as aliquots at −80 °C. This stock was used as parental CHIKV in all assays. The rabbit and mouse polyclonal antibodies against CHIKV E2 protein were generated by immunization of Japanese big-ear rabbits and BALB/C mice with SDS-PAGE purified CHIKV E2 protein, respectively.

2.2. Plasmid Construction

The viral RNA was extracted from the parental virus with QIAamp Viral RNA Mini Kit (Qiagen, Hilden, Germany) and used for reverse transcription PCR (RT-PCR). By using the SuperScript III One-Step RT-PCR System (Invitrogen), six cDNA fragments covering the complete viral genome of CHIKV were amplified from extracted viral RNA with six pairs of primers listed in Table 1, and then cloned into pACYC177 at PsiI and BamHI sites, yielding six subclones called CHIKV-A, CHIKV-B, CHIKV-C, CHIKV-D, CHIKV-E and CHIKV-F. Construct of CHIKV-A contained the sequence ranging from the 5′-end of the genome to the nucleotide position 2490 with a T7 promoter located before the 5′-end that was designed in primer CHIKV-5′UTR′-PsiI-F (Table 1). Constructs of CHIKV-B, CHIKV-C, CHIKV-D and CHIKV-E contained the sequences from nucleotide position 1877 to 4087, 4103 to 6802, 6604 to 9308 and 8713 to 10727, respectively. Construct of CHIKV-F included the sequence from nucleotide position 10,196 to the 3′-end with a poly(A) tail. The six subclones were assembled step-by-step into a full-length cDNA clone of pACYC-CHIKV following the scheme of the cloning strategy in Figure 2A.

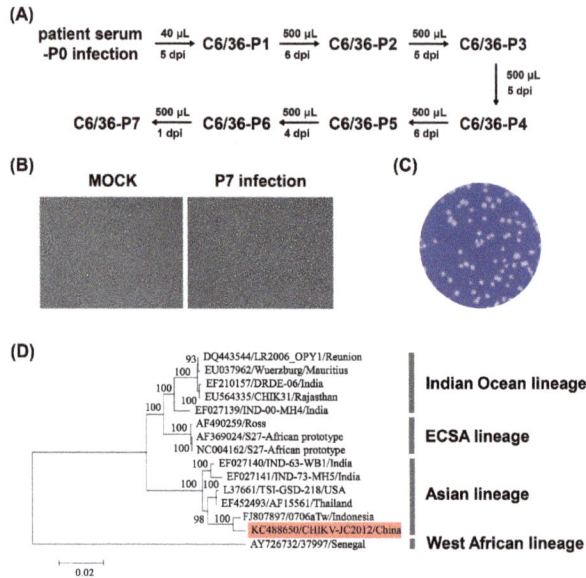

Figure 1. Isolation and characterization of CHIKV from clinical human cases. (**a**) the flow chart of virus isolation on C6/36 cells from human serum; (**b**) the CHIKV strain that was passaged for seven rounds on C6/36 cells showed apparent CPE on BHK-21 cells; (**c**) plaque morphology of the CHIKV strain on BHK-21 cells on the four days post-inoculation; and (**d**) phylogenetic analyses of CHIKV genome sequences using the neighbor-joining method. The newly isolated CHIKV strain is highlighted in red. ECSA lineage = the East, Central and South African lineage.

Table 1. Primer sequences.

Primer	Sequence
CHIKV-5'UTR'-PsiI-F	cgcTTATAATAATACGACTCACTATAGatggctgcgtgagacacacg
CHIKV-nsP2-BamHI-R(2490)	cgcGGATCCctcaccaaggcgatcaaggc
CHIKV-nsP2-PsiI-F(1877)	cgcTTATAACGTACGATGGCCGAGTCCTAGTG
CHIKV-nsP3-BamHI-R(4087)	cgcGGATCCgatatccatgcgtttacccggtac
CHIKV-nsP3-PsiI-F(4103)	cgcTTATAAgatatcGCGAAGAACGATGAAGAG
CHIKV-nsP4-BamHI-R(6802)	cgcGGATCCgcattaaggcggtaagcgca
CHIKV-nsP4-PsiI-F(6604)	cgcTTATAACTTGGCAACAGCGTACCTATG
CHIKV-E2-BamHI-R(9308)	cgcGGATCCtgcatgtcacatttgccagag
CHIKV-nsP2-PsiI-F(8713)	cgcTTATAACATGATTGGACCAAGCTGCG
CHIKV-E1-BamHI-R(10727)	cgcGGATCCggtgctgtgctgcagcgacg
CHIKV-E1-PsiI-F(10196)	cgcTTATAAGCCTACCTGATTACAGC
CHIKV-3'UTR-BamHI-R	cgcGGATCCCACTAGTTTTTTTTTTTTTTTTTTTTTTTTTTTTTTTTTTTTTgaaatatt
CHIKV-AscI-PacI-2A-F	TAAATACCAATCAGCCATAcggcgcgccaagACATACgccttaattaatCAGCTGTTGAATTTTGACCTTCTCAA
CHIKV-AscI-PacI-2A-R	GGTTGGGATAAACTCCATTGGCCCAGGGTTGGACTCGACGTCTCCCGCCAGCTTGAGAAGGTCAAAATTCAACA
CHIKV-PacI-2SG promoter-F	gccttaattaatGTCATAACCTTGTACGG
CHIKV-2SG promoter-C-F	AATCAGCCATAATGGAGTTTATCCCAACC
CHIKV-2SG promoter-C-R	ATAAACTCCATTATGGCTGATTGGTATTT
eGFP-AscI-F	ttggcgcgccatggtgagcaagggcgaggag
PacI-eGFP-stop-R	TccttaattaaCTActtgtacagctcgtccatgcc
CHIKV-7376-F	gaagtgcagggtatatcag
CHIKV-8498-R	gatgcttgtagcagctgat
CHIKV-9923-F	TGAGCGTCGGTGCCCAC
CHIKV-10003-R	GAGTCTTATACCGTACTCCCACCGT

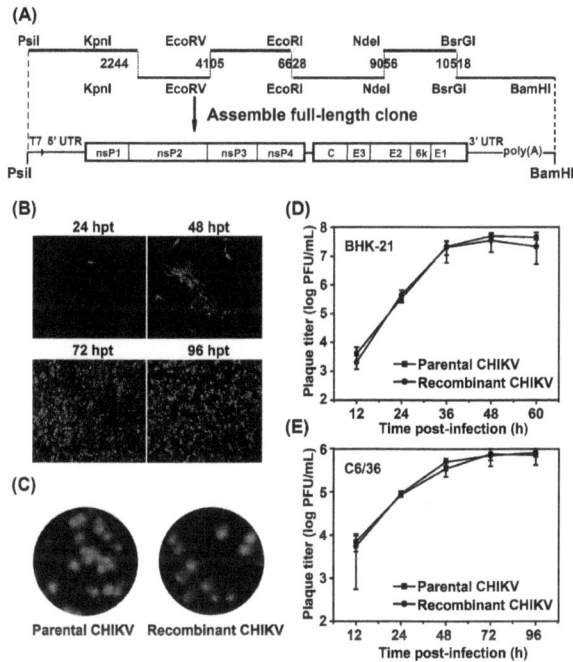

Figure 2. Construction and characterization of the full-length cDNA clone of CHIKV. (**a**) six cDNA fragments represented by thick lines were synthesized from genomic RNA through RT–PCR to cover the complete CHIKV genome and unique restriction sites as well as their nucleotide numbers are shown. Genome organization is depicted below. Individual fragments were assembled to form the full-length cDNA clone of CHIKV (pACYC-CHIKV). The complete CHIKV cDNA is under the control of T7 promoter elements for in vitro transcription. The numbers are the nucleotide positions based on the CHIKV genome sequences we identified in this study (KC488650); (**b**) IFA of viral protein expression in BHK-21 cells transfected with the full-length CHIKV RNA transcript. The transfected cells were analyzed by IFA at the indicated time points post-transfection; (**c**) plaque morphology of parental and recombinant CHIKV on BHK-21 cells; (**d,e**) comparison of the growth kinetics of recombinant and parental CHIKV in BHK-21 and C6/36 cells, respectively. The virus growth curves were compared at MOI = 0.1 on both cells. Three independent experiments were performed in duplicate, and the representative data were presented.

The eGFP-CHIKV reporter virus cDNAs were constructed in multiple steps according to two different strategies as depicted in Figure 3A. Details for construction were as followed. In the first strategy, a cassette containing the AscI/PacI cleavage sites and FMDV 2A in reading frame of the structural polyprotein were introduced to pACYC-CHIKV by overlap PCR. Then, the eGFP gene was amplified and inserted into AscI and PacI sites to generate CHIKV reporter virus. In the second strategy, an expression cassette expressing a repeated sub-genomic (sg) promoter and eGFP were inserted between the nonstructural nsP4 gene and the 5′-terminal of structural genes. All primers used for PCRs were listed in Table 1. All constructs were confirmed by sequencing.

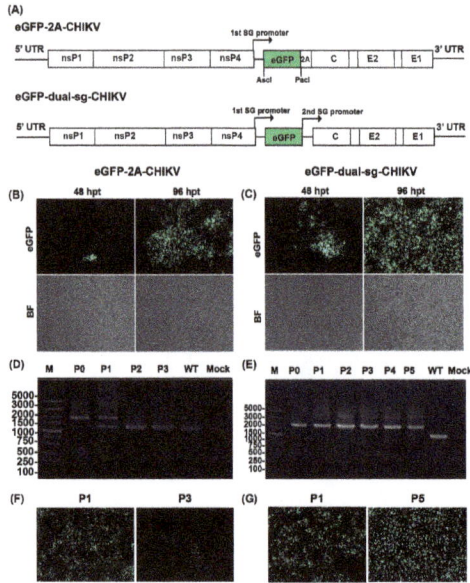

Figure 3. Construction and stability of CHIKV reporter viruses. (a) schematic of the construction of two versions of CHIKV reporter virus (eGFP-2A-CHIKV and eGFP-dual-sg-CHIKV). An infectious cDNA clone of pACYC-CHIKV was used as a backbone for the construction of CHIKV reporter viruses; (b,c) eGFP expressions in cells transfected with different versions of CHIKV reporter RNA transcripts. The data for eGFP-2A-CHIKV and eGFP-dual-sg-CHIKV were present on left and right panels, respectively, and thereafter. The expression of eGFP in transfected BHK-21 cells was analyzed by fluorescent microscopy at the indicated time points post-transfection. The lower panels were the images visualized under differential interference contrast microscopy; (d,e) detection of the eGFP gene during virus passage. Viral RNAs were extracted from culture supernatants of the indicated passages, respectively. RT–PCR was performed with primer pairs to cover the complete eGFP gene region. The resulting RT–PCR products were resolved by 1% agarose gel electrophoresis; and (f,g) detection of eGFP expression in BHK-21 cells infected with different passages of reporter viruses. eGFP expressions were observed under fluorescent microscope at 48 hpi. Both RT-PCR and eGFP expression results indicated that the reporter gene still maintained in eGFP-dual-sg-CHIKV. The eGFP-dual-sg-CHIKV was designated as eGFP-CHIKV.

2.3. RNA Transcription and Transfection

Genome-length and reporting viral RNAs were prepared from the BamHI-linearized cDNA plasmids through in vitro transcription using mMESSENGER mMACHINE T7 Kit (Ambion, Austin, TX, USA). The RNAs were transfected into BHK-21 cells with DMRIE-C (Invitrogen). After transfection, supernatants were collected at different time points. The culture medium containing viruses were aliquoted and stored at $-80\,^\circ$C for the next experiments.

2.4. Plaque Assay

BHK-21 cells were seeded into 24-well plates at a density of 1×10^5 cells per well one day before plaque assay. A series of 1:10 dilutions were made by mixing 15 μL of virus sample with 135 μL of DMEM. Then, 100 μL of each dilution were added to individual wells of 24-well plates containing confluent BHK-21 cells. The plates were incubated at 37 $^\circ$C with 5% CO_2 for 1 h before the layer of 2% methyl cellulose was added. After 3 days of incubation at 37 $^\circ$C with 5% CO_2, the cells were fixed with 3.7% formaldehyde and then stained with 1% crystal violet. Plaque morphology and numbers were recorded after washing the plates with tap water.

2.5. Immunofluorescence Assay (IFA)

BHK-21 cells transfected with the CHIKV genome-length RNA were seeded on Chamber Slide (Nalge Nunc, Rochester, NY, USA). At 24, 48, 72 and 96 h post-transfection (hpt), the transfected cells were collected, washed with PBS and fixed by cold ($-20\,^\circ$C) 5% acetic acid in acetone for 15 min at room temperature. The fixed cells were washed with PBS and incubated with CHIKV E2 polyclonal serum from rabbit (1:250 dilution with PBS) for 1 h. After washing with PBS, the cells were then incubated with goat anti-rabbit IgG conjugated with Texas-Red (Proteintech, Wuhan, China) at a 1:125 dilution with PBS at room temperature for another hour. The cells on the slide were mounted with 90% glycerol and examined under a fluorescent microscope. The fluorescent signal images were taken at $200\times$ magnification with a NIKON upright fluorescence microscope (Tokyo, Japan).

2.6. Antiviral Assay of eGFP-CHIKV by Ribavirin

One day prior to infection, BHK-21 cells were seeded in 12-well plates at a density of 1×10^5 cells per well. Then, the cells were inoculated with 0.2 mL of diluted recombinant CHIKV and eGFP-CHIKV virus (at a multiplicity of infection of 0.05) and incubated at 37 $^\circ$C with 5% CO_2 for 1 h. Subsequently, the culture was replaced with fresh medium containing different concentrations of ribavirin. After incubation at 37 $^\circ$C for 48 h, the supernatants were collected and viral titers were quantified by plaque assay. Antiviral activity of ribavirin was expressed as the 50% effective concentration (EC_{50}) and calculated as described previously [23].

2.7. Neutralization Assay Based on eGFP-CHIKV Reporter Virus

The neutralizing activities of the serum samples from CHIKV-infected patients and mice were determined using eGFP-CHIKV reporting virus in BHK-21 cells in triplicates. Briefly, BHK-21 cells were seeded in 24-well plates at a density of 8×10^4 cells per well. The following day, eGFP-CHIKV were used in the NT assay at an MOI of 0.05. eGFP-CHIKV were incubated with 4-fold serial dilutions of heat-inactivated patient sera (starting at 1:10 dilution) and mouse sera or CHIKV E2 polyclonal antibody (starting at 1:20 dilution) for 1 h at 37 °C before adding the mixture to the monolayer of BHK-21 cells in the 24-well plates. Simultaneously, the serum from health person was used as a negative control. After incubation for another 1 h at 37 °C, the inoculum was removed. Cells were washed with PBS and fresh medium was added to each well. We examined the level of neutralization activity by measurement of eGFP fluorescence intensity at 48 h post-infection (hpi) using both fluorescent microscopy and microplate fluorimeter. The percentage of infectivity was calculated as: %Infectivity = (fluorescence intensity from serum samples / fluorescence intensity from negative control samples). The experiments were independently repeated three times.

2.8. Plaque Reduction Neutralization Tests (PRNTs)

PRNT assay was carried out to detect and quantify the presence of neutralizing antibodies in the human serum samples as previously described [24]. In brief, about 100 PFU of infectious WT CHIKV were incubated with 100 µL of 4-fold serial dilutions of heat-inactivated serum samples (starting at 1:40 dilution) at 37 °C for 1 h. The virus-serum mixture was then inoculated to BHK-21 cell monolayer cultured in 12-well plates and incubated at 37 °C for 1 h with rocking every 15 min. Next, the supernatant was removed, and the layer of 2% methyl cellulose was added. After further incubation at 37 °C with 5% CO_2 for 2 days, the cells were fixed and plaque numbers were recorded as described above (section 2.4). WT CHIKV alone without serum incubation was served as negative control. The percentage of infectivity was calculated as: %Infectivity = (number of plaques from serum samples / number of plaques from negative control).

2.9. Real-Time RT-PCR Analysis

To detect and quantify the genome copy numbers of WT CHIKV, SYBR Green based real time RT-PCR analysis was used. Briefly, 100 µL of WT CHIKV corresponding to MOI of 1 were mixed with 100 µL heat-inactivated serum samples (1:40 dilution). After being incubated at 37 °C for 1 h, the virus-serum mixture was added to BHK-21 cell monolayer in 12-well plates. After 1 h of incubation at 37 °C, unbound virus was removed by three washes with cold PBS. The total cellular RNAs were extracted using Trizol reagent (TaKaRa, Dalian, China). Real time RT-PCR was performed using one step SYBR green PrimeScript PLUS RT-PCR kit (TaKaRa). The primer set was designed to amplify a 81 bp-long region within the E2-6K-E1 gene as described previously [25]. The primer sequences were CHIKV-9923-F (forward) and CHIKV-10003-R (reverse) (Table 1). Amplification conditions were as follows: 42 °C for 5 min, 95 °C for 10 s and 40 cycles of 95 °C for 5 s, 60 °C for 34 s. A melting curve analysis was performed to verify the authenticity of the amplification after each reaction. In vitro transcribed RNA were used to establish standard curve and the genomic RNA copies were determined from the respective C_T values based on the standard curve.

2.10. Statistical Analysis

The Student's t-test and ANOVA test were used to determine if there were significant differences ($p < 0.05$) for all tested viruses. The statistical analyses were performed in IBM SPSS Statistics v18.0 (Chicago, IL, USA).

3. Results

3.1. Isolation and Characterization of CHIKV

The CHIKV strain (KC488650) used for the construction of infectious clone was isolated from a clinical sample as depicted in Figure 1A. The isolated CHIKV replicated efficiently and caused typical cytopathic effects (CPE) on BHK-21 cells (Figure 1B) with a high viral titer of approximately 1.5×10^7 PFU/mL determined by plaque assay (Figure 1C). The viral complete genome sequencing and the phylogenetic analysis indicated that this newly isolated strain belonged to the Asian lineage of CHIKV. Our isolate was sister to the strain 0706aTw isolated from Indonesia in 2007 (Figure 1D) and two isolates shared high identity of their nucleotide sequences up to 99.4%.

3.2. Construction and Characterization of of the Full-Length CHIKV cDNA Clone

The full-length CHIKV cDNA clone, pACYC-CHIKV, was assembled with six subclones through distinct restriction cut sites as indicated in Figure 2A. A T7 promoter was engineered at the 5′-end of viral genome sequence for in vitro transcription. The recombinant CHIKV genomic RNA was in vitro transcribed from CHIKV cDNA clone, followed by linearization with BamHI. In vitro transcribed viral RNA was transfected into BHK-21 cells and IFA was firstly used to demonstrate viral replication by detecting viral E2 protein expression (Figure 2B). Increasing number of IFA-positive cells expressing viral E2 protein was observed from 24 to

96 hpt and almost 100% cells were IFA-positive at 72 and 96 hpt. Then, plaque morphologies were compared between the recombinant and parental viruses. As shown in Figure 2C, the plaque size of recombinant virus is similar to that of parental virus. Consistently, indistinguishable viral growth kinetics at MOI of 0.1 were observed between recombinant and parental viruses in either BHK-21 or C6/36 cells (Figure 2D,E). Overall, our data demonstrated that CHIKV infectious clone was constructed successfully and recombinant virus was infectious in both mosquito and mammalian cells. The recombinant CHIKV derived from infectious clone is designated as WT CHIKV in this paper.

3.3. Construction of the CHIKV Reporter Virus with eGFP

Two strategies were used to construct eGFP-CHIKV reporter viruses as shown in Figure 3A. The first strategy is that eGFP gene was inserted before the structure protein of capsid (C). In order to release the eGFP from capsid, FMDV 2A for autocleavage was introduced between them. The construct was designated as eGFP-2A-CHIKV. A dual subgenomic (sg) promoter method was used based on the second strategy to construct reporter virus that was called eGFP-dual-sg-CHIKV. Additional sg promoter was used for eGFP cassette expression that was inserted before structure genes.

Equal amounts of in vitro transcribed genome-length RNAs of eGFP-2A-CHIKV and eGFP-dual-sg-CHIKV were transfected into BHK-21 cells to compare their viral replication capacity. The expression levels of eGFP were monitored under a fluorescent microscope for these two constructs of reporter viruses. As demonstrated in Figure 3B,C, eGFP-dual-sg-CHIKV produced more positive eGFP cells than eGFP-2A-CHIKV at each time point post-transfection. Furthermore, more CPE were also observed in transfected cells for eGFP-dual-sg-CHIKV (Figure 3C), which indicated a replication advantage of eGFP-dual-sg-CHIKV over eGFP-2A-CHIKV (Figure 3B).

3.4. Characterization of the CHIKV Reporter Virus

After demonstrating that both reporter viruses are replication-competent, their stabilities were tested in cell culture. The viruses in supernatants from reporter virus RNA transfected BHK-21 cells were defined as P0 passage and were used for blinding passage on BHI-21 cells. The viruses from each passage were defined as P1 to P5 passages. Viral RNAs were first extracted from P0 to P3 passages of eGFP-2A-CHIKV and P0 to P5 passages of eGFP-dual-sg-CHIKV, respectively. RT-PCRs were performed to amplify the fragment between nsP4 and capsid that covers the region of the inserted reporter gene using primers CHIKV-7376-F and CHIKV-8498-R (Table 1). Different sizes of bands should be detected from RT-PCR products of WT (~1.5 Kb) and reporter viruses (~2 Kb) (Figure 3D,E). For eGFP-2A-CHIKV, reporter genes began to lose from P1 passage as indicated by two RT-PCR products of both 1.5 Kb and 2 Kb fragments, and were completely lost at the P2/P3 passage (Figure 3D). For eGFP-dual-sg-CHIKV, only specific 2 kb RT-PCR products were observed from P0 to P5 passages, which indicated eGFP-dual-sg-CHIKV reporter virus is much more stable than eGFP-2A-CHIKV (Figure 3E). Consistent with RT-PCR results, a high eGFP expression level was detected in BHK-21 cells infected with P5 of eGFP-dual-sg-CHIKV, but only sporadic eGFP positive cells were observed in P3 of eGFP-2A-CHIKV infected cells (Figure 3F,G). Our results demonstrated that eGFP-dual-sg-CHIKV reporter virus was stable and designated as eGFP-CHIKV in the following study.

Plaque morphology and viral growth curve were compared between WT CHIKV and eGFP-CHIKV reporter virus. As measured by plaque assay, eGFP-CHIKV could replicate efficiently although its viral titer was about 10-fold lower than that of WT CHIKV (Figure 4A). eGFP-CHIKV displayed a smaller plaque size than WT CHIKV (Figure 4B). To determine whether the level of eGFP expression is correlated with the efficiency of viral replication, different MOIs of eGFP-CHIKV were used to infect naïve BHK-21 cells, and the numbers of eGFP-positive cells were recorded at different time points post-infection. A time- and dose-dependent increase in the number of eGFP-positive cells was observed (Figure 4C), which indicated that eGFP expression levels could be used to monitor and quantify viral replication. We used MOI of 0.05 for the following experiments as almost 100% eGFP-positive cells were observed at 48 hpi. Overall, the results show that the eGFP-CHIKV reporter virus is replication-competent, stable and infectious.

3.5. Inhibitory Effects of Ribavirin on eGFP-CHIKV Reporter Virus

It has been demonstrated that ribavirin could inhibit CHIKV replication [26,27]. To further confirm the correlation between eGFP expression and viral replication, we assessed the antiviral activity of ribavirin using eGFP-CHIKV reporter virus. The eGFP expression levels and virus titers were determined at different concentrations of ribavirin that did not cause significant cytotoxic effects. The maximum concentration of ribavirin was 10 μg/mL and 2-fold serial dilutions were used in this study. The eGFP expressions of infected cells were reduced in a dose-dependent manner (Figure 5A), which was consistent with the reduction of viral titers (Figure 5B). Furthermore, the EC_{50} of ribavirin for eGFP-CHIKV was 1.03 μg/mL (4.22 μM), which was similar to that of WT CHIKV, i.e., $EC_{50} = 1.32$ μg/mL (5.41 μM) (Figure 5C). The results further confirmed that eGFP expression could be used to surrogate the replication of WT CHIKV. In addition, it was also suggested that the eGFP-CHIKV reporter virus system could be used for screening potential anti-CHIKV inhibitors.

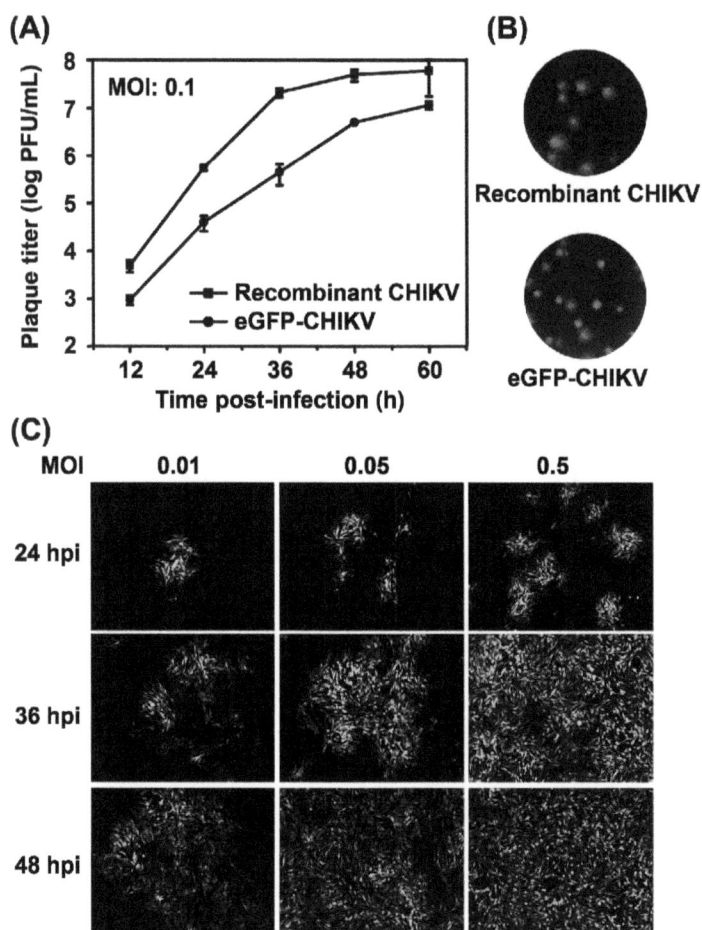

Figure 4. Characterization of eGFP-CHIKV reporter viruses. (**a**) comparison of the growth kinetics of recombinant WT CHIKV and eGFP-CHIKV in BHK-21 cells. The representative data were presented from three independent experiments; (**b**) plaque morphology of recombinant WT CHIKV and eGFP-CHIKV on BHK-21 cells; and (**c**) correlation between eGFP expression and different MOIs infection. BHK-21 cells were infected with eGFP-CHIKV at the indicated MOI and eGFP expression were observed at 24, 36 and 48 hpi.

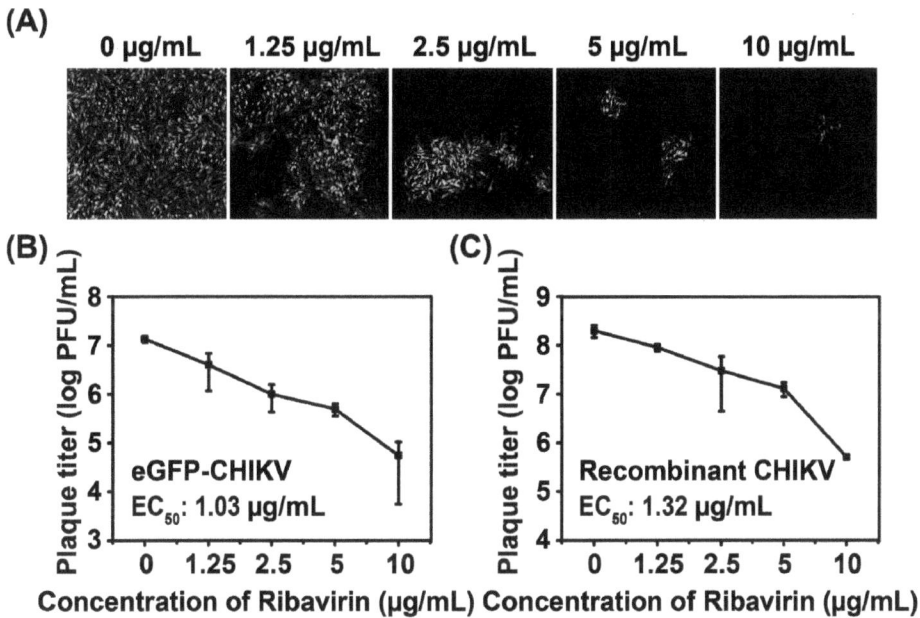

Figure 5. Antiviral activity of ribavirin on eGFP-CHIKV reporter viruses. (a) detection of eGFP expression by fluorescent microscope at different concentrations of ribavirin; (b) viral titer reduction assay of CHIKV with different concentrations of ribavirin. BHK-21 cells were infected with eGFP-CHIKV at MOI of 0.05 and treated with various concentrations of ribavirin. The viral titers were quantified by plaque assay at 48 hpi; and (c) viral titer reduction assay of WT CHIKV at different concentrations of ribavirin. The experiments were performed in triplicate, and the representative data were presented.

3.6. Confirmation of Neutralization Assay Using eGFP-CHIKV with CHIKV Patients' Sera

Neutralization assay with eGFP-CHIKV were firstly evaluated by using human sera from two CHIKV-infected patients (#34 and #81), which had been confirmed to be positive against CHIKV. The eGFP-CHIKV was mixed with the respective serum dilutions and incubated in BHK-21 cells. At 48 hpi, the numbers of eGFP-positive cells were recorded to determine the infectivity rates. As shown in Figure 6A, neutralizing activities were observed after incubations with CHIKV sera (#34 and #81) at different dilutions. There was little or no effect observed when eGFP-CHIKV was incubated with the negative control of a healthy human serum (Figure 6A), demonstrating that neutralization is specifically induced by the sera against CHIKV. Furthermore, the two CHIKV-positive human sera samples significantly reduced the percentage of infectivity of eGFP-CHIKV reporter viruses in a dose-dependent manner with quantification of eGFP positive cells (Figure 6B). In order to assess the efficacy of eGFP-CHIKV on neutralizing assay, the conventional PRNT assay was conducted. Similarly, there was an excellent correlation between the plaque reduction and the dilutions of neutralizing human sera (Figure 6C). Both dilution 1:40 of #34 and #81 sera could inhibit approximately 90% of infectious CHIKV in PRNT assay (Figure 6C), which was consistent with that of neutralization assay based on eGFP-CHIKV (Figure 6B). The percentages of infectivity of WT CHIKV was lower than 50% when sera #34 and #81 were diluted to 1:160 and 1:640 (Figure 6C). No significant differences in inhibition were observed between dilution 1:10240 of these two human sera and negative human serum (ANOVA test, $p > 0.05$). As control, the changes in CHIKV genome copies following the co-incubation of human sera and WT CHIKV in fusion/entry step were monitored by real-time RT-PCR analysis. The CHIKV genome copies can be decreased by nearly 22-fold and 9-fold when 1:40 dilutions of #34 and #81 sera were used, respectively (Figure 6D). We thus concluded that the ability of serum neutralization could be measured by eGFP expression with eGFP-CHIKV reporter viruses that are visible and suitable for the neutralization test.

3.7. Neutralization Assay with Sera from Mice Using eGFP-CHIKV

The serum samples collected from two mice immunized with formalin-inactivated CHIKV (#1 and #3 sera) and one mouse immunized with SDS-PAGE purified CHIKV E2 protein (anti-E2 serum) were analyzed with the NT assay. The eGFP-CHIKV was mixed with 4-fold serially diluted serum and incubated with cells for 48 h. All pre-immunization sera were negative for anti-CHIKV neutralizing antibodies and one of representative data at 1:20 dilution was shown in Figure 7A as a negative control. A dose dependence of eGFP positive cells were observed for both #1 and #3 sera from dilution 1:20 to 1:320 (Figure 7A). In contrast, there was only a minor effect on eGFP expression for anti-E2 serum at 1:20 dilution. To further confirm the reactivity of the tested sera against CHIKV, an IFA (Figure 7B) was performed using CHIKV-infected cells. All three sera (#1, #3 and anti-E2 sera) were found to be reactive against CHIKV as IFA positive cells were observed in CHIKV infected cells. At the same time, no positive cells were found in mock infected cells for all tested sera, which excluded the possibility of false positive results. Additionally, the antibody titers of #1, #3 and anti-E2 sera were assessed by ELISA. The coating antigen for ELISA was purified recombinant E2 protein. The results revealed that anti-E2 sera yielded the maximum antibody titers. Overall, the results demonstrated that our NT assay based on eGFP-CHIKV could easily differentiate neutralizing serum/antibody against CHIKV from non-neutralizing serum/antibody with E2 protein.

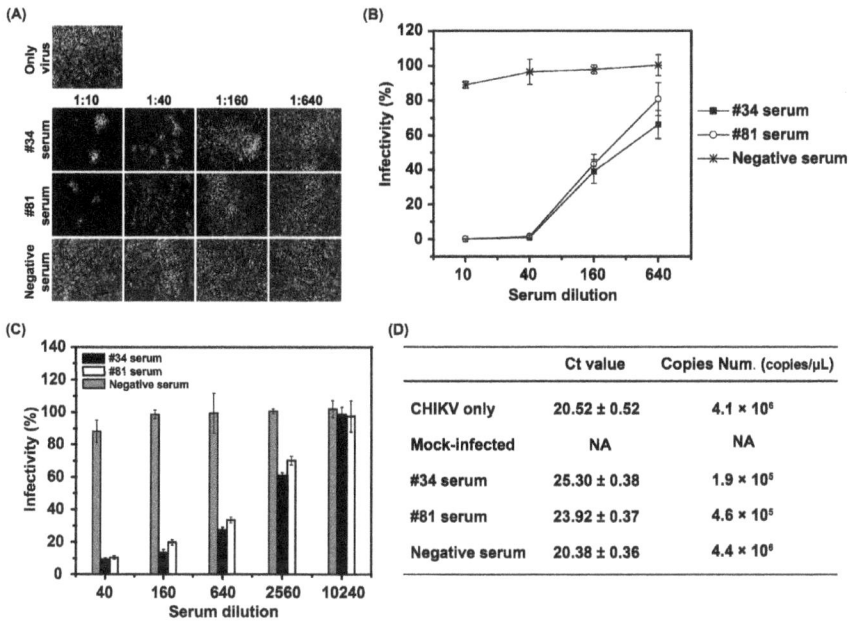

Figure 6. Confirmation of the availability of eGFP-CHIKV in neutralization assay with the serum samples from CHIKV-infected patients. (**a**) detection of eGFP expression by fluorescent microscope with different dilutions of human sera. The neutralization assay was performed in a 12-well plate. Serially 4-fold diluted human sera were incubated with eGFP-CHIKV at an MOI of 0.05 for 1 h at 37 °C before adding the mixture to the monolayer of BHK-21 cells in the 24-well plates. The readout of eGFP expression was recorded at 48 hpi. The serum from health person was used as a negative control; (**b**) quantification of neutralization assay with increasing serum dilution. The percentage inhibition of infectivity was normalized to eGFP-CHIKV infection without serum incubation by quantification of eGFP expression levels; (**c**) neutralizing activity of the human sera against WT CHIKV based on PRNT assay. WT CHIKV was incubated with 4-fold dilutions of individual sera before infecting BHK-21 cells. The percentage inhibition of infectivity for each dilution was normalized to WT CHIKV infection alone by counting the number of plaque forming units (PFU); (**d**) detection of CHIKV genome copy number in BHK-21 cells treated with virus or virus-antiserum mixture by real-time RT-PCR. The results are positive if Ct value ⩽34, otherwise negative if Ct value >34. The genome copy numbers of #34 and #81 were significantly decreased than those of control samples (*t*-test, $p < 0.05$). CHIKV only = BHK-21 cells were infected by WT CHIKV without serum incubation; mock-infected = BHK-21 cells were neither infected by WT CHIKV nor incubated with serum; NA = data not available. All experiments were performed in triplicate, and the representative data were presented.

227

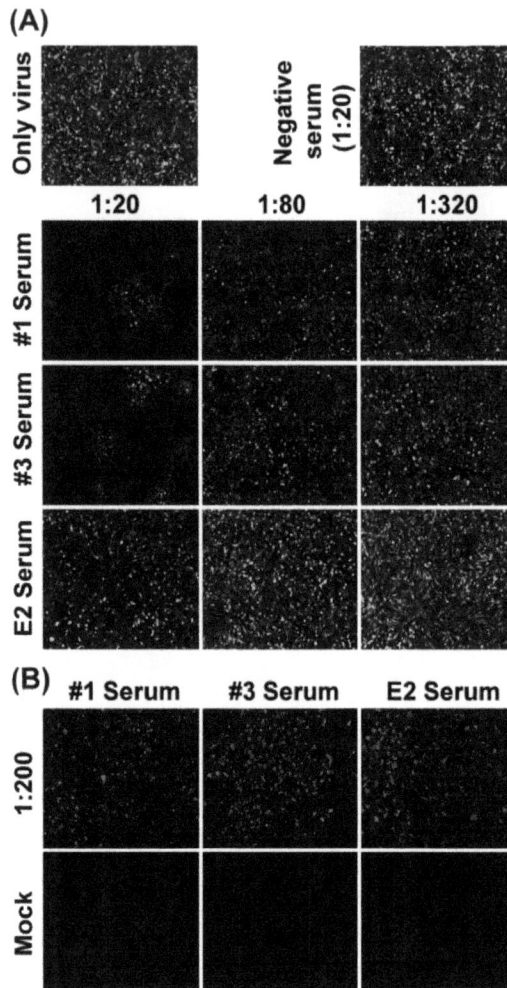

Figure 7. Neutralization assay with sera from mice using eGFP-CHIKV. (a) detection of eGFP expression by fluorescent microscope with different dilutions of sera from mice immunized with different antigens. #1 and #3 sera were from two mice immunized with formalin-inactivated CHIKV, and anti-E2 serum was obtained from one mouse immunized with SDS-PAGE purified CHIKV E2 protein. NT assay was performed as described for CHIKV-infected patient sera; (b) the reactivity of the tested sera against CHIKV through an IFA. All three sera (#1, #3 and E2 sera) were diluted at 1:200. Mock and CHIKV infected BHK-21 cells were used to performed IFA.

4. Discussion

CHIKV, a member of the *Alphavirus* genus, has re-emerged as a major threat to global public health and there is an urgent need for continued research into the epidemiology, pathogenesis, prevention, and treatment of CHIKV infections [2]. In recent years, significant progress has been made in our understanding of the replication mechanism of alphavirus [28,29]. These mechanism studies further prompted development of reverse genetic for alphavirus [4,30–35]. For CHIKV, different infectious clones and reporter viruses have been also developed to follow up viral replication both in vitro and in vivo [27,31,36–38]. In general, the CHIKV strain La Réunion (Indian Ocean lineage, Figure 1D) isolated from infected humans is commonly used for reporter virus system and functional research [37–39]. In our study, the new infectious clone was derived from a Chinese human isolate (strain CHIKV-JC2012) that belongs to the Asian lineage (Figure 1D). As we know, lineage-specific adaptive mutations could accumulate in CHIKVs from different genetic lineages, which may affect the infectivity range and epidemic dynamics [7,40]. Thus, recombinant and reporter CHIKVs derived from different lineages could expand the range and diversity of CHIKV related analyses and applications.

In this study, we developed an NT assay using the eGFP-CHIKV reporter virus to evaluate neutralizing activity of sera/antibody against CHIKV. On one hand, we constructed an infectious clone of CHIKV and a eGFP-CHIKV reporter virus based on the isolate in our lab (strain CHIKV-JC2012) (Figure 1). The recombinant CHIKV and reporter CHIKV are replication-competent and infectious (Figure 2). The expression level of eGFP correlated with viral replication, which enabled us to perform neutralization assay using eGFP-CHIKV to quantify neutralizing antibody/serum against CHIKV. We used this reporter virus to detect the neutralizing activities of the serum samples from two CHIKV-infected patients, and demonstrated that the replication of eGFP-CHIKV was inhibited in a dose-dependent manner. On the other hand, the conventional PRNT assay and real-time RT-PCR analysis were conducted to further confirm the suitability and reliability of eGFP-CHIKV based NT assay. Although the PRNT exhibited more efficient ability to inhibit the infectious CHIKV, we speculated that the initial amounts of CHIKV particles could account for the difference of performance between PRNT and eGFP-CHIKV based NT assay (MOI, 0.0005 vs. 0.05) as less viruses are likely to be more efficiently reduced by neutralizing antibody. Overall, the newly developed NT assay based on eGFP-CHIKV reporter viruses is suitable for neutralization test.

In our assay, we confirmed in a previous study that inactivated CHIKV represents a good immunogen for inducing neutralizing antibodies (Figure 7A) comparing with denaturing recombinant E2 protein although mice immunized with denaturing E2 protein produced highest antibody titer by ELISA. However, we cannot exclude the possibility that these antibodies from denaturing E2 protein immunized-mice are against the same denatured epitope with the coating antigen of ELISA, while inactivated CHIKV-immunized sera are likely against other conformational epitopes. Nonetheless, our results demonstrated that, although ELISAs are very fast and easy to perform for measuring and deciding the absence or presence of antibodies, ELISAs could not differentiate neutralizing and non-neutralizing antibodies. In summary, NT assay is the best way to evaluate virus-neutralizing antibodies biologically that will produce accurate and specific results.

When eGFP-CHIKV reporter viruses are used for NT assays, there are some advantages compared with VRP and pseudotyped lentiviral vectors [21,22]. Firstly, the production of eGFP-CHIKV reporter viruses is technically easy as the reporter viruses are quite stable and infectious. The growth curve of eGFP-CHIKV is comparable with that of WT CHIKV. Large amounts of eGFP-CHIKV could be obtained from infected cells once the reporter viruses were recovered from cells transfected with transcribed recombinant RNA. In contrast, a new cycle of transfection is required each time to get new batches of CHIKV VRPs or CHIKV-lentiviral-vectors, which are tedious and technically difficult. Secondly, the assays are not expensive, as no additional reagents are needed. The readouts of the NT assay based on eGFP-CHIKV reporter viruses were performed with a fluorescence microscope by direct observation of eGFP expression, which is much easier than conventional methods. Lastly, eGFP-CHIKV also has the high throughput potential by using high content assays with quantification of eGFP positive cells. It will omit any further plate processing procedures, such as substrate addition to produce different signals for readout, cell fixation, plate washing or cell lysis, which will further decrease the cost and complexity of the screening process. Moreover, the eGFP-CHIKV reporter virus derived from an Asian lineage could provide opportunities for the comparison study of antiviral activity of neutralizing antibodies and compounds among different lineages in a high-throughput manner. In this regard, the relationships between virulence, resistance, and viral fitness of CHIKV strains from distinct phylogenetic groups could be further investigated.

Comparing NT assays using the CHIKV VRP and pseudotyped lentiviral vectors, one disadvantage for eGFP-CHIKV is the biosafety issue as CHIKV is classified as a biosafety level 3 (BSL-3) pathogen. This problem may be resolved by using attenuated vaccine candidates to create eGFP-CHIKV reporter viruses. Recently, various live-attenuated CHIKV vaccine candidates have been developed through inserting an IRES (internal ribosome entry site) sequence into the genome of CHIKV [14] or deleting a large part of the gene encoding nsP3 or the entire gene encoding 6K [13]. These live vaccine candidates may help develop new versions of CHIKV reporter viruses to increase safety in the future.

5. Conclusions

In conclusion, the eGFP-CHIKV reporter virus constructed in our study is stable and efficient for the detection and quantification of CHIKV replication. The neutralization assay based on this reporter virus is demonstrated to be rapid and specific for evaluating neutralizing antibodies of viruses. Importantly, the eGFP-CHIKV reporter virus derived from a new Asian strain, which was rarely reported previously, could facilitate the systematic studies among different lineages of CHIKV in a high-throughput manner.

Acknowledgments: We thank the Core Facility and Technical Support, Wuhan Institute of Virology, and the Wuhan Key Laboratory on Emerging Infectious Diseases and Biosafety for helpful support during the course of the work. This work was supported by the National Basic Research Program of China (Grants 2012CB518904) and the National Natural Science Foundation of China (Grant No. 81572003).

Author Contributions: Bo Zhang conceived and designed the experiments; Cheng-Lin Deng, Si-Qing Liu, Lin-Lin Xu, Xiao-Dan Li, Pan-Tao Zhang and Peng-Hui Li performed the experiments; Cheng-Lin Deng, Si-Qing Liu and Xiao-Dan Li analyzed the data; Dong-Gen Zhou, Cheng-Feng Qin, Hong-Ping Wei and Zhi-Ming Yuan contributed reagents and materials; Bo Zhang, Han-Qing Ye and Si-Qing Liu wrote the paper.

Conflicts of Interest: The authors declare no conflict of interest. The founding sponsors had no role in the design of the study; in the collection, analyses, or interpretation of data; in the writing of the manuscript, and in the decision to publish the results.

Abbreviations

CHIKV	Chikungunya virus
mAbs	monoclonal antibodies
NT	neutralization
MOI	multiplicity of infection
eGFP	enhanced green fluorescent protein
Gluc	Gaussia luciferase

References

1. Burt, F.J.; Rolph, M.S.; Rulli, N.E.; Mahalingam, S.; Heise, M.T. Chikungunya: A re-emerging virus. *Lancet* **2012**, *379*, 662–671.
2. Morrison, T.E. Reemergence of Chikungunya Virus. *J. Virol.* **2014**, *88*, 11644–11647.
3. Cruz, D.J.M.; Bonotto, R.M.; Gomes, R.G.B.; da Silva, C.T.; Taniguchi, J.B.; No, J.H.; Lombardot, B.; Schwartz, O.; Hansen, M.A.E.; Freitas-Junior, L.H. Identification of novel compounds inhibiting Chikungunya virus-induced cell death by high throughput screening of a kinase inhibitor library. *PLoS Neglect. Trop. Dis.* **2013**, *7*, e2471.
4. Ziegler, S.A.; Nuckols, J.; McGee, C.E.; Huang, Y.-J.S.; Vanlandingham, D.L.; Tesh, R.B.; Higgs, S. In vivo imaging of Chikungunya virus in mice and *Aedes* mosquitoes using a *Renilla* luciferase clone. *Vector Borne Zoonotic Dis.* **2011**, *11*, 1471–1477.

5. Weaver, S.C.; Reisen, W.K. Present and future arboviral threats. *Antivir. Res.* **2010**, *85*, 328–345.

6. Josseran, L.; Paquet, C.; Zehgnoun, A.; Caillere, N.; le Tertre, A.; Solet, J.-L.; Ledrans, M. Chikungunya disease outbreak, Reunion island. *Emerg. Infect. Dis.* **2006**, *12*, 1994–1995.

7. Tsetsarkin, K.A.; Vanlandingham, D.L.; McGee, C.E.; Higgs, S. A single mutation in Chikungunya virus affects vector specificity and epidemic potential. *PLoS Pathog.* **2007**, *3*, e201.

8. Dash, A.; Bhatia, R.; Sunyoto, T.; Mourya, D. Emerging and re-emerging arboviral diseases in Southeast Asia. *J. Vector Borne Dis.* **2013**, *50*, 77–84.

9. Zheng, K.; Li, J.; Zhang, Q.; Liang, M.; Li, C.; Lin, M.; Huang, J.; Li, H.; Xiang, D.; Wang, N.; et al. Genetic analysis of chikungunya viruses imported to mainland China in 2008. *Virol. J.* **2010**, *7*, 1–6.

10. Wu, D.; Zhang, Y.; Zhouhui, Q.; Kou, J.; Liang, W.; Zhang, H.; Monagin, C.; Zhang, Q.; Li, W.; Zhong, H.; et al. Chikungunya virus with E1-A226V mutation causing two outbreaks in 2010, Guangdong, China. *Virol. J.* **2013**, *10*, 1–9.

11. Zhang, Q.; He, J.; Wu, D.; Wang, Z.; Zhong, X.; Zhong, H.; Ding, F.; Liu, Z.; Wang, S.; Huang, Z.; et al. Maiden outbreak of Chikungunya in Dongguan City, Guangdong Province, China: Epidemiological characteristics. *PLoS ONE* **2012**, *7*, e42830.

12. García-Arriaza, J.; Cepeda, V.; Hallengärd, D.; Sorzano, C.Ó.S.; Kümmerer, B.M.; Liljeström, P.; Esteban, M. A novel poxvirus-based vaccine, MVA-CHIKV, is highly immunogenic and protects mice against Chikungunya infection. *J. Virol.* **2014**, *88*, 3527–3547.

13. Hallengärd, D.; Kakoulidou, M.; Lulla, A.; Kümmerer, B.M.; Johansson, D.X.; Mutso, M.; Lulla, V.; Fazakerley, J.K.; Roques, P.; le Grand, R.; et al. Novel attenuated Chikungunya vaccine candidates elicit protective immunity in C57BL/6 mice. *J. Virol.* **2014**, *88*, 2858–2866.

14. Roy, C.J.; Adams, A.P.; Wang, E.; Plante, K.; Gorchakov, R.; Seymour, R.L.; Vinet-Oliphant, H.; Weaver, S.C. Chikungunya vaccine candidate is highly attenuated and protects nonhuman primates against telemetrically monitored disease following a single dose. *J. Infect. Dis.* **2014**, *209*, 1891–1899.

15. Tretyakova, I.; Hearn, J.; Wang, E.; Weaver, S.; Pushko, P. DNA vaccine initiates replication of live attenuated Chikungunya virus in vitro and elicits protective immune response in mice. *J. Infect. Dis.* **2014**, *209*, 1882–1890.

16. Piper, A.; Ribeiro, M.; Smith, K.M.; Briggs, C.M.; Huitt, E.; Nanda, K.; Spears, C.J.; Quiles, M.; Cullen, J.; Thomas, M.E.; et al. Chikungunya virus host range E2 transmembrane deletion mutants induce protective immunity against challenge in C57BL/6J mice. *J. Virol.* **2013**, *87*, 6748–6757.

17. Wang, E.; Kim, D.Y.; Weaver, S.C.; Frolov, I. Chimeric Chikungunya viruses are nonpathogenic in highly sensitive mouse models but efficiently induce a protective immune response. *J. Virol.* **2011**, *85*, 9249–9252.

18. Selvarajah, S.; Sexton, N.R.; Kahle, K.M.; Fong, R.H.; Mattia, K.-A.; Gardner, J.; Lu, K.; Liss, N.M.; Salvador, B.; Tucker, D.F.; et al. A neutralizing monoclonal antibody targeting the acid-sensitive region in Chikungunya virus E2 protects from disease. *PLoS Neglect. Trop. Dis.* **2013**, *7*, e2423.

19. Smith, S.A.; Silva, L.A.; Fox, J.M.; Flyak, A.I.; Kose, N.; Sapparapu, G.; Khomandiak, S.; Ashbrook, A.W.; Kahle, K.M.; Fong, R.H.; et al. Isolation and characterization of broad and ultrapotent human monoclonal antibodies with therapeutic activity against Chikungunya virus. *Cell Host Microbe* **2015**, *18*, 86–95.

20. Kumar, M.; Sudeep, A.B.; Arankalle, V.A. Evaluation of recombinant E2 protein-based and whole-virus inactivated candidate vaccines against Chikungunya virus. *Vaccine* **2012**, *30*, 6142–6149.

21. Gläsker, S.; Lulla, A.; Lulla, V.; Couderc, T.; Drexler, J.F.; Liljeström, P.; Lecuit, M.; Drosten, C.; Merits, A.; Kümmerer, B.M. Virus replicon particle based Chikungunya virus neutralization assay using Gaussia luciferase as readout. *Virol. J.* **2013**, *10*, 1–10.

22. Kishishita, N.; Takeda, N.; Anuegoonpipat, A.; Anantapreecha, S. Development of a pseudotyped-lentiviral-vector-based neutralization assay for Chikungunya virus infection. *J. Clin. Microbiol.* **2013**, *51*, 1389–1395.

23. Zhang, C.-H.; Ma, W.-Q.; Yang, Y.-L.; Wang, H.-M.; Dong, F.-T.; Huang, Z.-X. Median effective effect-site concentration of sufentanil for wake-up test in adolescents undergoing surgery: A randomized trial. *BMC Anesthesiol.* **2015**, *15*, 1–4.

24. Sim, A.C.N.; Lin, W.; Tan, G.K.X.; Sim, M.S.T.; Chow, V.T.K.; Alonso, S. Induction of neutralizing antibodies against dengue virus type 2 upon mucosal administration of a recombinant *Lactococcus lactis* strain expressing envelope domain III antigen. *Vaccine* **2008**, *26*, 1145–1154.

25. Zheng, K.; Ding, G.-Y.; Zhou, H.-Q.; Xie, X.-M.; Li, X.-B.; Shi, Y.-X.; Su, J.-K.; Huang, J.-C. Rapid detection of Dengue virus and Chikungunya virus by multiplexreal-time RT-PCR assay with an internal control. *Chin. J. Zoonoses* **2013**, *29*, 242–247. (In Chinese)

26. Briolant, S.; Garin, D.; Scaramozzino, N.; Jouan, A.; Crance, J.M. In vitro inhibition of Chikungunya and Semliki Forest viruses replication by antiviral compounds: Synergistic effect of interferon-α and ribavirin combination. *Antivir. Res.* **2004**, *61*, 111–117.

27. Scholte, F.E.M.; Tas, A.; Martina, B.E.E.; Cordioli, P.; Narayanan, K.; Makino, S.; Snijder, E.J.; van Hemert, M.J. Characterization of synthetic Chikungunya viruses based on the consensus sequence of recent E1–226V isolates. *PLoS ONE* **2013**, *8*, e71047.

28. Schwartz, O.; Albert, M.L. Biology and pathogenesis of Chikungunya virus. *Nat. Rev. Microbiol.* **2010**, *8*, 491–500.

29. Weaver, S.C.; Osorio, J.E.; Livengood, J.A.; Chen, R.; Stinchcomb, D.T. Chikungunya virus and prospects for a vaccine. *Expert Rev. Vaccines* **2012**, *11*, 1087–1101.

30. Frolova, E.; Gorchakov, R.; Garmashova, N.; Atasheva, S.; Vergara, L.A.; Frolov, I. Formation of nsP3-specific protein complexes during Sindbis virus replication. *J. Virol.* **2006**, *80*, 4122–4134.

31. Kümmerer, B.M.; Grywna, K.; Gläsker, S.; Wieseler, J.; Drosten, C. Construction of an infectious Chikungunya virus cDNA clone and stable insertion of mCherry reporter genes at two different sites. *J. Gen. Virol.* **2012**, *93*, 1991–1995.

32. Patterson, M.; Poussard, A.; Taylor, K.; Seregin, A.; Smith, J.; Peng, B.-H.; Walker, A.; Linde, J.; Smith, J.; Salazar, M.; et al. Rapid, non-invasive imaging of alphaviral brain infection: Reducing animal numbers and morbidity to identify efficacy of potential vaccines and antivirals. *Vaccine* **2011**, *29*, 9345–9351.

33. Phillips, A.T.; Stauft, C.B.; Aboellail, T.A.; Toth, A.M.; Jarvis, D.L.; Powers, A.M.; Olson, K.E. Bioluminescent imaging and histopathologic characterization of WEEV neuroinvasion in outbred CD-1 mice. *PLoS ONE* **2013**, *8*, e53462.

34. Poussard, A.; Patterson, M.; Taylor, K.; Seregin, A.; Smith, J.; Smith, J.; Salazar, M.; Paessler, S. In vivo imaging systems (IVIS) detection of a neuro-invasive encephalitic virus. *J. Vis. Exp.* **2012**.

35. Tamberg, N.; Lulla, V.; Fragkoudis, R.; Lulla, A.; Fazakerley, J.K.; Merits, A. Insertion of EGFP into the replicase gene of Semliki Forest virus results in a novel, genetically stable marker virus. *J. Gen. Virol.* **2007**, *88*, 1225–1230.

36. Delogu, I.; Pastorino, B.; Baronti, C.; Nougairède, A.; Bonnet, E.; de Lamballerie, X. In vitro antiviral activity of arbidol against Chikungunya virus and characteristics of a selected resistant mutant. *Antivir. Res.* **2011**, *90*, 99–107.

37. Tsetsarkin, K.; Higgs, S.; McGee, C.E.; Lamballerie, X.D.; Charrel, R.N.; Vanlandingham, D.L. Infectious clones of Chikungunya virus (La Reunion isolate) for vector competence studies. *Vector Borne Zoonotic Dis.* **2006**, *6*, 325–337.

38. Vanlandingham, D.L.; Tsetsarkin, K.; Hong, C.; Klingler, K.; McElroy, K.L.; Lehane, M.J.; Higgs, S. Development and characterization of a double subgenomic Chikungunya virus infectious clone to express heterologous genes in *Aedes aegypti* mosquitoes. *Insect Biochem. Mol. Biol.* **2005**, *35*, 1162–1170.

39. Sun, C.; Gardner, C.L.; Watson, A.M.; Ryman, K.D.; Klimstra, W.B. Stable, high-level expression of reporter proteins from improved alphavirus expression vectors to track replication and dissemination during encephalitic and arthritogenic disease. *J. Virol.* **2014**, *88*, 2035–2046.

40. Tsetsarkin, K.A.; Chen, R.; Leal, G.; Forrester, N.; Higgs, S.; Huang, J.; Weaver, S.C. Chikungunya virus emergence is constrained in Asia by lineage-specific adaptive landscapes. *Proc. Natl. Acad. Sci. USA* **2011**, *108*, 7872–7877.

Newcastle Disease Virus as a Vaccine Vector for Development of Human and Veterinary Vaccines

Shin-Hee Kim and Siba K. Samal

Abstract: Viral vaccine vectors have shown to be effective in inducing a robust immune response against the vaccine antigen. Newcastle disease virus (NDV), an avian paramyxovirus, is a promising vaccine vector against human and veterinary pathogens. Avirulent NDV strains LaSota and B1 have long track records of safety and efficacy. Therefore, use of these strains as vaccine vectors is highly safe in avian and non-avian species. NDV replicates efficiently in the respiratory track of the host and induces strong local and systemic immune responses against the foreign antigen. As a vaccine vector, NDV can accommodate foreign sequences with a good degree of stability and as a RNA virus, there is limited possibility for recombination with host cell DNA. Using NDV as a vaccine vector in humans offers several advantages over other viral vaccine vectors. NDV is safe in humans due to host range restriction and there is no pre-existing antibody to NDV in the human population. NDV is antigenically distinct from common human pathogens. NDV replicates to high titer in a cell line acceptable for human vaccine development. Therefore, NDV is an attractive vaccine vector for human pathogens for which vaccines are currently not available. NDV is also an attractive vaccine vector for animal pathogens.

Reprinted from *Viruses*. Cite as: Kim, S.-H.; Samal, S.K. Newcastle Disease Virus as a Vaccine Vector for Development of Human and Veterinary Vaccines. *Viruses* **2016**, *8*, 183.

1. Introduction

Infectious diseases have been emerging and reemerging over millennia [1]. Human immunodeficiency virus (HIV), severe acute respiratory syndrome coronavirus (SARS-CoV), and the most recent 2009 pandemic H1N1 influenza virus are only a few of many examples of emerging infectious pathogens in the modern world [2]. Each of these diseases has global societal and economic impact related to unexpected illnesses and deaths, as well as interference with travel, business, and daily activities. To overcome emerging, reemerging, as well as stable infectious diseases, the demand for development of efficient vaccines has greatly increased. Historically, live attenuated vaccines have provided the most effective protection against viral infection and disease [3]. However, there have been safety concerns with the risk of reversion to the wild-type pathogen phenotype as shown with some traditional live attenuated vaccines such as the polio vaccine. Furthermore, development of live attenuated vaccines has not been successful for many important pathogens. On the other hand, inactivated vaccines are generally not very effective and require a high containment laboratory for cultivation of highly virulent pathogens. Also, there is a risk of incomplete inactivation for inactivated vaccines. Therefore, there is a need for an alternative approach for development of vaccines.

Replicating viral vector vaccines offer a live vaccine approach without requiring involvement of the complete pathogen or cultivation of the pathogen [4]. Replicating viral vectors have the ability to synthesize the foreign antigen intracellularly and induce humoral, cellular, and mucosal immune responses. Specifically, vectored vaccines can have advantages for (i) viruses for which a live attenuated vaccine might not be feasible (i.e., HIV); (ii) viruses that do not grow well in vitro (i.e., human papillomavirus, hepatitis C virus, and norovirus); (iii) highly pathogenic viruses that present safety challenges during vaccine development (i.e., SARS-CoV and Ebola virus); (iv) viruses that lose infectivity due to physical instability (i.e., respiratory syncytial virus (RSV)); and (v) viruses that can exchange genes with circulating viruses (i.e., coronaviruses, influenza viruses, and enteroviruses) [4]. A vectored vaccine can be rapidly engineered against a newly emerging pathogenic virus by inserting the gene of the protective antigen of the virus into the genome of the viral vector. In general, the magnitude of the immune response to live viral vector vaccines is substantially greater and broader than that induced by vaccines based on subunit proteins or inactivated viruses. Furthermore, manufacturing of vectored vaccines against highly pathogenic viruses do not require a high level of biosafety containment laboratories.

Newcastle disease virus (NDV) is a fast-replicating avian virus that is prevalent in all species of birds [5]. In most avian species, NDV infections do not result in disease. In chickens, NDV causes a highly contagious respiratory and neurologic disease, leading to severe economic losses in the poultry industry worldwide [6]. NDV strains vary widely in virulence. Based on the severity of the disease in chickens, NDV strains are classified into three pathotypes: lentogenic strains which cause mild or asymptomatic infections that are restricted to the respiratory tract; mesogenic strains which are of intermediate virulence; and velogenic strains which cause systemic infections with high mortality [5]. Naturally occurring low-virulent NDV strains, such as LaSota and B1, are widely used as live attenuated vaccines to control Newcastle disease in poultry. Although NDV primarily infects avian species, many non-avian species have also been shown to be naturally or experimentally susceptible to infection. The advent of a reverse genetics system to manipulate the genome of NDV not only allowed us to study the molecular biology and pathogenesis of NDV but also to develop NDV as a vaccine vector against diseases of humans and animals. NDV vector has several advantages over other replicating viral vectors.

Avirulent NDV strains are highly safe in avian and non-avian species. NDV replicates well in vivo and induces a robust immune response. In contrast to adeno, herpes, and pox virus vectors whose genome encodes a large number of proteins, NDV encodes only seven proteins and is thus less competition for immune responses between vector proteins and the expressed foreign antigen. NDV replicates in the cytoplasm, does not integrate into the host cell DNA, and does not establish persistent infection. Recombination involving NDV is extremely rare. NDV has a modular genome that facilitates genetic manipulation. NDV infects via the intranasal route and therefore induces both mucosal and systemic immune responses. A wide range of NDV strains exists that can be used as vaccine vectors. NDV-vectored vaccine can also be used as a "differentiating infected from vaccinated animals" (DIVA) vaccine. In this review article, we have reviewed the biology of NDV, development of reverse genetic systems for generation of NDV-vectored vaccines, and use of NDV vector for development of human and veterinary vaccines.

2. Biology of Newcastle Disease Virus (NDV)

NDV is a member of the genus *Avulavirus* in the family *Paramyxoviridae* [5]. NDV virions are pleomorphic, but mostly spherical with a diameter of 100 nm. The virion is enveloped with a bilayer lipid membrane. The genome of NDV is a non-segmented, negative-sense, single-stranded RNA of 15,186 to 15,198 nucleotides containing six transcriptional units (3'-N-P-M-F-HN-L-5') (Figure 1). The genome encodes a nucleocapsid protein (N), a phosphoprotein (P), a matrix protein (M), a fusion protein (F), a hemagglutinin-neuraminidase protein (HN), and a large polymerase protein (L). An additional protein called the V protein is produced by RNA editing of the P gene. The beginning and end of each gene contain control sequences, known as gene-start (GS) and gene-end (GE), respectively. The viral RNA-dependent RNA polymerase begins transcription at the 3' end of the genomic RNA, in a sequential manner by a stop-start mechanism [5]. The re-initiation of transcription at the GS is not perfect, thus leading to a gradient of mRNA abundance with high levels of mRNA transcription located at the 3' end. The genome length of NDV must be an even multiple of six for efficient virus replication following the "rule of six" [5].

In NDV, the HN and F proteins are the two integral membrane proteins. The HN protein is responsible for attachment of the virion to sialic acid containing cell surface receptors. The F protein mediates entry of the virus into the host cell by fusion of the viral envelope to the plasma membrane. The F protein is synthesized as a precursor (F0) that is cleaved by host cell protease into two biologically active F1 and F2 subunits. Cleavage of the F protein is a pre-requisite for virus entry and cell-to-cell fusion. The amino acid sequence at the F protein cleavage site has been identified as the primary determinant of virulence [7,8]. Virulent NDV strains have multibasic residues that conform to the preferred cleavage site of the intracellular protease furin present in most cell types. In contrast, avirulent NDV strains typically contain one or two basic residues at the F protein cleavage site and are delivered to the plasma membrane in an uncleaved form for cleavage by extracellular proteases, thus restricting viral replication to the respiratory and enteric tracts where secreted proteases for cleavage are available.

NDV genomic RNA

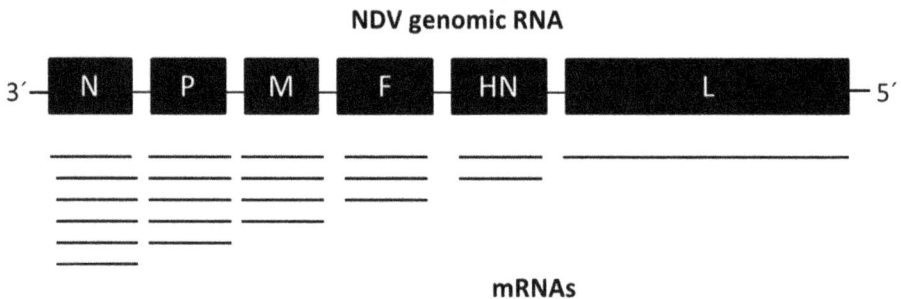

mRNAs

Figure 1. Genome organization and transcription scheme of Newcastle disease virus (NDV).

3. Construction of NDV-Vectored Vaccines

Infectious NDV can be recovered entirely from cloned cDNA by transfecting cultured cells with plasmids encoding the viral components of a functional nucleocapsid, full-length antigenomic RNA, and the major proteins involved in replication and transcription, i.e., the N, P, and L proteins under the control of bacteriophage T7 RNA polymerase promoter [5] (Figure 2). This method, which is also known as reverse genetics technique, is now available for all three pathotypes of NDV strains [9–12]. In general, a foreign gene flanked by NDV GS and GE sequences is inserted into a 3′ non-coding region of an NDV genome as an additional transcription unit. Due to a polar gradient transcription, foreign genes are expressed more efficiently when placed closer to 3′ end of the genome. Although a foreign gene can be placed between any two genes of NDV, the insertion site between the P and M genes has been found optimal for efficient expression of the foreign protein and replication of NDV [13–15]. The insertion of a foreign gene into NDV genome increases its genome length and gene number and often has a growth retardation effect on virus replication in vitro and in vivo [16]. NDV accommodates foreign genes (at least 4.5 kb in length) with a good degree of stability [5]. A single NDV vector can also express at least three different foreign genes.

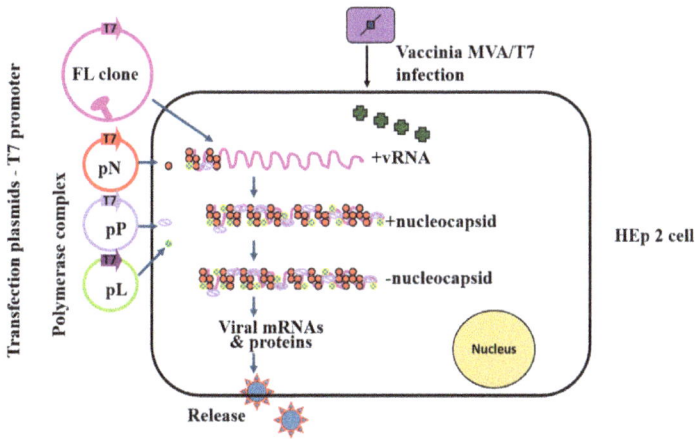

Figure 2. Plasmid-based recovery of recombinant NDV. HEp-2 cells are cotransfected with the antigenome plasmid and expression plasmids encoding the N, P and L proteins of NDV. The T7 RNA polymerase is provided by the recombinant vaccinia MVA/T7 strain.

Avirulent NDV strains LaSota and B1 are commonly used as vaccine vectors because of their proven track records of safety. Mesogenic and velogenic NDV strains are not used as vaccine vectors because they are virulent in chickens. In an experimental study, the mesogenic strain Beaudette C (BC) was evaluated as a vaccine vector in nonhuman primates [17]. Strain BC replicated to a higher titer and induced a substantially higher level of antibody response compared to strain LaSota, indicating it would be an effective vaccine vector.

4. NDV as a Vaccine Vector for Human Use

NDV has several advantages for use as a vaccine vector in humans. NDV is safe in humans, due to a natural host range restriction. In nonhuman primates, the intranasal and intratracheal inoculation of African green and rhesus monkeys with $10^{6.5}$ plaque-forming units (PFU) per site of NDV did not cause any disease symptoms and its replication was restricted to the respiratory tract [17]. In humans, infection by NDV appears to be limited and benign based on both anecdotal observations with bird handlers and in clinical studies using NDV as an oncolytic agent [5]. According to a clinical study for NDV as an oncolytic agent in humans, intravenous administration of 10^{10} PFU of NDV to humans was safe without causing adverse effects [18]. NDV shares only a low level of amino acid sequence identity with known human paramyxoviruses and are antigenically distinct from common human and animal pathogens, and thus would not be affected by preexisting immunity in humans. NDV infects via the intranasal route and has been shown to induce humoral and cellular immune responses both at the mucosal and systemic levels in murine and nonhuman primate models [13,17,19–26]. NDV is a strong stimulator of the host immune response, thus providing an adjuvant effect. The use of avirulent pathotypes of NDV in humans prevents the possibility of accidental spread of a virulent virus strain from treated patients to birds. NDV grows to high titers not only in embryonated eggs (10^9 PFU/mL) but also in Vero cells (10^8 PFU/mL), which is acceptable for human vaccine development. In fact, NDV has been used to express protective antigens of various human pathogens and has shown promising results in nonhuman primates. NDV-vectored vaccines for several human pathogens are discussed as follows (Table 1).

The potential of recombinant NDV strain B1 as an effective vaccine vector for humans was first evaluated by expressing an influenza virus (A/WSN/33) hemagglutinin (HA) protein [13]. The expressed HA protein was incorporated into virions and appeared to be cleaved, indicating that the HA protein was accessible to proteolytic enzymes. In vitro growth kinetics and pathogenicity test in embryonated chicken eggs indicated attenuation of the recombinant NDV. Intravenous administration of mice induced higher titers of antibody to influenza virus HA than intraperitoneal administration. Further, immunized mice by the intravenous route were completely protected against a lethal dose of influenza virus, suggesting that NDV can be a safe and effective vaccine vector for possible use in mammalian and avian species.

The potential of NDV as a vaccine vector for use in humans was first determined in nonhuman primates. Two NDV strains LaSota and BC were evaluated as vaccine vectors in nonhuman primates by inserting the HN protein of human parainfluenza virus type 3 (HPIV3) as a protective antigen [17]. Two doses of immunization with NDV strains confirmed their restricted replication in African green monkeys (NDV-BC and NDV-LS) and in rhesus monkeys (NDV-BC only). However, the serum antibody response following the second dose exceeded that observed with HPIV3 infection, even though HPIV3 replicated much more efficiently than NDV in these animals. This is the first study to demonstrate efficacy of NDV-vectored vaccine in nonhuman primates.

Table 1. NDV-vectored vaccines against human pathogens.

Pathogen	Disease	Antigen	NDV Strain	Animal Model	Route of Inoculation *	Dose	Ref.
Influenza A H1N1	Respiratory	HA	B1	Mouse	iv, ip	3×10^7 PFU; two doses	[13]
HPAIV H5N1	Respiratory	HA	LaSota	African green monkey	in	2×10^7 PFU; two doses	[21]
HPIV3	Respiratory	HN	LaSota BC	African green and rhesus monkeys	in, it	$10^{6.5}$ PFU; two doses	[17]
EBOV	Hemorrhagic fever	GP	LaSota	Rhesus monkeys	in, it	Two doses	[19]
SARS-CoV	Respiratory	S	LaSota	African green monkeys	in, it	Two doses	[20]
SIV	simian AIDS	Gag	B1	Mouse	iv, ip, in	5×10^7 PFU; one or two doses	[27]
HIV	AIDS	Gag	B1	Mouse	in	5×10^5 PFU; two doses	[14]
HIV	AIDS	Env	LaSota	Guinea pig	in, im	10^6 PFU; two or three doses	[28]
HIV	AIDS	Env Gag	LaSota	Guinea pig	in	10^6 PFU; two or three doses	[29]
HIV	AIDS	Env (gp160, gp120)	LaSota	Guinea pig	in	10^6 PFU; three doses	[30]
HRSV	Respiratory	F	B1	Mouse	in	$10^{5.5}$ PFU; one dose	[31]
NiV	Encephalitis	F/G	LaSota	Mouse Pig	im	10^9 EID$_{50}$; two doses	[32]
Borrelia burgdorferi	Lyme	BmpA OspC	LaSota	Hamster	in, im, ip	10^6 PFU	[33]
NoV	Gastroenteritis	VP1	LaSota BC	Mouse	in	10^6 EID$_{50}$; two doses	[25]

* im: intramuscular, in: intranasal, ip: intraperitoneal, it: intratracheal, iv: intravenous, EID$_{50}$: 50% embryo infective dose.

Ebola virus (EBOV) causes severe hemorrhagic fever in humans with a fatality rate of up to 88% (species *Zaire ebolavirus*) of infected individuals [19]. Due to the limitation of inactivated vaccines, viral vectors based on common human pathogens have been used for EBOV vaccine. To overcome the high seroprevalence against vectors based on common human pathogens in the adult human population, recombinant NDV strain LaSota expressing the EBOV GP envelope protein was generated to evaluate its potential as a vaccine for EBOV [19]. Following one intranasal and intratracheal inoculation of rhesus monkeys with NDV/GP, titers of EBOV-specific antibodies and serum EBOV-neutralizing antibodies, were undetectable or low compared to those induced by HPIV3/GP. However, a second immunization resulted in a substantial boost in serum immunoglobulin (Ig) G enzyme-linked immunosorbent assay (ELISA) titers, yet the titers remained lower than those induced by a second dose of HPIV3/GP. In contrast, the ELISA IgA titers in respiratory tract secretions and the serum EBOV-neutralizing antibody titers were equal to those induced after the second dose of HPIV3/GP, showing that the efficacy of NDV vector can be comparable to that of HPIV3 vector by prime-boosting vaccination [23].

NDV was evaluated as a vaccine vector for another important emerging pathogen, the severe acute respiratory syndrome-associated coronavirus (SARS-CoV) [20]. Two NDV vectors were constructed: mesogenic strain BC (NDV-BC) and lentogenic strain LaSota in which the F protein cleavage sequence was modified to that of strain BC (NDV-VF) [20]. These NDV vectors were engineered to express the SARS-CoV spike S glycoprotein, the major protective antigen. Two dose immunizations of African green monkeys induced a robust neutralizing antibody response, resulting in reduction of virus shedding after challenge with SARS-CoV (10^6 50 % Tissue Culture Infective Dose (TCID$_{50}$)). Specifically, immunization with NDV-VF vector resulted in SARS-CoV titers of a 5-fold, 61-fold, and 236-fold reduction in nasal turbinate, trachea, and lung, respectively, compared with the control animals. The NDV-BC vector was even more effective, with average reductions in viral titer of 13-fold, 276-fold, and 1102-fold in the nasal turbinate, trachea, and lung, respectively. This study demonstrated the safety and protective efficacy of NDV as a topical respiratory vaccine vector for SARS-CoV.

The use of viral vectors expressing selected HIV antigens has been a promising vaccine strategy. The potential of NDV-vectored vaccine against HIV infection was first evaluated by generating recombinant NDV expressing simian immunodeficiency virus (SIV) Gag protein (rNDV/SIVgag) [27]. The vaccine virus induced Gag-specific cellular immune responses in mice. Among intravenous, intraperitoneal, and intranasal immunization routes, intranasal administration induced the strongest protective immune response against a surrogate challenge virus (rVac/SIVgag) following a booster immunization with recombinant influenza viruses expressing immunogenic portions of SIV Gag. Specifically, this heterologous vaccination approach resulted in approximately, a 10^6-fold reduction in rVac/SIVgag titers at day 5 after challenge compared to titers of control mice injected with phosphate-buffered saline (PBS). The magnitude of the protective immune response also correlated with the levels of cellular immune responses to Gag. These results suggest that NDV vector can be a suitable candidate vaccine against HIV.

The HIV Gag and Env proteins have been expressed by NDV vector [14,28–30]. The expression level of Gag protein was optimized using different insertion sites in the NDV genome. It was found that the codon-optimized Gag inserted between the P and M genes of NDV induced the highest level of protein expression and an enhanced immune response against HIV Gag in mice [14]. In another study, expression of gp160 Env protein by NDV vector LaSota also induced systemic and mucosal antibody responses in guinea pigs [28]. Priming/boosting by the intranasal route was more immunogenic than by the intramuscular route. Further, coexpression of gp160 Env and p55 Gag by vector LaSota enhanced both Env-specific and Gag-specific immune responses in guinea pigs [29]. This approach was efficient in inducing cellular and protective immune responses to challenge with vaccinia viruses expressing HIV-1 Env and Gag in mice. These results suggest that vaccination with a single NDV vector coexpressing Env and Gag represents a promising strategy to enhance immunogenicity and protective efficacy against HIV. In addition, heterologous prime (NDV expressing gp160) and boosting (purified gp120 protein) approach induced high neutralizing antibody titer in guinea pigs [30]. These findings suggest that vaccination with multiple HIV antigens in combination can broaden antiviral immune responses.

Respiratory syncytial virus (RSV) is a major cause of severe lower respiratory tract disease in infants and elderly [31]. The development of an effective vaccine against RSV is a high priority. In order to develop a vector vaccine against RSV, NDV strain B1 was used to express the fusion glycoprotein of RSV [31]. NDV was chosen as a viral vector because of its ability to induce a strong interferon (IFN)-α/β response. The RSV F protein was more immunogenic when presented by NDV-F than by live RSV, and this correlated with an increased ability of NDV to activate antigen-presenting cells in vitro and to induce high levels of IFN-α/β in vivo. RSV F-specific, CD8+ memory T cells were present in greater numbers in NDV-F-primed BALB/c mice than in animals previously infected with RSV. Consequently, NDV vaccine virus provided protection from RSV challenge. This study also highlights the adjuvant effect of NDV vector mediated by the potent IFN induction.

NDV has also been used as a vector to express the immunogens of a bacterial pathogen. Lyme borreliosis is a prevalent vector-borne disease in the United States, Europe and parts of Asia. NDV was used to express the basic membrane protein A (BmpA) and the outer surface protein C (OspC) of the Lyme disease pathogen *Borrelia burgdorferi* [33]. C3H or Balb/C mice that were immunized intranasally with the NDV vectors mounted vigorous serum antibody responses against the NDV vector, but failed to mount a robust response against either the intracellular or extracellular forms of BmpA or OspC. In contrast, a single immunization of hamsters with the NDV vectors via the intranasal, intramuscular, or intraperitoneal route resulted in rapid and rigorous antibody responses against the BmpA and OspC. Challenged with *B. burgdorferi* (10^8 cells/animal), immunization with vector-expressing BmpA provided a reduction of the pathogen load in the joints. This study showed the potential of NDV as a vaccine vector against bacterial pathogens.

Nipah virus (NiV) is a deadly emerging zoonotic pathogen that causes fatal encephalitis in humans and pigs [32]. The glycoprotein (G) and fusion protein (F) are two major NiV surface glycoproteins that stimulate protective immune responses. NDV strain LaSota expressing the NiV G and F proteins (rLa-NiVG and rLa-NiVF, respectively) were evaluated for their immunogenicity in mice and in pigs [32]. Following the second dose of immunization, rLa-NiVG and rLa-NiVF induced NiV-specific neutralizing antibodies in mice and long-lasting neutralizing antibodies in pigs (at least for 21 weeks). This study also showed that rLa-NiVG induced higher levels of neutralizing antibodies than rLa-NiVF. Although the protective efficacy of the vaccines was not evaluated in this study, the vaccine viruses showed the potential to be used for protecting humans and animals against NiV infection.

Norovirus (NoV) is the most frequent cause of viral gastroenteritis in people of all ages [34]. The inability of NoV to grow in the cell culture system has greatly hindered development of effective vaccines. To circumvent this obstacle, virus-like particles (VLPs) produced by the baculovirus expression system have been commonly used as NoV vaccine candidates. As a live vaccine vector, LaSota and modified BC strains were used to express the capsid protein (VP1) of NoV strain VA387 (GII.4) and Norwalk virus (GI.1) [25,26]. For the modified BC vector, the multibasic cleavage site sequence of the F gene was changed to that of strain LaSota. The NoV-expressed VP1 protein formed VLPs in cell culture and in allantoic fluid of embryonated chicken eggs. The modified BC-vectored vaccine induced higher levels of serum, cellular, and mucosal immune responses than the baculovirus-expressed VLPs in mice. These results suggested that NDV has great potential for developing a live NoV vaccine. Alternatively, VLPs produced in large quantities in embryonated eggs or in cell culture by NDV can be a cost-effective method for producing a VLP-based vaccine for humans. This study also has implications for development of NDV-vectored vaccines for other non-cultivable pathogens of humans.

5. NDV as a Vaccine Vector for Veterinary Use

NDV-vectored vaccines have been evaluated in several animal species (i.e., chicken, cattle, sheep, cat, mouse, pig, and dog) for veterinary use [35] (Table 2).

NDV is a natural vaccine vector for poultry pathogens. Live attenuated NDV vaccines are widely used all over the world. Therefore, an NDV vector carrying the protective antigen of another avian pathogen can be used as a bivalent vaccine. Such a vaccine will be economical for poultry farmers.

As a bivalent vaccine, the NDV strain LaSota was first used to express the host-protective immunogen VP2 of infectious bursal disease virus (IBDV), a birnavirus, which causes a highly immunosuppressive disease in chickens [36]. The protective efficacy of LaSota-expressing VP2 protein was evaluated by challenging vaccinated chickens with a highly virulent NDV strain Texas GB or a virulent IBDV variant strain. Vaccination with rLaSota/VP2 provided 90% protection against NDV and IBDV. Booster immunization induced higher levels of antibody responses against both NDV and IBDV and conferred complete protection against both viruses. These results indicate that the recombinant NDV can be used as a vaccine vector for other avian pathogens.

Infectious laryngotracheitis is a major respiratory disease in chickens and caused by infectious laryngotracheitis virus (ILTV), a herpes virus [47]. Bivalent NDV-vectored vaccines against ILTV have been developed to improve the safety of current live attenuated ILTV vaccines [38,48]. The protective efficacy of NDVs expressing the three major ILTV surface glycoproteins, namely, gB, gC, and gD was evaluated against ILTV infection in chickens [38]. Particularly, rNDV-expressing gD induced the highest level of neutralizing antibodies among the tested vaccine candidates and completely protected chickens against the challenge of virulent ILTV and NDV, showing its potential as a bivalent vaccine. This protective efficacy of rNDV gD vaccine was attributed to high levels of envelope incorporation and cell surface expression of gD compared to gB and gC. In another study, LaSota viruses expressing gB and gD of ILTV were generated and vaccination of chickens with the two viruses conferred protection against virulent ILTV and NDV challenges [48]. In addition, LaSota with gB showed the protection of commercial broilers against clinical disease. Discrepancy of vaccine efficacy between these two studies could be due to the different levels of gB and gD expressions by NDV vectors and experimental conditions.

Table 2. NDV-vectored vaccines against veterinary pathogens.

Pathogen	Disease	Antigen	NDV Strain	Animal Model	Route of Inoculation *	Dose	Ref.
IBDV	Immunosuppressive disease	VP2	LaSota	Chicken	o	10^4 EID$_{50}$; two doses	[36]
CDV	Canine distemper	F/HN	LaSota	Mink	im	10^9 EID$_{50}$; two doses	[37]
ILTV	Respiratory	gB, gC, gD	LaSota	Chicken	on	10^6 TCID$_{50}$	[38]
IBV	Respiratory	S2	LaSota	Chicken	o	$10^{9.8}$ EID$_{50}$; two doses	[39]
RVFV	Rift Valley fever	Gn	LaSota	Calf	im	10^7 TCID$_{50}$; two doses	[40]
RVFV	Rift Valley fever	Gn, Gc	LaSota	Mouse; Lamb	im	$10^{5.3}$TCID$_{50}$; $10^{7.3}$ TCID$_{50}$; two doses	[41]
BHV-1	Respiratory	gD	LaSota	Calf	in	10^6 PFU; one dose	[42]
RV	Rabies	G	LaSota	Cat; Dog	im	$10^{9.8}$ EID$_{50}$; $10^{8.3}$ EID$_{50}$; three doses	[43]
HPAIV H5N1	Respiratory	HA	LaSota	Chicken	in	10^6 EID$_{50}$; one dose	[44,45]
HPAIV H7N2	Respiratory	HA	B1	Chicken	o	10^6 EID$_{50}$; one dose	[46]

* o: ocular, on: oculonasal, im: intramuscular, in: intranasal, it: intratracheal, EID$_{50}$: 50% embryo infectious dose, TCID$_{50}$: 50% tissue culture infective dose.

Infectious bronchitis virus (IBV), a coronavirus, is an important avian pathogen, causing respiratory disease in broilers and poor egg production in breeders and layers worldwide [49]. The spike polypeptide S2 gene was expressed by LaSota (rLS/IBV.S2) [39]. The vaccine virus effectively elicited hemagglutination inhibition antibodies against NDV and protected chickens against lethal challenge with virulent strain NDV/CA02. IBV heterotypic protection was assessed using a prime-boost approach with a commercially available attenuated IBV Massachusetts (Mass)-type vaccine. Chickens primed ocularly with rLS/IBV.S2 and boosted with Mass were completely protected against challenge with a virulent Ark-type strain. The protective efficacy of this heterologous vaccination was similar to that of priming and boosting with Mass (Mass + Mass). Based on clinical signs, both vaccinated groups appeared equally protected against challenge compared to unvaccinated challenged chickens. In shedding of challenge virus in the trachea, viral RNA was detected in 50% of rLS/IBV.S2 + Mass-vaccinated chickens while chickens vaccinated with Mass + Mass and unvaccinated challenged controls showed 84% and 90% incidence of IBV RNA detection, respectively. These results demonstrate the potential of NDV-vectored vaccine for IBV infection.

NDV vector has also been used for the prevention of economically important livestock diseases. Rift Valley fever virus (RVFV), a bunyavirus, causes recurrent large outbreaks in humans and in livestock [50]. NDV expressing the RVFV structural glycoproteins Gn was generated (NDFL-Gn) [40]. Immunization of calves via the intranasal route elicited no detectable antibody responses, whereas intramuscular immunization elicited antibodies against both NDV and the Gn protein. In general, the titers of RVFV-neutralizing antibodies were modest, varying from 8 to 32. To improve the efficacy of NDV-vectored vaccine, Gn was coexpressed with another glycoprotein Gc [41], which resulted in the formation of VLPs and subsequent release from the producing cells. A homologous prime-boost vaccination of mice with this vaccine virus induced neutralizing antibodies and provided complete protection from a lethal RVFV challenge. The immunogenicity of the vaccine virus was further evaluated in lamb, the main target species of RVFV. A single intramuscular vaccination induced neutralizing antibodies, and this response was significantly boosted by a second vaccination. Although coexpression of the Gn and Gc induced a good immune response, protective efficacy of this vaccine needs to be further evaluated.

Bovine herpesvirus-1 (BHV-1) is a major cause of respiratory tract diseases in cattle. Since modified live BHV-1 vaccines can cause latent infection in immunized animals, NDV expressing the glycoprotein D (gD) of BHV-1 was generated as a vectored vaccine [42]. A single intranasal and intratracheal inoculation of calves with NDV elicited mucosal and systemic antibodies specific to BHV-1. Challenge with BHV-1 showed reduced virus shedding and clinical signs in immunized calves compared to unimmunized claves. In addition, the titers of serum antibodies specific to BHV-1 were higher in immunized animals compared to unimmunized animals, indicating that the vaccines primed for secondary responses. This indicates that NDV can be used as a vaccine vector in bovines, and BHV-1 gD may be useful as a mucosal vaccine against BHV-1 infection. However, vaccination might require augmentation by a second dose or the inclusion of additional BHV-1 antigens.

Rabies virus (RV), a rhabdovirus, causes a fatal neurologic disease in humans and in animals [43]. To generate an effective, safe, and affordable rabies vaccine, NDV strain LaSota expressing the rabies virus glycoprotein G (rL-RVG) was evaluated. The safety of rL-RVG vaccine virus was confirmed in cats and dogs. Intramuscular vaccination with rL-RVG induced strong and long-lasting protective neutralization antibody responses against rabies virus in dogs and cats. Although three doses of vaccination were conducted, the second dose induced the highest levels of immune responses in both cats and dogs. Vaccination dose of 10 50% embryo infective dose (EID) completely protected dogs from challenge after one year. This study demonstrated protective efficacy of NDV-vectored vaccine against rabies in dogs. This vaccine may also have potential use in high-risk human individuals to control rabies virus infections.

Canine distemper virus (CDV), a morbillivirus, infects many carnivores and cause several high-mortality disease outbreaks [37]. The current CDV live vaccine cannot be safely used in some exotic species, such as mink and ferret. NDV strain LaSota expressing envelope glycoproteins, hemagglutinin (H, rLa-CDVH) and fusion protein (F, rLa-CDVF), were generated as vaccine candidates. In immunized minks, rLa-CDVH induced higher titers of neutralization antibodies against CDV than rLa-CDVF neutralizing antibodies. Further, rLa-CDVH provided complete protection against virulent CDV challenge during the four weeks of observation. In contrast, all animals immunized with rLa-CDVF developed clinical signs of distemper and virus shedding. This study suggested that recombinant NDV expressing the H protein of CDV is a safe and efficient candidate vaccine against CDV in mink. The efficacy of rLa-CDVH virus also needs to be evaluated in other host carnivore species.

6. NDV-Vectored Vaccines against Highly Pathogenic Avian Influenza Virus

Highly pathogenic avian influenza virus (HPAIV) is an economically important pathogen of poultry worldwide. The outbreaks involving H5N1 or H7N7 influenza viruses resulted in lethal infections in poultry and the death of a limited number of people [51]. Therefore, vaccination of poultry against HPAIV could play an important role in reducing virus shedding and raising the threshold for infection and transmission [46]. However, development of vaccines against HPAIV has been hampered due to poor immunogenicity of the virus [52]. Furthermore, inactivated vaccines are not commonly used because of the high cost due to the requirement of enhanced biosafety level 3 containment and the difficulty in "differentiating infected from vaccinated animals" (DIVA). The use of live attenuated influenza viruses as vaccines in avian or mammalian species can also raise a major biosafety concern, because the vaccine viruses may become virulent through mutation or genetic reassortment with circulating strains. Alternatively, NDV can be an ideal vaccine vector for development of an avian influenza vaccine. NDV infects via the intranasal route and therefore induces both local and systemic immune responses at the respiratory tract [5]. Therefore, it provides a convenient platform for rapid, efficient, and economical immunization. In fact, NDV has been most commonly used as a vaccine vector against AIV. Protective efficacy of NDV-vectored vaccines has been evaluated and verified by many different vaccination studies [21,22,44–46,53,54].

For the generation of vaccines, a major protective antigen, hemagglutinin (HA) of HPAIV has been placed between the P and M genes or between the F and HN genes in lentogenic NDV strains LaSota or B1. To address a safety concern, an NDV-vectored vaccine was further generated by replacing the polybasic cleavage site in HPAIV HA with that from a low-pathogenicity strain of influenza virus [22]. In addition, the HA gene has been modified to enhance its expression levels by NDV. Specifically, elimination of an NDV transcription termination signal-like sequence located within the HA open reading frame of H5 enhanced expression levels of HA protein by NDV and completely protected chickens after challenge with a lethal dose of velogenic NDV or highly pathogenic AIV, respectively [44]. In addition, the ectodomain of an H7N7 or H5N1 avian influenza virus HA was fused with the transmembrane and cytoplasmic domains derived from the F protein of NDV [46,53]. This approach resulted in enhanced incorporation of the foreign protein into virus particles and the protection of chickens against both HPAIV and a highly virulent NDV. These studies also demonstrated that NDV can be used to generate a bivalent vaccine.

Although use of avirulent NDV vectors has been effective in protecting chickens against clinical disease and mortality, some studies also found virus shedding in chickens after challenge with HPAIV [45]. To enhance the replication of vaccine virus, attenuated mesogenic NDV strain BC has been generated by changing the multibasic cleavage site sequence of the F protein to the dibasic sequence of strain LaSota [54]. Additionally, the BC, F, and HN proteins were modified in several ways to enhance virus replication. The modified BC-based vectors replicated better than LaSota vector, and expressed higher levels of HA protein and provided complete protection against challenge virus shedding, suggesting its potential to be safely used as a vaccine vector.

For effective human vaccines against HPAIV, the immunogenicity of NDV expressing the HA of H5N1 was evaluated in African green monkeys by the intranasal route of administration [21]. Two doses of NDV-vectored vaccine

(2×10^7 PFU) induced a high titer of H5N1 HPAIV-neutralizing serum antibodies in all of the immunized monkeys. Moreover, a substantial mucosal IgA response was induced in the respiratory tract, which can potentially reduce or prevent transmission of the virus during an outbreak or a pandemic. The intranasal route of administration is also advantageous for needle-free immunization and is thus suitable for mass immunization. The protective efficacy of vaccine viruses was evaluated in African green monkeys by the intranasal/intratracheal route or by the aerosol route of administration [22]. Each of the vaccine constructs was highly restricted for replication, with only low levels of virus shedding detected in respiratory secretions. All groups developed high levels of neutralizing antibodies against homologous (A/Vietnam/1203/04) and heterologous (A/egret/Egypt/1162-NAMRU3/06) strains of HPAIV and were protected against challenge with 2×10^7 PFU of homologous HPAIV. This study demonstrated that needle-free, highly attenuated NDV-vectored vaccines were immunogenic and protective in a nonhuman primate model of HPAIV infection.

7. Conclusions

Newcastle disease virus (NDV) is an attractive vaccine vector for both human and animal pathogens. The live attenuated vaccine strains used as vaccine vectors have a proven track record of safety and efficacy. NDV vectors not only induce robust humoral and cellular immune responses but also induce mucosal immune response. Therefore, NDV can be a vector of choice for mucosal immunization. The ability of NDV to infect a wide variety of non-avian species makes it a potential vector for other animals. NDV is also a promising vaccine vector for use in humans. One advantage is that most humans do not have pre-existing immunity to NDV. NDV-vectored vaccines have also become available commercially (i.e., H5N1 HPAIV vaccine for poultry).

Author Contributions: S.H.K. and S.K.S. wrote the manuscript.

Conflicts of Interest: The authors declare no conflict of interest.

References

1. Morens, D.M.; Fauci, A.S. Emerging infectious diseases: Threats to human health and global stability. *PLoS Pathog.* **2013**, *9*, e1003467.
2. Fauci, A.S.; Morens, D.M. The perpetual challenge of infectious diseases. *N. Engl. J. Med.* **2012**, *366*, 454–461.
3. Belshe, R.B.; Edwards, K.M.; Vesikari, T.; Black, S.V.; Walker, R.E.; Hultquist, M.; Kemble, G.; Connor, E.M.; CAIV-T Comparative Efficacy Study Group. Live attenuated versus inactivated influenza vaccine in infants and young children. *N. Engl. J. Med.* **2007**, *356*, 685–696.

4. Bukreyev, A.; Collins, P.L. Newcastle disease virus as a vaccine vector for humans. *Curr. Opin. Mol. Ther.* **2008**, *10*, 46–55.

5. Samal, S.K. Newcastle disease and related avian paramyxoviruses. In *The Biology of Paramyxoviruses*; Samal, S.K., Ed.; Caister Academic Press: Norfolk, UK, 2011; pp. 69–114.

6. Alexander, D.J. Newcastle disease and other avian paramyxoviruses. *Rev. Sci. Tech.* **2000**, *19*, 443–462.

7. Peeters, B.P.H.; de Leeuw, O.S.; Koch, G.; Gielkens, A.L.J. Rescue of Newcastle disease virus from cloned cDNA: Evidence that cleavability of the fusion protein is a major determinant for virulence. *J. Virol.* **1999**, *73*, 5001–5009.

8. Panda, A.; Huang, Z.; Elankumaran, S.; Rockemann, D.D.; Samal, S.K. Role of fusion protein cleavage site in the virulence of Newcastle disease virus. *Microb. Pathog.* **2004**, *36*, 1–10.

9. Krishnamurthy, S.; Huang, Z.; Samal, S.K. Recovery of a virulent strain of Newcastle disease virus from cloned cDNA: Expression of a foreign gene results in growth retardation and attenuation. *Virology* **2000**, *278*, 168–182.

10. Huang, Z.; Krisnamurthy, S.; Panda, A.; Samal, S.K. High-level expression of a foreign gene from the 3′ proximal first locus of a recombinant Newcastle disease virus. *J. Gen. Virol.* **2001**, *82*, 1729–1736.

11. Paldurai, A.; Kim, S.H.; Nayak, B.; Xiao, S.; Collins, P.L.; Samal, S.K. Evaluation of the contributions of the individual viral genes to Newcastle disease virulence and pathogenesis. *J. Virol.* **2014**, *88*, 8579–8596.

12. Xiao, S.; Nayak, B.; Samuel, A.; Paldurai, A.; Kanabagattebasavarajappa, M.; Collins, P.L.; Samal, S.K. Generation by reverse genetics of an effective, stable, live-attenuated Newcastle disease virus vaccine based on a currently circulating, highly virulent Indonesian strain. *PLoS ONE* **2012**, *7*, e52751.

13. Nakaya, T.; Cros, J.; Park, M.S.; Nakaya, Y.; Zheng, H.; Sagrera, A.; Villar, E.; García-Sastre, A.; Palese, P. Recombinant Newcastle disease virus as a vaccine vector. *J. Virol.* **2001**, *75*, 11868–11873.

14. Carnero, E.; Li, W.; Borderia, A.V.; Moltedo, B.; Moran, T.; García-Sastre, A. Optimization of human immunodeficiency virus Gag expression by Newcastle disease virus vectors for the induction of potent immune responses. *J. Virol.* **2009**, *83*, 584–597.

15. Zhao, H.; Peeters, B.P.H. Recombinant Newcastle disease virus as a viral vector: Effect of genomic location of foreign gene on gene expression and virus replication. *J. Gen. Virol.* **2013**, *84*, 781–788.

16. Bukreyev, A.; Skiadopoulos, M.H.; Murphy, B.R.; Collins, P.L. Nonsegmented negative-strand viruses as vaccine vectors. *J. Virol.* **2006**, *80*, 10293–10306.

17. Bukreyev, A.; Huang, Z.; Yang, L.; Elankumaran, S.; Claire, M.; Murphy, B.R.; Samal, S.K.; Collins, P.L. Recombinant Newcastle disease virus expressing a foreign viral antigen is attenuated and highly immunogenic in primates. *J. Virol.* **2005**, *79*, 13275–13284.

18. Freeman, A.I.; Zakay-Rones, Z.; Gomori, J.M.; Linetsky, E.; Rasooly, L.; Greenbaum, E.; Rozenman-Yair, S.; Panet, A.; Libson, E.; Irving, C.S.; et al. Phase I/II trial of intravenous NDV-HUJ oncolytic virus in recurrent glioblastoma multiforme. *Mol. Ther.* **2006**, *13*, 221–228.

19. Bukreyev, A.; Rollin, P.E.; Tate, M.K.; Yang, L.; Zaki, S.R.; Shieh, W.J.; Murphy, B.R.; Collins, P.L.; Sanchez, A. Successful topical respiratory tract immunization of primates against Ebola virus. *J. Virol.* **2007**, *81*, 6379–6388.

20. DiNapoli, J.M.; Kotelkin, A.; Yang, L.; Elankumaran, S.; Murphy, B.R.; Samal, S.K.; Collins, P.L.; Bukreyev, A. Newcastle disease virus, a host range-restricted virus, as a vaccine vector for intranasal immunization against emerging pathogens. *Proc. Natl. Acad. Sci. USA* **2007**, *104*, 9788–9793.

21. DiNapoli, J.M.; Yang, L.; Suguitan, A.; Elankumaran, S.; Dorward, D.W.; Murphy, B.R.; Samal, S.K.; Collins, P.L.; Bukreyev, A. Immunization of primates with a Newcastle disease virus-vectored vaccine via the respiratory tract induces a high titer of serum neutralizing antibodies against highly pathogenic avian influenza virus. *J. Virol.* **2007**, *81*, 11560–11568.

22. DiNapoli, J.M.; Nayak, B.; Yang, L.; Finneyfrock, B.W.; Cook, A.; Andersen, H.; Torres-Velez, F.; Murphy, B.R.; Samal, S.K.; Collins, P.L.; et al. Newcastle disease virus-vectored vaccines expressing the hemagglutinin or neuraminidase protein of H5N1 highly pathogenic avian influenza virus protect against virus challenge in monkeys. *J. Virol.* **2010**, *84*, 1489–1503.

23. Dinapoli, J.M.; Yang, L.; Samal, S.K.; Murphy, B.R.; Collins, P.L.; Bukreyev, A. Respiratory tract immunization of non-human primates with a Newcastle disease virus-vectored vaccine candidate against Ebola virus elicits a neutralizing antibody response. *Vaccine* **2010**, *29*, 17–25.

24. Ge, J.; Deng, G.; Wen, Z.; Tian, G.; Wang, Y.; Shi, J.; Wang, X.; Li, Y.; Hu, S.; Jiang, Y.; et al. Newcastle disease virus-based live attenuated vaccine completely protects chickens and mice from lethal challenge of homologous and heterologous H5N1 avian influenza viruses. *J. Virol.* **2007**, *81*, 150–158.

25. Kim, S.H.; Chen, S.; Jiang, X.; Green, K.Y.; Samal, S.K. Newcastle disease virus vector producing human norovirus-like particles induces serum, cellular, and mucosal immune responses in mice. *J. Virol.* **2014**, *88*, 9718–9727.

26. Kim, S.H.; Chen, S.; Jiang, X.; Green, K.Y.; Samal, S.K. Immunogenicity of Newcastle disease virus vectors expressing Norwalk virus capsid protein in the presence or absence of VP2 protein. *Virology* **2015**, *484*, 163–169.

27. Nakaya, Y.; Nakaya, T.; Park, M.S.; Cros, J.; Imanishi, J.; Palese, P.; García-Sastre, A. Induction of cellular immune responses to simian immunodeficiency virus Gag by two recombinant negative-strand RNA virus vectors. *J. Virol.* **2004**, *78*, 9366–9375.

28. Khattar, S.K.; Samal, S.; Devico, A.L.; Collins, P.L.; Samal, S.K. Newcastle disease virus expressing human immunodeficiency virus type 1 envelope glycoprotein induces strong mucosal and serum antibody responses in Guinea pigs. *J. Virol.* **2011**, *85*, 10529–10541.

29. Khattar, S.K.; Manoharan, V.; Bhattarai, B.; LaBranche, C.C.; Montefiori, D.C.; Samal, S.K. Mucosal immunization with Newcastle disease virus vector coexpressing HIV-1 Env and Gag proteins elicits potent serum, mucosal, and cellular immune responses that protect against vaccinia virus Env and Gag challenges. *mBio* **2015**, *6*, e01005.

30. Khattar, S.K.; Devico, A.L.; Labranche, C.C.; Panda, A.; Montefiori, D.C.; Samal, S.K. Enhanced immune responses to HIV-1 envelope elicited by a vaccine regimen consisting of priming with Newcastle disease virus expressing HIV gp160 and boosting with gp120 and SOSIP gp140 proteins. *J. Virol.* **2016**, *90*, 1682–1686.

31. Martinez-Sobrido, L.; Gitiban, N.; Fernandez-Sesma, A.; Cros, J.; Mertz, S.E.; Jewell, N.A.; Hammond, S.; Flano, E.; Durbin, R.K.; Garcia-Sastre, A.; et al. Protection against respiratory syncytial virus by a recombinant Newcastle disease virus vector. *J. Virol.* **2006**, *80*, 1130–1139.

32. Kong, D.; Wen, Z.; Su, H.; Ge, J.; Chen, W.; Wang, X.; Wu, C.; Yang, C.; Chen, H.; Bu, Z. Newcastle disease virus-vectored Nipah encephalitis vaccines induce B and T cell responses in mice and long-lasting neutralizing antibodies in pigs. *Virology* **2012**, *432*, 327–335.

33. Xiao, S.; Kumar, M.; Yang, X.; Akkoyunlu, M.; Collins, P.L.; Samal, S.K.; Pal, U. A host-restricted viral vector for antigen-specific immunization against Lyme disease pathogen. *Vaccine* **2011**, *29*, 5294–5303.

34. Hall, A.J.; Lopman, B.A.; Payne, D.C.; Patel, M.M.; Gastañaduy, P.A.; Vinjé, J.; Parashar, U.D. Norovirus disease in the United States. *Emerg. Infect. Dis.* **2013**, *19*, 1198–1205.

35. Duan, Z.; Xu, H.; Ji, X.; Zhao, J. Recombinant Newcastle disease virus-vectored vaccines against human and animal infectious diseases. *Future Microbiol.* **2015**, *10*, 1307–1323.

36. Huang, Z.; Elankumaran, S.; Yunus, A.S.; Samal, S.K. A recombinant Newcastle disease virus expressing VP2 protein of infectious Bursal disease virus protects against NDV and IBDV. *J. Virol.* **2004**, *78*, 10054–10063.

37. Ge, J.; Wang, X.; Tian, M.; Gao, Y.; Wen, Z.; Yu, G.; Zhou, W.; Zu, S.; Bu, Z. Recombinant Newcastle disease viral vector expressing hemagglutinin or fusion of canine distemper virus is safe and immunogenic in minks. *Vaccine* **2015**, *33*, 2457–2462.

38. Kanabagatte Basavarajappa, M.; Kumar, S.; Khattar, S.K.; Gebreluul, G.T.; Paldurai, A.; Samal, S.K. A recombinant Newcastle disease virus (NDV) expressing infectious laryngotracheitis virus (ILTV) surface glycoprotein D protects against highly virulent ILTV and NDV challenges in chickens. *Vaccine* **2014**, *32*, 3555–3563.

39. Toro, H.; Zhao, W.; Breedlove, C.; Zhang, Z.; Yub, Q. Infectious bronchitis virus S2 expressed from recombinant virus confers broad protection against challenge. *Avian Dis.* **2014**, *58*, 83–89.

40. Kortekaas, J.; Dekker, A.; de Boer, S.M.; Weerdmeester, K.; Vloet, R.P.; de Wit, A.A.; Peeters, B.P.; Moormann, R.J. Intramuscular inoculation of calves with an experimental Newcastle disease virus-based vector vaccine elicits neutralizing antibodies against Rift Valley fever virus. *Vaccine* **2010**, *28*, 2271–2276.

41. Kortekaas, J.; de Boer, S.M.; Kant, J.; Vloet, R.P.M.; Antonis, A.F.G.; Moormann, R.J.M. Rift Valley fever virus immunity provided by a paramyxovirus vaccine vector. *Vaccine* **2010**, *28*, 4394–4401.

42. Khattar, S.K.; Collins, P.L.; Samal, S.K. Immunization of cattle with recombinant Newcastle disease virus expressing bovine herpesvirus-1 (BHV-1) glycoprotein D induces mucosal and serum antibody responses and provides partial protection against BHV-1. *Vaccine* **2010**, *28*, 3159–3170.

43. Ge, J.; Wang, X.; Tao, L.; Wen, Z.; Feng, N.; Yang, S.; Xia, X.; Yang, C.; Chen, H.; Bu, Z. Newcastle disease virus-vectored rabies vaccine is safe, highly immunogenic, and provides long-lasting protection in dogs and cats. *J. Virol.* **2011**, *85*, 8241–8252.

44. Veits, J.; Wiesner, D.; Fuchs, W.; Hoffmann, B.; Granzow, H.; Starick, E.; Mundt, E.; Schirrmeier, H.; Mebatsion, T.; Mettenleiter, T.C.; et al. Newcastle disease virus expressing H5 hemagglutinin gene protects chickens against Newcastle disease and avian influenza. *Proc. Natl. Acad. Sci. USA* **2006**, *103*, 8197–8202.

45. Nayak, B.; Rout, S.N.; Kumar, S.; Khalil, M.S.; Fouda, M.M.; Ahmed, L.E.; Earhart, K.C.; Perez, D.R.; Collins, P.L.; Samal, S.K. Immunization of chickens with Newcastle disease virus expressing H5 hemagglutinin protects against highly pathogenic H5N1 avian influenza viruses. *PLoS ONE* **2009**, *4*, e6509.

46. Park, M.S.; Steel, J.; Garcia-Sastre, A.; Swayne, D.; Palese, P. Engineered viral vaccine constructs with dual specificity: Avian influenza and Newcastle disease. *Proc. Natl. Acad. Sci. USA* **2006**, *103*, 8203–8208.

47. McGeoch, D.J.; Dolan, A.; Ralph, A.C. Toward a comprehensive phylogeny for mammalian and avian herpesviruses. *J. Virol.* **2000**, *74*, 10401–10406.

48. Zhao, W.; Spatz, S.; Zhang, Z.; Wen, G.; Garcia, M.; Zsak, L.; Yu, Q. Newcastle disease virus (NDV) recombinants expressing infectious laryngotracheitis virus (ILTV) glycoproteins gB and gD protect chickens against ILTV and NDV challenges. *J. Virol.* **2014**, *88*, 8397–8406.

49. Chu, V.C.; McElroy, L.J.; Chu, V.; Bauman, B.E.; Whittaker, G.R. The avian coronavirus infectious bronchitis virus undergoes direct low-pH-dependent fusion activation during entry into host cells. *J. Virol.* **2006**, *80*, 3180–3188.

50. Pepin, M.; Bouloy, M.; Bird, B.H.; Kemp, A.; Paweska, J. Rift Valley fever virus (Bunyaviridae: Phlebovirus): An update on pathogenesis, molecular epidemiology, vectors, diagnostics and prevention. *Vet. Res.* **2010**, *41*, 61.

51. OIE (World Organisation for Animal Health) Avian Influenza. Available online: http://www.oie.int/fileadmin/Home/eng/Health_standards/tahm/2.03.04_AI.pdf(2012) (accessed on 30 July 2013).

52. Treanor, J.J.; Campbell, J.D.; Zangwill, K.M.; Rowe, T.; Wolff, M. Safety and immunogenicity of an inactivated subvirion influenza A (H5N1) vaccine. *N. Engl. J. Med.* **2006**, *354*, 1343–1351.

53. Nayak, B.; Kumar, S.; DiNapoli, J.M.; Paldurai, A.; Perez, D.R.; Collins, P.L.; Samal, S.K. Contributions of the avian influenza virus HA, NA, and M2 surface proteins to the induction of neutralizing antibodies and protective immunity. *J. Virol.* **2010**, *84*, 2408–2420.

54. Kim, S.H.; Paldurai, A.; Xiao, S.; Collins, P.L.; Samal, S.K. Modified Newcastle disease virus vectors expressing the H5 hemagglutinin induce enhanced protection against highly pathogenic H5N1 avian influenza virus in chickens. *Vaccine* **2014**, *32*, 4428–4435.

Recombinant Ranaviruses for Studying Evolution of Host–Pathogen Interactions in Ectothermic Vertebrates

Jacques Robert and James K. Jancovich

Abstract: Ranaviruses (*Iridoviridae*) are large DNA viruses that are causing emerging infectious diseases at an alarming rate in both wild and captive cold blood vertebrate species all over the world. Although the general biology of these viruses that presents some similarities with poxvirus is characterized, many aspects of their replication cycles, host cell interactions and evolution still remain largely unclear, especially in vivo. Over several years, strategies to generate site-specific ranavirus recombinant, either expressing fluorescent reporter genes or deficient for particular viral genes, have been developed. We review here these strategies, the main ranavirus recombinants characterized and their usefulness for in vitro and in vivo studies.

Reprinted from *Viruses*. Cite as: Robert, J.; Jancovich, J.K. Recombinant Ranaviruses for Studying Evolution of Host–Pathogen Interactions in Ectothermic Vertebrates. *Viruses* **2016**, *8*, 187.

1. Introduction

Ranavirus pathogens are causing emerging infectious diseases in many ectothermic vertebrate species worldwide [1,2]. The increase in prevalence and in the range of infected hosts from fish and amphibians to reptiles is alarming for biodiversity in the wild as well as for aquaculture and international trade of animals [3–5]. Importantly, ranaviruses are capable of crossing species barriers of numerous ectothermic vertebrates, suggesting that these pathogens possess potent immune evasion strategies [6–8]. Furthermore, although some species are highly susceptible to ranaviruses, others are more resistant and can serve as asymptomatic carriers for viral dissemination [9,10].

The World Organization for Animal Health (OIE) lists ranaviruses that infect amphibians as notifiable pathogens, which means that international trade of live amphibians and related products should require health certifications to be applied according to OIE standards. In addition, recent warnings about the international trade of amphibians in the dissemination of this disease highlight the importance of these pathogens [11]. However, movement of ranavirus host species continues and with that the increased probability of further pathogen dissemination. As ranaviruses continue to influence wild and cultured cold-blooded vertebrate populations, it is crucial to better understand the molecular biology driving these infectious agents.

Ranaviruses are large, icosahedral, double-stranded DNA viruses that belong to the family *Iridoviridae*, which is part of the monophyletic group of nucleocytoplasmic large DNA viruses (NCLDV) that also includes *Poxviridae* [12]. Ranaviruses possess large genomes, ranging between 105 and 155 kilobase pairs in size and encoding 95 to 162 open reading frames (ORFs; [12]). Despite their growing economic and ecological significance, ranaviruses have not been as extensively studied as other families of large double-stranded DNA (dsDNA) viruses (e.g., *Poxviridae* and *Herpesviridae*), and their mechanisms of replication, infection, and pathogenesis are still poorly understood.

Based on early studies using *Frog virus* 3 (FV3), which is the type species of the genus [13], and more recent work with other ranaviruses, the general outlines of the ranavirus replication cycle are known. Similar to poxvirus, no cellular receptor critical for viral entry has been identified to date for these promiscuous viruses that can infect many different cell types in vitro and in vivo [14]. In fact, ranaviruses can multiply at 32 °C or lower even in mammalian and avian cells, although they cannot replicate at 37 °C [15].

A distinctive feature of FV3 and other ranaviruses is that a viral envelope added onto virions budding from the plasma membrane of infected cells is not required for infectivity. As such naked virus particles released during cell lysis are infectious. However, naked viral DNA is not infectious, which suggests that virion-associated proteins are required for infection. Viral entry into the cytoplasm is thought to occur by receptor-mediated endocytosis for enveloped virions or by release of viral DNA into the cytoplasm by naked virus [12,16,17].

In contrast to poxviruses that replicate only in the cytoplasm, ranaviruses present the peculiarity of replicating both in the nucleus and in the cytoplasm of infected cells. After viral entry, early viral transcription and a first round of unit length genome replication take place within the nucleus using the host DNA polymerase. At later stages, when the viral DNA polymerase is produced, genomes are transported into the cytoplasm and serve as templates for concatemer formation. Ranavirus genomic DNA is both circularly permuted and terminally redundant [18], leading to a genome map that is circular, whereas the actual molecule is still linear [19]. In addition, ranavirus genomes are highly methylated, a feature unique among animal viruses. This is another marked difference, particularly among genera within the family *Iridoviridae*, as ranavirus genomes are highly modified with greater than 20% of its cytosine CpG sequences methylated [20]. While the function of viral methylation has not been conclusively determined, the high level of methylation suggests an important evolutionary advantage for this group of viruses.

Ranavirus virions are assembled in cytoplasmic assembly sites that contain viral DNA and various virus-encoded proteins. Newly formed ranavirus particles can be found free within the cytoplasm, accumulated in para-crystalline arrays or bud from the plasma membrane and acquire an envelope. Recently, work with Singapore grouper iridovirus, a ranavirus related to FV3, suggests that ranavirus maturation and envelope formation can also occur intracellularly using cytoplasmic vesicles [21]. Mature naked and enveloped virions released help spread the viral infection locally and into the environment. In many host species, ranaviruses can infect almost every tissue type in a host and median lethal dose (LD_{50}) studies provide supportive evidence of the infectious power of ranavirus pathogens [22].

To date, as many as 19 iridovirus genomes have been sequenced [1,12,23–25]. However, precise functions of most ranaviral genes are still unknown. For example, only about one-third of FV3's ~100 genes have been assigned putative functions and this mostly by sequence homology to known viral and host genes. As a result, there is a limited knowledge of the host–pathogen interactions between ranaviruses and their cold-blooded vertebrate hosts. Because of the increasing role of ranaviruses in amphibian declines and disease outbreaks among commercially important amphibian and fish species, it is imperative to gain better insight into the determinants of virulence encoded by these emerging pathogens as well as to be able to trace viral infection in vivo.

Here, we review recent progress in generating recombinant ranaviruses with a special emphasis on the characterization of putative virulence or immune evasion genes and using expression of reporter fluorescent gene under viral promoter to trace viral infection.

2. *Ambystoma Tigrinum Virus* (ATV)

The *Ambystoma Tigrinum Virus* (ATV) was first isolated from tiger salamanders in southern Arizona during an epizootic in the San Rafael Valley in 1995 and the genome sequenced [26]. Interestingly, ATV encodes around 95 ORFs, yet less than half of these ORFs have a predicted function.

The first recombinant ATV (rATV) has the viral gene for the homolog of the eukaryotic translation initiation factor 2 α (vIF2α) replaced with the selectable marker neomycin [27]. This rATV containing the *neoR* gene was generated by homologous recombination between viral DNA and a linear DNA recombination cassette. The recombination cassette was constructed by overlapping PCR; it contained 1200 nucleotides (nt) of up- and down-stream homologous sequences flanking an ATV promoter and 200 nt of sequence upstream of the immediate early *ICP-18* ATV gene fused to the *neoR* gene. rATV can be selected, isolated and purified because of the sensitivity of wild-type ATV (wtATV) to neomycin. Cells are infected with wtATV and then transfected with the linear recombination cassette. The use of this linear recombination cassette DNA molecule allows for the generation of a recombinant virus lacking the target gene, in this case the 57R ORF that is neomycin resistant. Neomycin resistance can only be propagated in progeny virions if two recombination events take place—one with the left (i.e., upstream) flanking sequence and a second with the right (i.e., downstream) flanking sequence. Subsequent rounds of selection inhibit wtATV growth, while promoting rATV growth. Therefore, recombinant virus can be enriched for, and eventually, isolated in pure culture and characterization of viral replication in the absence of the gene can be carried out both in vitro and in vivo. In ATV, the deletion of the *vIF2α* gene (ORF 57R) resulted in modulation of protein kinase PKZ, a molecule resembling protein kinase R (PKR) in fish and amphibians [28–30]. PKR is the interferon inducible double-stranded RNA protein kinase R that phosphorylates the eukaryotic translation initiation factor 2-α (eIF2α), thereby inhibiting protein synthesis [31–33]. The deletion of 57R in ATV resulted in degradation of PKZ in transiently transfected fathead minnow cells (FHM) and sensitivity to FHM interferon [27]. In addition, ATVΔ57R showed reduced pathogenesis in tiger salamanders providing support that this gene enhances viral replication and pathogenesis. Interestingly, when the ATV 57R gene was inserted into the vaccinia virus (VACV) *E3L* locus, a gene that modulates PKR activation by binding and sequestering dsRNA produced by the virus [34,35], it could not rescue the deletion of *E3L* in vivo and suggested a role in interferon sensitivity in vitro [36].

Recently, a simplified, reliable and standardized process to identify essential and non-essential genes in ATV by homologous recombination has been developed [37]. This process uses a linear DNA molecule expressing a fusion protein containing green fluorescent protein (GFP) and neomycin resistance (referred to as GNR) that is expressed using a cytomegalovirus (CMV) promoter (cassette referred to as CMV-GNR) to replace ORFs of interest in the viral genome. If the target ORF is non-essential, green, neomycin resistant plaques can be isolated and purified (Figure 1A), and subsequently, if the gene is essential, a rATV cannot be easily generated. Notably, the use of an ectopic promoter diminishes the risk of recombination between the transgene and other ranaviral promoters. This approach can be used to scan for essential and non-essential genes in ATV and can easily be adapted for use in other ranavirus systems. To simplify and accelerate assembly, we have designed a PCR based amplification and cloning strategy using adaptor sequences (Figure 1B). This methodology standardizes assembly and generation of mutant virus.

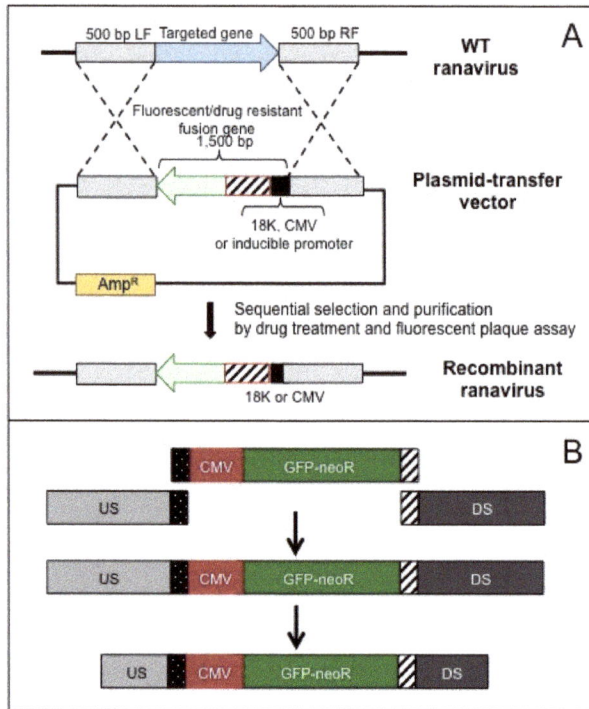

Figure 1. (**A**) Schematic for generating ranavirus recombinant by site-specific integration of a selection cassette. This cassette consists of a fluorescent reporter gene fused to a drug resistance gene by a short linker that is under the control of a ranaviral or ectopic promoter. This cassette is flanked by a left and right sequence portion (500 bp) of the targeted site and is cloned into a convenient bacterial plasmid. Cells are transfected with the construct using lipofectin and then infected with *wt* ranavirus to generate homologous recombination. The selection is performed sequentially by virus replication in the presence the drug and then by isolation of fluorescent plaques. (**B**) Schematic representation of the standardized process to generate a recombination cassette for *Ambystoma tigrinum virus* (ATV). Primers are designed to amplify the neomycin resistance (GNR) cassette as well as approximately 1000 nt of the upstream (US) and downstream (DS) flanking sequences for each target open reading frame (ORF). Adapter sequence added to the 3′ end of the US sequence and the 5′ end of CMV promoter. In addition, a different adapter is added to the 3′ end of the cytomegalovirus promoter (CMV-GNR) cassette and the 5′ end of the DS sequence. A standardized overlapping PCR protocol assembles the recombination cassette that is then agarose gel purified and re-amplified using primers that truncate the US and DS sequences t. This PCR product is then used to generate a recombinant virus.

3. *Frog virus 3* (FV3)

FV3, the main member and the type species of the genus *Ranavirus* [13], was originally isolated from the native North American leopard frog *Rana pipiens*. FV3 and FV3-like viruses are now found worldwide infecting different genera and species of amphibians, fish and reptiles, making it a potentially serious global threat to ectothermic vertebrates.

Initial attempts to identify genes essential for FV3 replication involved the isolation of a number of temperature-sensitive mutants [38–40]. More recent alternative ways to elucidate the function of several viral genes used transient knock down of viral gene by antisense morpholino oligonucleotides [41] or small interfering RNA (siRNA) [42]. However, the random nature of temperature sensitive mutants and the inability to readily perform knock down in vivo, have limited the usefulness of these approaches, especially if one wishes to target virulence genes that are usually non-essential for viral replication in vitro. Thus, to advance the characterization of ranaviral genes, the development of a stable, reliable and efficient site-specific mutagenesis methodology similar to that used with other DNA virus such as poxvirus, is paramount.

Although a traceable fluorescent gene reporter such as GFP is often used to screen and isolate recombinant virus ([43]; see next section), the relatively high ranavirus recombination rate could make the insertion of this gene unstable. Therefore, similar to the strategy used for ATV, a more robust, dual-selection system consisting of the puromycin-resistance gene fused to the gene for enhanced green fluorescent protein, PuroR/GFP was chosen [44]. This method is based on the use of plasmid constructs in which the PuroR/GFP gene is located downstream from the highly active 18K immediate-early promoter and the cassette flanked by several hundred nucleotides derived from sequences immediately adjacent to the genes of interest. Rather than fish or amphibian cell line, this method takes advantage of a mammalian cell line, the baby hamster kidney (BHK) cell line that supports culture at 30 °C necessary for FV3 replication. BHK cells do not generate interferon response and the higher temperature accelerates the production of high titer virus.

Typically, BHK cells are infected with wtFV3 and transfected two hours later with the PuroR/GFP plasmid. Virus harvested two days later is subjected to four rounds of selection in the presence of 50 μg/mL puromycin, followed by 4–5 rounds of selection for plaques expressing GFP (Figure 1A). This procedure has proved highly effective in isolating KO mutants. It is noteworthy that puromycin selection is also important to ensure that the cassette is not lost during large-scale production of recombinant virus.

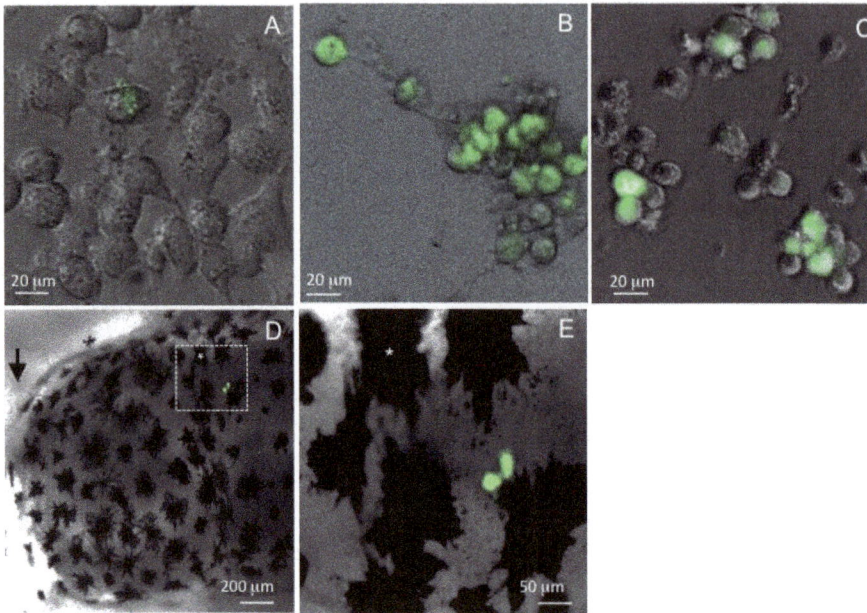

Figure 2. Detection of FV3-GFP knock-in mutant expressing GFP reporter under the control of the immediate early 18K promoter during infection in vitro in mammalian cell lines and in vivo in *X. laevis* tadpoles. (**A**) BHK cells at 2 h post-infection at permissive (30 °C) temperature; (**B**) mouse BV2 macrophage-like microglial cells at 24 h post-infection at permissive (30 °C) temperature; (**C**) mouse sertoli macrophage TM4 at 24 h post-infection at non-permissive (37 °C) temperature; and (**D,E**) midbrain view of pre-metamorphic tadpole brain at 1 day post-infection at low (**D**) and higher (**E**) magnification. (*) Indicates the same melanophore in panel D and E. Images are composite of phase contrast and fluorescence for cells (**A–C**) and of bright field and fluorescence of the whole-mounted tadpole, taken under a Leica DMIRB inverted fluorescence microscope and Infinity 2 digital camera (objectives ×5/×10/×20; Zeiss). Digital images were analyzed and processed by ImageJ software.

3.1. GFP Knock-in FV3

As a first FV3 recombinant, the PuroR/GFP cassette was integrated into a noncoding genomic region to serve as a control for the presence of the puromycin resistance and *EGFP* gene. This knock-in recombinant virus (FV3-GFP) has revealed useful for various studies. FV3-GFP replicate as well as wt-FV3 in BHK cells [44]. Because EGFP expression is under the control of the immediate-early promoter 18K, it should be one of the first viral genes produced upon infection and serves as a good marker of early stages of infection. In cell culture for example, we can detect GFP signal as early as two hours post-infection in a few BHK cells (Figure 2A).

GFP fluorescence can be detected in other mammalian cells infected with FV3-GFP when adapted to grow at 30 °C including HeLa and murine macrophage lines like microglial BV2 cells (Figure 2B). In addition, GFP can even be detected in mouse cells infected with FV3-GFP at the non-permissive temperature of 37 °C, which is known to prevent FV3 replication 24 hours post-infection (Figure 2C). This suggests that at the non-permissive temperature FV3 is not only able to bind and infect mammalian cells but is also able to express at least some of its immediate-early genes [45].

More importantly, FV3-GFP can be used to trace infection in vivo. In *Xenopus laevis* tadpoles at one day post-infection, GFP signal is detected in the kidney, which is the main site of viral replication. In addition, some GFP signal is also detectable in the brain of tadpoles infected with this recombinant (Figure 2D,E). This observation confirms our recent report that the blood-brain barrier (BBB) is compromised during FV3 infection in tadpole, which results in the dissemination of FV3 in the brain [46].

3.2. FV3 KO Mutants.

By analogy to poxviruses and other NCLDV, where more than two dozen viral gene products play critical roles in the evasion of host immunity and virulence, it is postulated that ranaviruses such as FV3 should encode multiple virulence genes. Some putative FV3 virulence or immune evasion genes can be inferred based on homology with genes of eukaryotes or other viruses. However, it is likely that, among the numerous FV3 ORFs that are conserved across ranaviruses but do not share any significant sequence similarity with other viruses or eukaryotes, there are additional, potentially novel, virulence or immune evasion genes. To date four FV3 putative virulence/immune evasion genes have been successfully knocked-out and partially characterized:

1. A viral homolog of the cellular translation factor eIF2α (vIF2α) that is an antagonist of PKR [28]. The vIF2α gene is present and conserved among ranaviruses but is truncated in FV3 and lacks the PKR N-terminal binding and central helicase domains [44].
2. A Caspase Activation and Recruitment Domain (CARD) motif-containing ranavirus gene (vCARD) postulated to interfere with CARD domains containing pro-apoptotic, pro-inflammatory and/or interferon responsive molecules [47,48].
3. A putative ranavirus homolog of β-hydroxysteroid dehydrogenase (vβHSD) that similar to poxviruses may play a role in dampening host immune responses [49,50].
4. An immediate-early gene, 18K, of unknown function but conserved among ranaviruses

Each FV3 KO mutant was examined in vitro to control for the correct insertion of the puromycin/EGFP cassette at the targeted site and to rule out any contamination by wtFV3. Importantly, compared to wtFV3, all FV3 KO mutants replicate as efficiently in vitro in the mammalian BHK cell line and the fish fat head minnow (FMH) cell line. In contrast, infection of susceptible *Xenopus laevis* (*X. laevis*) tadpoles reveals that the replication and virulence of these FV3 KO mutants is affected in vivo leading to lower levels of mortality than wtFV3. To obtain further insight into the putative virulence/immune evasion mechanisms of these viral genes, the *X. laevis* A6 kidney cell line was used. Unlike the non-amphibian BHK cell line, which does not have a type I interferon response, the replication of 3 FV3 KO mutants (ΔvCARD-, ΔvβHSD- and ΔvIF2α-FV3) were significantly reduced compared to wtFV3 in the *X. laevis* A6 cell line. This is interesting because type I interferon response is elicited by FV3 in A6 cells [51]. Increased sensitivity to amphibian type I IFN was further shown for ΔvCARD- and ΔvIF-2α-FV3 by pretreating A6 cells with *X. laevis* recombinant type I IFN [52], suggesting a role in modulating the cellular interferon response by these FV3 genes.

Further insight into the respective role of these genes in overcoming or subverting host cell defenses by FV3 was obtained by looking at apoptosis induction in A6 cells [52]. Since, interaction of protein with CARD domains is critical in the apoptosis pathway and thus could be interfered by a viral CARD-like gene, it was postulated that disruption of this gene would have an impact on apoptosis during infection. Indeed, ΔvCARD-FV3 triggered significantly more apoptosis than wtFV3 in A6 cells, but surprisingly FV3 with a deletion of the truncated vIF2α (ΔvIF2α-FV3) was also less effective in preventing apoptosis, whereas apoptosis induced by ΔvβHSD-FV3 was comparable to wtFV3. It is noteworthy that vCARD KO mutant induces as much type I IFN as wtFV3, which suggests that vCARD is probably not directly interfering with IFN-I synthesis, but rather may block IFN-induced apoptosis.

The role of the truncated vIF-2α gene in FV3 is enigmatic since it lacks the N-terminal PKR-binding and central helicase domains [44]. These domains have been shown to be critical in vIF2α proteins of ATV [27] and RCV-Z, [28] for counteracting the protein translational arrests mediated by PKR. The impaired replication and virulence of ΔvIF2α-FV3 in in vivo in tadpoles as well as its sensitivity to type I IFN and pro-apoptotic activity, all imply that albeit truncated gene is critical counteracting the programmed cell death induced by viral infection, presumably through the IFN response. It is possible that truncated FV3 vIF2α has adopted unique interactions with PKR, or that FV3 vIF2α may function by targeting a distinct and yet undefined cellular antiviral mechanism.

Although ΔvβHSD-FV3 did also show attenuation of virulence in vivo, which is suggestive of the implication of this gene in promoting FV3 infection, its characterization is in progress.

18K is the only gene characterized in FV3 that does not exhibit significant match with other genes outside ranaviruses. Given that it is not an essential gene and that it is expressed at very early stage of infection, it represents an ideal candidate for a novel virulence or immune evasion gene as evidenced by the dramatic growth impairment and reduced lethality in infected tadpole. Notably, FV3-Δ18K is more resistant to r*Xl*IFN prestimulation than the three other knockout (KO) mutants, whereas 18K deletion results in substantially increased apoptosis. Similar results were obtained by 18K knockdown with morpholino [41]. These findings suggest a distinctive function of 18K that may be to regulate timely FV3 gene expression and release.

4. Other Ranaviruses

In addition to ATV and FV3, other ranaviruses, especially those in Asia and Europe, have been used to generate mutants expressing foreign genes. In fact, one of the first ranaviruses expressing GFP was made in soft-shelled turtle iridovirus (STIV), an Asian ranavirus isolated from infected turtles [43]. This recombinant mutant virus was isolated by selecting for EGFP expressing plaques as the viral envelope protein, VP55, was fused to EGFP. Homologous recombination between a DNA construct containing the VP55-EGFP fusion construct and viral DNA in infected cells generated the knock-in virus, similar to the technique described above for FV3. This recombinant STIV initiated a wave of similar research with other ranavirus isolates from Asian host species.

Rana grylio virus (RGV) has been used to generate a number of knock-out recombinant viruses as well as recombinant virus that can specifically induce expression of the target gene. RGV was isolated from infected pig frogs (*Rana grylio*) and caused high levels of mortality in pig frogs in multiple locations throughout China [53,54]. RGV is genetically similar to FV3 [55] and recombinants deleting non-essential genes have been generated. RGV deleted of the viral envelope protein (ORF 53R) and the thymidine kinase (TK, 92R) was isolated by plaque purification using a plasmid vector recombination cassette that express only GFP [56]. RGV 53R and 92R are not required for growth in cell culture allowing isolation of these mutants without any drug selection. That is, green plaques can easily be observed and recombinant virus purified. Recently, this method of using only a visual marker was used to generate a double deletion RGV expressing EGFP in the TK (92R) locus and red fluorescent protein (RFP) in the deoxyuridine triphosphatase (dUTPase, DUT) (67R) [57]. All of the above RGV mutants were made using plasmid sequences containing DNA to delete the target ORF with a fluorescent marker. These mutants can easily be isolated if the target gene/ORF does not significantly influence viral replication in tissue culture cells. If the gene/ORF is even semi-essential, a single crossover recombination would insert the entire plasmid, including the GFP, without deleting the gene. Although it should be unstable, these events could be propagated in progeny virions with both the target gene and the selection gene being present in the genome. Therefore, confirmation of the site specific deletion is required to ensure the recombinant has not retained the gene while inserting the plasmid, and GFP, into the viral genome. The long-term stability of these mutants, especially during production of high titer viruses remains to be evaluated.

European sheatfish virus (ESV) is the only fish ranavirus used to express foreign genes. In this case, the dihydrofolate reductase gene (DHFR) was knocked-out by homologous recombination using GFP-neomycin resistance [58]. Generation of this mutant was performed using techniques similar to those described above. In this case, the ESV DHFR gene was shown not to play a role in viral replication or pathogenesis in a zebrafish model. This study highlights a fish model system to study ranavirus in vivo as zebrafish have been used to study iridovirus pathogen dynamics [59] and there are many molecular and genetic tools (e.g., transgenic lines with various cell types expressing fluorescent tracers) available that can be used to help characterize ranavirus–host interactions. Characterized ranavirus mutants to date are listed in Table 1.

Table 1. Recombinant Ranaviruses.

Virus	ORF	Predicted Function	Mutant Phenotype	Reporter Marker	Reference
FV3					
	26R	eIF2α homologue	antagonist of PKR; IFNˢ; increased apoptosis; reduced pathogenesis	EGFP-puromycin resistance	[44]
	82R	ICP-18	increased apoptosis; increased induction of type I IFN; reduced pathogenesis	EGFP-puromycin resistance	"
	52L	β-hydroxysteroid dehydrogenase homolog	tbd; reduced pathogenesis	EGFP-puromycin resistance	[52]
	64R	caspase activation & recruitment domain-containing (CARD) protein	IFNˢ; increased apoptosis; reduced pathogenesis	EGFP-puromycin resistance	"
ATV					
	57R	eIF2α homologue	antagonist of PKZ; reduced pathogenesis	neomycin resistance	[27]
	11R	unknown	essential gene	GFP-neomycin resistance	[37]
	25R	RNase III	degrades RNA	GFP-neomycin resistance	"
	40L	CARD-containing gene	tbd; see FV3 above	GFP-neomycin resistance	"
	53R	Unknown—essential	essential gene	GFP-neomycin resistance	"
	54R	unknown	tbd	GFP-neomycin resistance	"

Table 1. *Cont.*

Virus	ORF	Predicted Function	Mutant Phenotype	Reporter Marker	Reference
RGV					
	53R	viral envelope protein	green virus	EGFP	[56]
	92R	thymidine kinase (TK)	non-essential	EGFP	"
	53R	viral envelope protein	required for viral production; reduced growth when not expressed	IPTG inducible; EGFP	[60]
	2L	viral envelope protein	required for viral production; reduced growth when not expressed	IPTG inducible; EGFP	[61]
	92R67R	TK and deoxyuridine triphosphatase (dUTPase, DUT)		EGFP/RFP	[57]
ESV					
	114L	dihydrofolate reductase (DHFR)	non-essential	EGFP-neomycin resistance	[58]
STIV					
	VP55	viral envelope protein	green virus	EGFP-VP55 fusion	[43]

tbd = to be determined; IFN^s = interferon sensitivity; " =same reference as above.

5. Conclusions and Perspective

Although we have now several optimized methods to generate knock-out ranavirus mutants as well as in vitro and in vivo systems to assess the effects of viral gene loss-of-functions, several issues remain that can potentially hamper advancing investigation of these viruses.

A first critical issue is that compared to other DNA viruses, a method to obtain revertants to any KO ranavirus has yet to be developed. This is a critical approach to fully validate that the biological effects observed with knock-out ranavirus mutants are only due to the targeted gene and not by other possible genetic defects. One solution would be to sequence the genome of the recombinant virus to identify any second site mutations [62]. Another difficulty is the selection process to generate recombinants. To introduce a second construct in the genome would require the use of a different selection marker. In our view, a drug selection is important to enforce the stability of the site-specific gene disruption. As summarized above, there have been several reports of gene knock-in in other ranavirus species including RGV that have only used fluorescent reporter gene for selection [56,57,60,61]. Although such a methodology appears to work and may seem simpler, we think that drug selection is safer to minimize the risk of recombination and loss of the insertion during virus production. One possible solution would be to develop a transient dominant selection protocol as used for poxvirus recombinant selection [63]. This would allow for use of drug selection when making the recombinant and could include a fluorescent marker (i.e., GFP or RFP) to visualize the mutant but would not rely on the insertion of a drug selection marker into the virus. Once a recombinant is made and isolated in pure culture, a revertant virus could then be generated by recombination between the mutant virus and a DNA construct containing the gene, and perhaps a tag to identify the revertant, by loss of GFP/RFP during plaque purification.

Another issue concerns the possibility that critical virulence and/or immune evasion may reveal to be essential, and therefore could not be obtained by the existing methodology. Identification of these essential genes can provide valuable insight into their function. In such a case, characterization of these essential genes may require ectopic expression of viral genes or the generation of inducible expression mutants. In the later case, the use of conditional lethal mutants that allow for regulation of expression may help resolve this issue (see below).

As some viral genes likely play roles in both replication and virulence, attempts to knock-out this category of gene can adversely impact the ability to propagate these mutants in culture. Therefore, alternative approaches must be taken to understand gene function. For example, knock-out methodology can be modified to permit analysis of infection using the tetracycline controlled gene expression system [64]. This method permits to propagate virus in vitro (in the presence of the inducer), then use that virus to infect tadpoles (in the absence of inducer) and determine the impact of gene loss in vivo. This system has been used successfully to construct conditional lethal mutants in vaccinia virus and herpesvirus [64–68] and should be adaptable to ranavirus. An alternative method that uses the *E. coli lac* operon as has been successful in generating mutants in VACV and African swine fever virus (ASFV) [69–72]. Although promising, this technique will need to be further evaluated, as regulation of gene expression is leaky and not 100% efficient.

In comparison to other large DNA viruses such as poxviruses, our understanding of ranavirus gene function is extremely poor. Poxvirus recombination techniques to explore gene function were developed in the 1980s [73,74] and continue to advance today. Therefore, using techniques developed in heterologous systems like those from poxviruses will continue to help advance our progress in understanding the function of ranavirus putative virulence and immune evasion genes. In addition, it may be possible to utilize this well-defined heterologous system to understand ranavirus gene function [36,75]. This approach does have drawbacks, specifically: (i) the higher temperature infection parameters may not resemble the cold-blooded vertebrate host; (ii) requires having an identifiable phenotype; and (iii) orthologous gene swapping. In addition, it is possible that one of the many poxvirus genes may mask ranavirus gene function identification and characterization. Nevertheless, the poxvirus heterologous system is advantageous for multiple reasons: (i) extremely well characterized model viruses in all aspects of the viral life cycle in vitro and in vivo; (ii) easy to manipulate and faster growth rates; (iii) many molecular resources and well defined host–pathogen interactions; and (iv) mechanisms to generate revertant viruses. Therefore, this heterologous system may be useful as a complementary system to study ranaviral gene functions that have broader impacts.

While ranaviruses as expression vectors of foreign genes is relatively recent, significant strides have been made in a relative short time at developing technology that help uncover the function of the large number of genes within this group. In addition, newer powerful technology such as the clustered regularly interspaced short palindromic repeats/Cas9 CRISPR/Cas9 system that has been used to facilitate the generation of knock-out recombinant poxviruses [76,77] are likely applicable to ranaviruses. As we continue to characterize the function of ranavirus genes, we anticipate the continued development of technology, resources and methodology to help in our endeavors to uncover the molecular mysteries of these important pathogens.

Acknowledgments: We would like to thank Odalys Torres Luquis, Jing Wang and Francisco De Jesus Andino for their technical contribution. We would like also to thank Brian Ward for discussions and for critical reading of the manuscript. Murine macrophage cell lines have been kindly provided by Michael Elliott. This work was supported by R24-AI-059830 from the NIH and IOB-074271 from the NSF to JR and by 1-R15AI101889-01 from the NIH to JKJ.

Conflicts of Interest: The authors declare no conflict of interest.

References

1. Chinchar, V.G.; Hyatt, A.; Miyazaki, T.; Williams, T. Family Iridoviridae: Poor viral relations no longer. *Curr. Top. Microbiol. Immunol.* **2009**, *328*, 123–170.
2. Duffus, A.; Waltzek, T.; Stöhr, A.; Allender, M.; Gotesman, M.; Whittington, R.; Hick, P.; Hines, M.; Marschang, R. Distribution and Host Range of Ranaviruses. In *Ranaviruses: Lethal Pathogens of Ectothermic Vertebrates*; Gray, M.J., Chinchar, V.G., Eds.; Springer: New York, NY, USA, 2015; pp. 9–59.
3. Chinchar, V.G.; Yu, K.H.; Jancovich, J.K. The molecular biology of frog virus 3 and other iridoviruses infecting cold-blooded vertebrates. *Viruses* **2011**, *3*, 1959–1985.
4. Gray, M.J.; Miller, D.L.; Hoverman, J.T. Ecology and pathology of amphibian ranaviruses. *Dis. Aquat. Organ.* **2009**, *87*, 243–266.
5. Green, D.E.; Converse, K.A.; Schrader, A.K. Epizootiology of sixty-four amphibian morbidity and mortality events in the USA, 1996–2001. *Ann. N. Y. Acad. Sci.* **2002**, *969*, 323–339.
6. Johnson, A.J.; Pessier, A.P.; Wellehan, J.F.; Childress, A.; Norton, T.M.; Stedman, N.L.; Bloom, D.C.; Belzer, W.; Titus, V.R.; Wagner, R.; et al. Ranavirus infection of free-ranging and captive box turtles and tortoises in the United States. *J. Wildl. Dis.* **2008**, *44*, 851–863.
7. Mao, J.; Green, D.E.; Fellers, G.; Chinchar, V.G. Molecular characterization of iridoviruses isolated from sympatric amphibians and fish. *Virus Res.* **1999**, *63*, 45–52.

8.   Stohr, A.C.; Lopez-Bueno, A.; Blahak, S.; Caeiro, M.F.; Rosa, G.M.; Alves de Matos, A.P.; Martel, A.; Alejo, A.; Marschang, R.E. Phylogeny and differentiation of reptilian and amphibian ranaviruses detected in Europe. *PLoS ONE* **2015**, *10*, e0118633.

9.   Hoverman, J.T.; Gray, M.J.; Haislip, N.A.; Miller, D.L. Phylogeny, life history, and ecology contribute to differences in amphibian susceptibility to ranaviruses. *Ecohealth* **2011**, *8*, 301–319.

10. Teacher, A.G.; Garner, T.W.; Nichols, R.A. Evidence for directional selection at a novel major histocompatibility class I marker in wild common frogs (Rana temporaria) exposed to a viral pathogen (Ranavirus). *PLoS ONE* **2009**, *4*, e4616.

11. Garner, T.W.; Stephen, I.; Wombwell, E.; Fisher, M.C. The amphibian trade: bans or best practice? *Ecohealth* **2009**, *6*, 148–151.

12. Jancovich, J.; Steckler, N.; Waltzek, T. Ranavirus Taxonomy and Phylogeny. In *Ranaviruses: Lethal Pathogens of Ectothermic Vertebrates*; Gray, M.J., Chinchar, V.G., Eds.; Springer: New York, NY, USA, 2015; pp. 59–71.

13. Hyatt, A.D.; Gould, A.R.; Zupanovic, Z.; Cunningham, A.A.; Hengstberger, S.; Whittington, R.J.; Kattenbelt, J.; Coupar, B.E. Comparative studies of piscine and amphibian iridoviruses. *Arch. Virol.* **2000**, *145*, 301–331.

14. Moss, B. Poxvirus cell entry: How many proteins does it take? *Viruses* **2012**, *4*, 688–707.

15. Lopez, C.; Aubertin, A.M.; Tondre, L.; Kirn, A. Thermosensitivity of frog virus 3 genome expression: Defect in early transcription. *Virology* **1986**, *152*, 365–374.

16. Braunwald, J.; Nonnenmacher, H.; Tripier-Darcy, F. Ultrastructural and biochemical study of frog virus 3 uptake by BHK-21 cells. *J. Gen. Virol.* **1985**, *66*, 283–293.

17. Gendrault, J.L.; Steffan, A.M.; Bingen, A.; Kirn, A. Penetration and uncoating of frog virus 3 (FV3) in cultured rat Kupffer cells. *Virology* **1981**, *112*, 375–384.

18. Goorha, R.; Murti, K.G. The genome of frog virus 3, an animal DNA virus, is circularly permuted and terminally redundant. *Proc. Natl. Acad. Sci. USA* **1982**, *79*, 248–252.

19. Houts, G.E.; Gravell, M.; Granoff, A. Electron microscopic observations on early events of frog virus 3 replication. *Virology* **1974**, *58*, 589–594.

20. Willis, D.B.; Granoff, A. Frog virus 3 DNA is heavily methylated at CpG sequences. *Virology* **1980**, *107*, 250–257.

21. Liu, Y.; Tran, B.N.; Wang, F.; Ounjai, P.; Wu, J.; Hew, C.L. Visualization of Assembly Intermediates and Budding Vacuoles of Singapore Grouper Iridovirus in Grouper Embryonic Cells. *Sci. Rep.* **2016**, *6*, 18696.

22. Forzn, M.J.; Jones, K.M.; Vanderstichel, R.V.; Wood, J.; Kibenge, F.S.; Kuiken, T.; Wirth, W.; Ariel, E.; Daoust, P.Y. Clinical signs, pathology and dose-dependent survival of adult wood frogs, Rana sylvatica, inoculated orally with frog virus 3 Ranavirus sp., Iridoviridae. *J. Gen. Virol.* **2015**, *96*, 1138–1149.

23. Williams, T.; Barbosa-Solomieu, V.; Chinchar, V.G. A decade of advances in iridovirus research. *Adv. Virus Res.* **2005**, *65*, 173–248.

24. Jancovich, J.K.; Bremont, M.; Touchman, J.W.; Jacobs, B.L. Evidence for multiple recent host species shifts among the Ranaviruses (family Iridoviridae). *J. Virol.* **2010**, *84*, 2636–2647.

25. Tan, W.G.; Barkman, T.J.; Gregory Chinchar, V.; Essani, K. Comparative genomic analyses of frog virus 3, type species of the genus Ranavirus (family Iridoviridae). *Virology* **2004**, *323*, 70–84.

26. Jancovich, J.K.; Mao, J.; Chinchar, V.G.; Wyatt, C.; Case, S.T.; Kumar, S.; Valente, G.; Subramanian, S.; Davidson, E.W.; Collins, J.P.; et al. Genomic sequence of a ranavirus (family Iridoviridae) associated with salamander mortalities in North America. *Virology* **2003**, *316*, 90–103.

27. Jancovich, J.K.; Jacobs, B.L. Innate immune evasion mediated by the Ambystoma tigrinum virus eIF2{alpha} homologue. *J. Virol.* **2011**, *85*, 5061–5069.

28. Rothenburg, S.; Chinchar, V.G.; Dever, T.E. Characterization of a ranavirus inhibitor of the antiviral protein kinase PKR. *BMC Microbiol.* **2011**, *11*, 56.

29. Rothenburg, S.; Deigendesch, N.; Dey, M.; Dever, T.E.; Tazi, L. Double-stranded RNA-activated protein kinase PKR of fishes and amphibians: Varying the number of double-stranded RNA binding domains and lineage-specific duplications. *BMC Biol.* **2008**, *6*, 12.

30. Rothenburg, S.; Deigendesch, N.; Dittmar, K.; Koch-Nolte, F.; Haag, F.; Lowenhaupt, K.; Rich, A. A PKR-like eukaryotic initiation factor 2alpha kinase from zebrafish contains Z-DNA binding domains instead of dsRNA binding domains. *Proc. Natl. Acad. Sci. USA* **2005**, *102*, 1602–1607.

31. Metz, D.H.; Esteban, M. Interferon inhibits viral protein synthesis in L cells infected with vaccinia virus. *Nature* **1972**, *238*, 385–388.

32. Jagus, R.; Anderson, W.F.; Safer, B. The regulation of initiation of mammalian protein synthesis. *Prog. Nucleic Acid Res. Mol. Biol.* **1981**, *25*, 127–185.

33. Galabru, J.; Hovanessian, A. Autophosphorylation of the protein kinase dependent on double-stranded RNA. *J. Biol. Chem.* **1987**, *262*, 15538–15544.

34. Chang, H.W.; Watson, J.C.; Jacobs, B.L. The E3L gene of vaccinia virus encodes an inhibitor of the interferon-induced, double-stranded RNA-dependent protein kinase. *Proc. Natl. Acad. Sci. USA* **1992**, *89*, 4825–4829.

35. Watson, J.C.; Chang, H.W.; Jacobs, B.L. Characterization of a vaccinia virus-encoded double-stranded RNA-binding protein that may be involved in inhibition of the double-stranded RNA-dependent protein kinase. *Virology* **1991**, *185*, 206–216.

36. Jentarra, G.M.; Heck, M.C.; Youn, J.W.; Kibler, K.; Langland, J.O.; Baskin, C.R.; Ananieva, O.; Chang, Y.; Jacobs, B.L. Vaccinia viruses with mutations in the E3L gene as potential replication-competent, attenuated vaccines: Scarification vaccination. *Vaccine* **2008**, *26*, 2860–2872.

37. Aron, M.M.; Allen, A.G.; Kromer, M.; Galvez, H.; Vigiland, B.; Jancovich, J. Identification of essential and non-essential genes in Ambystoma tigrinum virus. *Virus Res.* **2016**, in press.

38. Chinchar, V.G.; Granoff, A. Temperature-sensitive mutants of frog virus 3: Biochemical and genetic characterization. *J. Virol.* **1986**, *58*, 192–202.

39. Goorha, R.; Willis, D.B.; Granoff, A.; Naegele, R.F. Characterization of a temperature-sensitive mutant of frog virus 3 defective in DNA replication. *Virology* **1981**, *112*, 40–48.

40. Goorha, R.; Dixit, P. A temperature-sensitive (TS) mutant of frog virus 3 (FV3) is defective in second-stage DNA replication. *Virology* **1984**, *136*, 186–195.

41. Sample, R.; Bryan, L.; Long, S.; Majji, S.; Hoskins, G.; Sinning, A.; Olivier, J.; Chinchar, V.G. Inhibition of iridovirus protein synthesis and virus replication by antisense morpholino oligonucleotides targeted to the major capsid protein, the 18 kDa immediate-early protein, and a viral homolog of RNA polymerase II. *Virology* **2007**, *358*, 311–320.

42. Whitley, D.S.; Sample, R.C.; Sinning, A.R.; Henegar, J.; Chinchar, V.G. Antisense approaches for elucidating ranavirus gene function in an infected fish cell line. *Dev. Comp. Immunol.* **2011**, *35*, 937–948.

43. Huang, Y.; Huang, X.; Cai, J.; Ye, F.; Guan, L.; Liu, H.; Qin, Q. Construction of green fluorescent protein-tagged recombinant iridovirus to assess viral replication. *Virus Res.* **2011**, *160*, 221–229.

44. Chen, G.; Ward, B.M.; Yu, K.H.; Chinchar, V.G.; Robert, J. Improved knockout methodology reveals that frog virus 3 mutants lacking either the 18K immediate-early gene or the truncated vIF-2alpha gene are defective for replication and growth in vivo. *J. Virol.* **2011**, *85*, 11131–11138.

45. Cordier, O.; Tondre, L.; Aubertin, A.M.; Kirn, A. Restriction of frog virus 3 polypeptide synthesis to immediate early and delayed early species by supraoptimal temperatures. *Virology* **1986**, *152*, 355–364.

46. De Jesús Andino, F.; Letitia, B.; Maggirwar, S.; Robert, J. Frog Virus 3 dissemination in the brain of tadpoles, but not in adult Xenopus, involves blood brain barrier dysfunction. *Sci. Rep.* **2016**.

47. Besch, R.; Poeck, H.; Hohenauer, T.; Senft, D.; Hacker, G.; Berking, C.; Hornung, V.; Endres, S.; Ruzicka, T.; Rothenfusser, S.; et al. Proapoptotic signaling induced by RIG-I and MDA-5 results in type I interferon-independent apoptosis in human melanoma cells. *J. Clin. Investig.* **2009**, *119*, 2399–2411.

48. Meylan, E.; Curran, J.; Hofmann, K.; Moradpour, D.; Binder, M.; Bartenschlager, R.; Tschopp, J. Cardif is an adaptor protein in the RIG-I antiviral pathway and is targeted by hepatitis C virus. *Nature* **2005**, *437*, 1167–1172.

49. Moore, J.B.; Smith, G.L. Steroid hormone synthesis by a vaccinia enzyme: A new type of virus virulence factor. *EMBO J.* **1992**, *11*, 1973–1980.

50. Sroller, V.; Kutinova, L.; Nemeckova, S.; Simonova, V.; Vonka, V. Effect of 3-beta-hydroxysteroid dehydrogenase gene deletion on virulence and immunogenicity of different vaccinia viruses and their recombinants. *Arch. Virol.* **1998**, *143*, 1311–1320.

51. Grayfer, L.; de Jesus Andino, F.; Robert, J. The amphibian (Xenopus laevis) type I interferon response to frog virus 3: new insight into ranavirus pathogenicity. *J. Virol.* **2014**, *88*, 5766–5777.

52. De Andino, F.J.; Grayfer, L.; Chen, G.; Chinchar, V.G.; Edholm, E.S.; Robert, J. Characterization of Frog Virus 3 knockout mutants lacking putative virulence genes. *Virology* **2015**, *485*, 162–170.

53. Zhan, Q.Y.; Xiao, F.; Li, Z.Q.; Gui, J.F.; Mao, J.; Chinchar, V.G. Characterization of an iridovirus from the cultured pig frog Rana grylio with lethal syndrome. *Dis. Aquat. Organ.* **2001**, *48*, 27–36.

54. Zhang, Q.Y.; Zhao, Z.; Xiao, F.; Li, Z.Q.; Gui, J.F. Molecular characterization of three Rana grylio virus (RGV) isolates and Paralichthys olivaceus lymphocystis disease virus (LCDV-C) in iridoviruses. *Aquaculture* **2006**, *251*, 1–10.

55. Lei, X.Y.; Ou, T.; Zhu, R.L.; Zhang, Q.Y. Sequencing and analysis of the complete genome of Rana grylio virus (RGV). *Arch. Virol.* **2012**, *157*, 1559–1564.

56. He, L.B.; Ke, F.; Zhang, Q.Y. Rana grylio virus as a vector for foreign gene expression in fish cells. *Virus Res.* **2012**, *163*, 66–73.

57. Huang, X.; Fang, J.; Chen, Z.; Zhang, Q. Rana grylio virus TK and DUT gene locus could be simultaneously used for foreign gene expression. *Virus Res.* **2016**, *214*, 33–38.

58. Martin, V.; Mavian, C.; Lopez Bueno, A.; de Molina, A.; Diaz, E.; Andres, G.; Alcami, A.; Alejo, A. Establishment of a Zebrafish Infection Model for the Study of Wild-Type and Recombinant European Sheatfish Virus. *J. Virol.* **2015**, *89*, 10702–10706.

59. Xu, X.; Zhang, L.; Weng, S.; Huang, Z.; Lu, J.; Lan, D.; Zhong, X.; Yu, X.; Xu, A.; He, J. A zebrafish (Danio rerio) model of infectious spleen and kidney necrosis virus (ISKNV) infection. *Virology* **2008**, *376*, 1–12.

60. He, L.B.; Gao, X.C.; Ke, F.; Zhang, Q.Y. A conditional lethal mutation in Rana grylio virus ORF 53R resulted in a marked reduction in virion formation. *Virus Res.* **2013**, *177*, 194–200.

61. He, L.B.; Ke, F.; Wang, J.; Gao, X.C.; Zhang, Q.Y. Rana grylio virus (RGV) envelope protein 2L: Subcellular localization and essential roles in virus infectivity revealed by conditional lethal mutant. *J. Gen. Virol.* **2014**, *95*, 679–690.

62. Alharbi, N.K.; Spencer, A.J.; Hill, A.V.; Gilbert, S.C. Deletion of Fifteen Open Reading Frames from Modified Vaccinia Virus Ankara Fails to Improve Immunogenicity. *PLoS ONE* **2015**, *10*, e0128626.

63. Wyatt, L.S.; Earl, P.L.; Moss, B. Generation of Recombinant Vaccinia Viruses. *Curr. Protoc. Microbiol.* **2015**, *39*, 14a.4.1–14a.4.18.

64. Hedengren-Olcott, M.; Hruby, D.E. Conditional expression of vaccinia virus genes in mammalian cell lines expressing the tetracycline repressor. *J. Virol. Methods* **2004**, *120*, 9–12.

65. Traktman, P.; Liu, K.; DeMasi, J.; Rollins, R.; Jesty, S.; Unger, B. Elucidating the essential role of the A14 phosphoprotein in vaccinia virus morphogenesis: construction and characterization of a tetracycline-inducible recombinant. *J. Virol.* **2000**, *74*, 3682–3695.

66. Rupp, B.; Ruzsics, Z.; Sacher, T.; Koszinowski, U.H. Conditional cytomegalovirus replication in vitro and in vivo. *J. Virol.* **2005**, *79*, 486–494.

67. Nichols, R.J.; Stanitsa, E.; Unger, B.; Traktman, P. The vaccinia virus gene I2L encodes a membrane protein with an essential role in virion entry. *J. Virol.* **2008**, *82*, 10247–10261.

68. Munoz, A.L.; Gadea, I.; Lerma, L.; Varela, L.; Torres, M.; Martin, B.; Garcia-Culebras, A.; Lim, F.; Tabares, E. Construction and properties of a recombinant pseudorabies virus with tetracycline-regulated control of immediate-early gene expression. *J. Virol. Methods* **2011**, *171*, 253–259.

69. Garcia-Escudero, R.; Andres, G.; Almazan, F.; Vinuela, E. Inducible gene expression from African swine fever virus recombinants: analysis of the major capsid protein p72. *J. Virol.* **1998**, *72*, 3185–3195.

70. Andres, G.; Garcia-Escudero, R.; Salas, M.L.; Rodriguez, J.M. Repression of African swine fever virus polyprotein pp220-encoding gene leads to the assembly of icosahedral core-less particles. *J. Virol.* **2002**, *76*, 2654–2666.

71. Suarez, C.; Gutierrez-Berzal, J.; Andres, G.; Salas, M.L.; Rodriguez, J.M. African swine fever virus protein p17 is essential for the progression of viral membrane precursors toward icosahedral intermediates. *J. Virol.* **2010**, *84*, 7484–7499.

72. Rodriguez, J.F.; Smith, G.L. Inducible gene expression from vaccinia virus vectors. *Virology* **1990**, *177*, 239–250.

73. Piccini, A.; Perkus, M.E.; Paoletti, E. Vaccinia virus as an expression vector. *Methods Enzymol.* **1987**, *153*, 545–563.

74. Falkner, F.G.; Moss, B. Transient dominant selection of recombinant vaccinia viruses. *J. Virol.* **1990**, *64*, 3108–3111.

75. Coupar, B.E.; Goldie, S.G.; Hyatt, A.D.; Pallister, J.A. Identification of a Bohle iridovirus thymidine kinase gene and demonstration of activity using vaccinia virus. *Arch. Virol.* **2005**, *150*, 1797–1812.

76. Yuan, M.; Zhang, W.; Wang, J.; Al Yaghchi, C.; Ahmed, J.; Chard, L.; Lemoine, N.R.; Wang, Y. Efficiently editing the vaccinia virus genome by using the CRISPR-Cas9 system. *J. Virol.* **2015**, *89*, 5176–5179.

77. Yuan, M.; Gao, X.; Chard, L.S.; Ali, Z.; Ahmed, J.; Li, Y.; Liu, P.; Lemoine, N.R.; Wang, Y. A marker-free system for highly efficient construction of vaccinia virus vectors using CRISPR Cas9. *Mol. Ther. Methods Clin. Dev.* **2015**, *2*, 15035.

Reporter-Expressing, Replicating-Competent Recombinant Arenaviruses

Luis Martínez-Sobrido and Juan Carlos de la Torre

Abstract: Several arenaviruses cause hemorrhagic fever (HF) disease in humans and pose an important public health problem in their endemic regions. To date, no Food and Drug Administration (FDA)-licensed vaccines are available to combat human arenavirus infections, and current anti-arenaviral drug therapy is limited to an off-label use of ribavirin that is only partially effective. The development of arenavirus reverse genetic approaches has provided investigators with a novel and powerful approach for the study of arenavirus biology including virus–host interactions underlying arenavirus induced disease. The use of cell-based minigenome systems has allowed examining the *cis*- and *trans*-acting factors involved in arenavirus replication and transcription, as well as particle assembly and budding. Likewise, it is now feasible to rescue infectious arenaviruses containing predetermined mutations in their genomes to investigate virus-host interactions and mechanisms of pathogenesis. The use of reverse genetics approaches has also allowed the generation of recombinant arenaviruses expressing additional genes of interest. These advances in arenavirus molecular genetics have also facilitated the implementation of novel screens to identify anti-arenaviral drugs, and the development of novel strategies for the generation of arenavirus live-attenuated vaccines. In this review, we will summarize the current knowledge on reporter-expressing, replicating-competent arenaviruses harboring reporter genes in different locations of the viral genome and their use for studying and understanding arenavirus biology and the identification of anti-arenaviral drugs to combat these important human pathogens.

Reprinted from *Viruses*. Cite as: Martínez-Sobrido, L.; de la Torre, J.C. Reporter-Expressing, Replicating-Competent Recombinant Arenaviruses. *Viruses* **2016**, *8*, 197.

1. Arenaviruses and Their Impact on Human Health

Arenaviruses cause chronic infections of rodents with a worldwide distribution [1]. Asymptomatically infected animals move freely in their natural habitat and may invade human dwellings. Humans are infected most likely through mucosal exposure to aerosols, or by direct contact between infectious materials and abrade skin. These infections are common and in some cases severe [1].

The family *Arenaviridae* consists currently of two genera: (1) *Mammarenavirus* and (2) *Reptarenavirus. Reptarenavirus* is a new genus that has been established to accommodate the distinct features of recently discovered snake arenaviruses [2–5]. Classification of the at least 25 recognized members of the genus *Mammarenavirus* into two distinct groups, Old World (OW) and New World (NW) arenaviruses [1], was originally based on serological cross-reactivity, but classification is still well supported by recent sequence-based phylogenetic studies [1]. Genetically, OW arenaviruses constitute a single lineage, while NW arenaviruses segregate into clades A, B, A/B, and C [1]. The OW arenavirus Lassa virus (LASV) is estimated to infect several hundred thousand individuals yearly in its endemic regions of West Africa, resulting in a high number of Lassa fever (LF) cases, a hemorrhagic fever (HF) disease associated with high morbidity and significant mortality [1,6–10]. Moreover, increased travel has led to the importation of cases of LF into the USA, Europe, Japan, and Canada [11–13]. Furthermore, recent studies indicating that LASV endemic regions are expanding [14] and the association of Lujo virus (LUJV) [15], a newly identified OW arenavirus, with an outbreak of HF in Southern Africa in 2008, has raised concerns about the emergence of novel HF OW arenaviruses outside their current known endemic regions. The NW arenavirus Junin virus (JUNV), endemic to the Pampas of Argentina, causes Argentine HF (AHF) with a high (15% to 30%) case-fatality rate [6] and places a population of about five million people at risk. Likewise, the NW arenaviruses Machupo virus (MACV) [16,17] and Chapare virus (CHPV) [18], Sabia virus [19,20] and Guanarito virus (GTOV) [21–23] are responsible for causing HF in Bolivia, Brazil and Venezuela, respectively. In addition, Whitewater Arroyo virus in the USA [24,25] and Ocozocoautla de Espinosa virus in Mexico [26] have been linked to sporadic cases of HF. Moreover, mounting evidence indicates that the worldwide-distributed prototypic arenavirus lymphocytic choriomeningitis virus (LCMV) is a neglected human pathogen of clinical significance, especially in cases of congenital infection [27–31]. In addition, LCMV poses a special threat to immunocompromised individuals, which has been recently illustrated by fatal cases of transplant-associated infections by LCMV [32–34]. OW arenaviruses LASV and LUJV and NW arenavirus JUNV, MACV, GTOV and CHPV have features that make them credible biodefense threats and have been included by the National Institute of Allergy and Infectious Diseases (NIAID) as Category A biological agents that pose a significant biodefense concern [35]. Concerns about human pathogenic arenavirus infections are aggravated by the lack of Food and Drug Administration (FDA)-licensed vaccines and antiviral drug treatment being limited to the use of ribavirin [36] that is only partially effective, associated with significant side effects and must be administered early, and intravenously, during viral infection for optimal efficacy [37–39]. Evidence indicates that with morbidity and mortality of LASV, and other HF arenaviruses, infections are associated with the failure of the host's innate immune response to restrict virus replication and to facilitate the initiation of an effective adaptive immune response [9]. Accordingly, viremia is a highly predictive factor for the outcome of LF patients [9]. Therefore, therapeutic interventions resulting in reduced virus load, without requiring virus clearance, are predicted to improve the infection outcome by promoting the recovery of appropriate host defense responses to control arenavirus multiplication and associated disease.

Studies with the prototypic member in the family, LCMV, have led to significant advances in both virology and immunology that have been shown to universally apply to other viral infections in humans, including virus-induced immunopathological disease, major histocompatibility complex restriction and mechanisms of virus-induced immunosuppression [40,41]. The outcome of LCMV infection on its natural host, the mouse, varies dramatically depending on the host age, genetic background, strain, and immunocompetence, as well as the route of infection and the specific viral strain and dose [40,41]. This provides researchers with a unique model system to investigate parameters that influence many aspects of virus–host interactions, including the heterogeneity of phenotypic manifestations often associated with infection by the same virus.

2. Arenavirus Genome Organization and Virion Structure

Arenaviruses are bi-segmented negative-sense, single-stranded, RNA viruses that belong to the *Arenaviridae* family (Figure 1) [1]. Each arenaviral segment encodes, using an ambisense coding strategy, two viral proteins in opposite orientation separated by a non-coding intergenic region (IGR) (Figure 1A) [1]. The large (L) segment encodes the viral RNA dependent RNA polymerase (RdRp) or L polymerase protein (Figure 1A, blue) involved in viral replication and gene transcription [1,42]. In the opposite direction, the L segment encodes the small really interesting new gene (RING) finger protein Z (Figure 1A, orange), which is the counterpart of the matrix (M) protein present in other negative-stranded (NS) RNA viruses, and the major driving force of arenavirus assembly and budding [1,43–45]. The small (S) segment encodes for the viral glycoprotein precursor (GPC) (Figure 1A, purple) that is co-translationally cleaved by signal peptidase to produce a stable 58 amino acid stable signal peptide (SSP) and GPC that is post-translationally processed by the cellular Site 1 Protease (S1P) to yield the two mature virion glycoproteins (GP1 and GP2) that form the spikes that decorate the virus surface and mediate receptor recognition and cell entry [1,46–48]. The S segment also encodes the viral nucleoprotein (NP) (Figure 1A, green) that encapsidates the viral RNA, and together with the polymerase L and the viral RNA, constitute the viral ribonucleoproteins (vRNPs), which are the minimal factors involved in arenavirus genome replication and gene transcription [1,49,50]. In addition, arenavirus NP mediates the incorporation of the vRNPs into mature infectious virions by interacting with Z [51]. NP has also been shown to counteract the cellular host type-I interferon (IFN-I) [52–57] and inflammatory [56,58] responses during viral infection. Cellular host-derived ribosomes are also incorporated into arenavirus virions, giving them a "sandy" appearance by electron microscopy [1]. This particular property is the origin of the family name, reference to the Latin word arena (sand) [1]. However, to date, the function and relevance of the host-derived ribosomes in the viral life cycle are unknown [1].

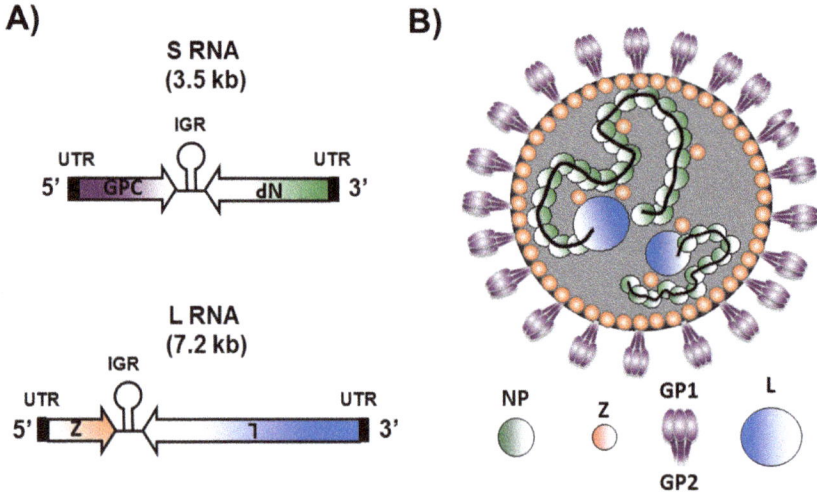

Figure 1. Arenavirus genome organization and virion structure. (**A**) genome organization: Arenaviruses are enveloped viruses with a single-stranded, bi-segmented RNA genome of negative polarity. Each of the two viral RNA genome segments uses an ambisense coding strategy to direct the synthesis of two viral polypeptides in opposite orientation. The Small (S) RNA segment (3.5 kb, top) encodes the viral glycoprotein precursor (GPC, purple) and nucleoprotein (NP, green). The Large (L) RNA segment (7.2 kb, bottom) encodes the RNA-dependent RNA polymerase (L, blue) and the small really interesting new gene (RING) finger protein (Z, orange); (**B**) virion structure: Arenaviruses are surrounded by a lipid bilayer containing the post-translationally processed viral glycoprotein involved in receptor binding (GP1) and viral cell entry (GP2). Underneath the lipid bilayer is a protein layer composed of the Z protein, which plays a major role in viral assembly and budding, and is the arenavirus counterpart of the matrix protein present in other enveloped negative-stranded (NS) RNA viruses. The core of the virus is made of a viral ribonucleoprotein (vRNP) complex, composed of the viral genome segments encapsidated by the viral NP. Incorporation of the vRNPs into newly nascent virions is mediated by NP-Z interaction. Associated with the vRNPs is the L polymerase protein that, together with NP, are the minimal components for viral genome replication and gene transcription.

3. Arenavirus Life Cycle

Arenavirus enveloped virions are pleomorphic but often spherical (Figure 1B). Arenavirus multiplication cycle occurs entirely in the cytoplasm of infected cells [1]. Homo-trimer complexes, consisting of the GP1 globular head and GP2 stalk region, form the spikes that decorate the surface of the arenavirus envelope [1,59] (Figure 1B). GP1, located at the top of the spike, mediates attachment of the virus particle to receptors located in the surface of the cell and is held in place by ionic interactions with the N-terminus of the transmembrane GP2 that forms the stalk of the spike [60]. Alpha-dystroglycan has been described as the main receptor for OW and NW clade C arenaviruses [61–63]. However, clade A, B, and A/B NW arenaviruses appear to use the human transferrin receptor protein 1 (TfR1) as the cellular receptor for viral entry [64]. Once bound to the surface of the cell, arenavirus enters the cell via receptor-mediated endocytosis [60,65]. The acidic environment of the late endosome induces a GP2 conformational change that promotes the fusion of viral and cell membranes [65], which releases the vRNPs into the cell cytoplasm where viral RNA replication and gene transcription occur [1].

Arenavirus gene transcription is mediated by the viral genome and antigenome promoters located within the untranslated regions (UTRs) at the 3' termini of viral RNA (vRNA) and complementary RNA (cRNA) species, respectively [1] (Figure 2). NP and L proteins, located at the 3' end of the S and L viral segments, are translated from mRNAs with antigenomic sense polarity transcribed directly from the vRNAs and, therefore, are the first arenaviral proteins encoded upon infection [1] (Figure 2A,B). Transcription termination is mediated by a secondary stem-loop structure within the IGR [1]. The IGR seems to also play a role in the packaging of infectious virions [66]. GPC and Z open reading frames (ORFs) are located, correspondingly, at the 5' end of the S and L genome segments and are translated from mRNA transcribed from the cRNAs [1] (Figure 2A,B). The cRNA species also serve as templates for the synthesis of nascent vRNAs [1]. Newly synthesized vRNAs are encapsidated by the viral NP to form the vRNP complexes and are packaged into progeny infectious virions by interaction of the viral Z [51]. Arenavirus assembly involves the interaction of the newly formed vRNP complexes with the GP1/GP2 complexes present in the membrane of infected cells, a process mediated by interaction with the Z protein [67]. Newly synthesized and assembled virions bud from infected cells, a process mediated by the Z protein [43,44,68,69].

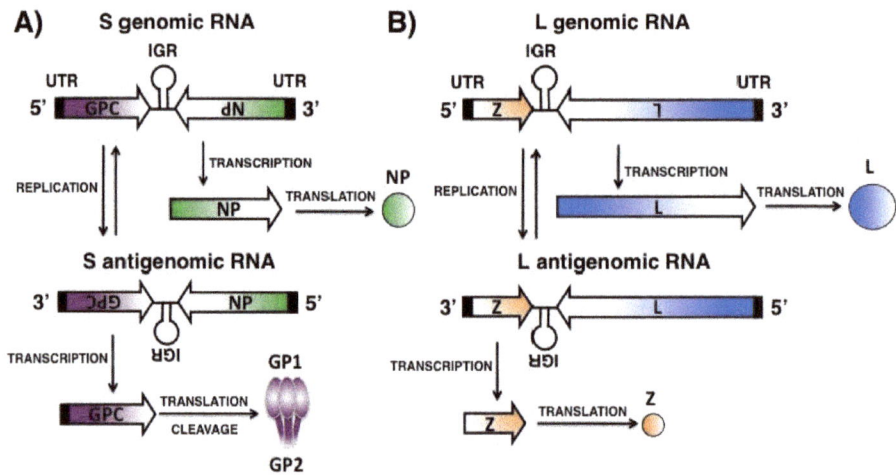

Figure 2. Arenavirus genome replication and gene transcription: The arenavirus replication cycle takes place entirely in the cytoplasm of infected cells. The L polymerase associated with the vRNPs initiates transcription from the viral promoter located within the untranslated region (UTR, black boxes) at the 3′ termini of the vRNAs. Primary transcription results in the synthesis of NP (**A**) and L (**B**) mRNAs from the S and L segments, respectively. Transcription termination is mediated by a secondary stem-loop structure formed by the intergenic region (IGR) found in both vRNA segments between each of the two viral genes. Subsequently, the virus polymerase L adopts a replicase mode and moves across the IGR to generate a copy of the full-length antigenome S and L vRNAs. The antigenomic RNA S and L segments serve as templates for the synthesis of GPC (**A**, S segment) and Z (**B**, L segment) mRNAs. The antigenomic RNA S and L segments also serve as templates for the amplification of the corresponding viral RNA genome species.

4. Current Strategies to Combat Human Arenavirus Infections

4.1. Arenavirus Vaccines

The live-attenuated vaccine strain Candid#1 strain of JUNV, has been shown to be effective at combating AHF in humans without causing serious adverse effects [6,70,71]. However, outside Argentina, Candid#1 is licensed only as an investigational new drug, and studies addressing the stability of its attenuation, long-term immunity, and safety, have not been conducted. Moreover, Candid#1 does not protect against LASV [1,6–10]. Despite significant efforts dedicated to the development of LASV vaccines, not a single LASV vaccine candidate has entered a clinical trial. Pre-clinical work with MOPV/LASV reassortant ML29, as well as recombinant vesicular stomatitis virus (VSV) and vaccinia virus expressing LASV antigens, has shown promising results in animal models, including non-human primates, of LASV infection [72]. However, the high prevalence of HIV within LASV-endemic regions in West Africa raises safety concerns about the use of VSV- or vaccinia-based platforms. Likewise, the mechanisms of ML29 attenuation remain poorly understood, and additional mutations in ML29 could result in enhanced virulence.

The recent development of reverse genetics systems for JUNV [73,74] and LASV [75,76] could facilitate the elucidation of the genetic determinants of JUNV and LASV virulence. This, in turn, should help with the design of safer live attenuated arenavirus vaccines and thereby minimize concerns related to reversion of virulence, establishment of persistent infection and vaccination of immunocompromised individuals with live-attenuated arenavirus vaccines.

4.2. Arenavirus Antiviral Drugs

In vitro and in vivo studies have documented the prophylactic and therapeutic value of the nucleoside analogue ribavirin (1-β-D-ribofuranosyl-1,2,4-triazole-3-carboxamide) against several arenaviruses [36]. Moreover, the drug has been shown to reduce significantly both morbidity and mortality associated with LASV infection in humans [36,77], and experimentally in MACV [78] and JUNV [79] infections, if given early in the course of clinical disease. The mechanisms by which ribavirin exerts its anti-arenaviral action are not fully understood and likely involve targeting different steps of the virus life cycle [80,81]. Recent evidence indicates that the nucleoside analogue can be used as a substrate by the RdRp of some riboviruses, leading to C to U and G to A transitions [82,83]. This mutagenic activity of ribavirin has been linked to its antiviral activity via lethal mutagenesis. However, the drug was also shown to strongly inhibit LCMV replication without exerting any noticeable mutagenic effect on the viral genome RNA [84]. Anemia and congenital disorders are two significant side effects associated with the use of ribavirin. In addition, oral administration is significantly less effective than intravenous administration, which poses logistic complications in regions with limited clinical infrastructure [37–39].

Several inhibitors of inosine-5′-monophosphate (IMP) dehydrogenase, as well as acyclic and carbocyclic adenosine analogue inhibitors of the S-adenosyl-L-homocysteine (SAH) hydrolase, have also been shown to have anti-arenavirus activity [36]. Likewise, the pyrimidine biosynthesis inhibitor A3, which exhibits broad-spectrum antiviral activity against negative- and positive-sense RNA viruses, retroviruses and DNA viruses [85], has been shown to be more efficient than ribavirin in controlling arenavirus multiplication in vitro. This inhibitory effect is due, at least in part, to its ability to interfere with viral RNA replication and transcription [86]. Moreover, since ribavirin and A3 target different metabolic pathways within the cell, they are excellent candidates for combination anti-arenaviral therapy to circumvent some limitations of monotherapy [86]. However, the antiviral effect of A3 against arenaviruses has not been evaluated in vivo.

Various sulfated polysaccharides and phenothiazines have been reported to have activity against several arenaviruses [36]. However, in general, these compounds displayed only modest and rather non-specific effects often associated with significant toxicity. Promising results have been shown with the broad-spectrum RdRp inhibitor favipiravir, a pyrazinecarboxamide derivative, which provided protection (20% survival) in a guinea pig model of fatal AHF [87,88]. Recently, cell-based screens have identified small molecules that prevent cell entry of NW [89] and OW [90,91] arenaviruses, and whose mechanism of action appear to be based on disruption of the pH-dependent fusion event mediated by GP2. These findings illustrate how complex chemical libraries, used in the context of appropriate screening assays, can be harnessed as a powerful tool to identify candidate antiviral drugs with highly specific activities. Towards this goal, the development of arenavirus reverse genetic systems and the use of reporter-expressing, replicating competent recombinant arenavirus represent an excellent platform for the identification of novel antivirals in high-throughput screening (HTS) approaches using libraries of small molecule compounds. Moreover, recent progress in the understanding of the molecular and cell biology of arenaviruses have opened new avenues for the identification of the steps in the life cycle targeted by the identified anti-arenavirus drugs.

4.3. Arenavirus Reverse Genetics

The development of reverse genetics systems to generate infectious recombinant arenaviruses from plasmid DNA has significantly advanced the investigation of arenavirus biology [1,42], including the characterization of *cis*-acting and the *trans*-acting factors that control each of the different steps of the arenavirus infectious life cycle [92–95]. Similarly, the generation of recombinant arenaviruses with predetermined mutations in their genomes has facilitated the identification and functional characterization of viral determinants of pathogenesis and associated disease in validated animal models of infection [53,96–99]. Likewise, implementation of arenavirus reverse genetics has allowed researchers to study arenavirus–host interactions [94,100,101], the role of NP in the inhibition of the IFN-I response [53,98], and the potential generation of novel live-attenuated arenavirus vaccine candidates and arenavirus-based vaccine vectors [95,100,102–105]. Advances in arenavirus molecular genetics have also led to the development of screening strategies to identify and characterize novel anti-arenavirus drugs that target specific steps of the virus life cycle [91,100]. Additionally, the use of arenavirus reverse genetics have permitted the generation of single-cycle infectious, reporter-expressing, recombinant arenaviruses, which can only replicate in GPC-expressing complementing cell lines [106,107]. These single-cycle arenavirus platforms have provided a new experimental approach to the study of some aspects of the biology of highly pathogenic arenaviruses (e.g., neutralizing antiviral responses and identification of inhibitors of GPC-mediated cell entry) without needing special biosafety conditions [106,107], which are required to study HF-causing members in the family [108].

Originally established for the prototyped member in the family, LCMV [109,110], plasmid-based arenavirus reverse genetics techniques have been extended to the OW LASV [75,76,111] and LUJV [112] arenaviruses, and NW JUNV [73,74], Pinchinde virus (PICV) [104,113] and MACV [114] arenaviruses. Both T7 RNA polymerase [73,75,104,109,111,113] and RNA Pol-I [74,95,100–103,110] based systems have been successfully used to launch the intracellular synthesis of the S and L RNA genome or antigenome species. These antigenome species are subsequently replicated and transcribed by the virus L and NP, the minimal viral *trans*-acting factors required for viral genome replication and gene transcription, encoded by RNA Pol-II dependent promoter protein plasmids [74,95,100–103,110]. The transcriptional activity of the RNA Pol-I exhibits species specificity, which prevents direct rescue of recombinant arenaviruses in human cells using the murine Pol-I based system [100]. This barrier was solved by the implementation of human Pol-I promoters to drive vRNA expression, which allowed for the generation of recombinant OW (LCMV) and NW (Candid#1) arenaviruses from human 293T and FDA-approved Vero cell lines [100,101]. More recently, the ability to successfully generate recombinant arenaviruses has been reduced to two plasmids by combining within the same plasmid, Pol-I-driven vRNA with Pol-II-driven protein constructs [100]. The benefit of performing arenavirus rescues using a two-plasmid approach is to increase successful co-transfection of cells that are poorly transfected, such as Vero cells, with the goal of vaccine implementation [100].

5. Reporter-Expressing Recombinant Arenaviruses

5.1. Recombinant Tri-Segmented (r3) Arenaviruses

Several approaches have been used to successfully generate recombinant NS RNA viruses expressing foreign genes. These include the use of dicistronic genome segments containing internal promoters [115,116], the use of internal ribosome entry sites (IRES) [117,118], and the use of virus-specific packaging signals within vRNA segments [119,120]. Although a viable strategy in other NS segmented RNA viruses, these approaches were unsuccessful in yielding recombinant arenaviruses encoding foreign genes [103]. Successful rescue of r3 arenavirus packaging two S and one L segments into mature, infectious virions have been described for the OW arenavirus LCMV [95,100,101,103,105] and the NW arenaviruses JUNV [74,100,101] and PICV [104] (Figure 3A). Within this approach, the S segment is altered to replace one of the viral-encoded proteins (e.g., GPC and NP) by a foreign reporter gene (RG) (Figure 4A,B) [95,100,101,103–105]. The physical separation of the GPC and NP proteins into two different S segments (S1 and S2) represents a strong selective pressure to maintain a virus capable of packaging one L segment and two S segments [95,100,101,103–105]. The ability of arenaviruses to package two S segments had been suggested based on genetics [121] and structural [122,123] analysis. Moreover, because of the stability of the r3 arenaviruses, these findings suggest that production of infectious arenavirus particles containing two S and one L segments are a common event [1,103]. Importantly, each of the S segments can direct expression of a RG, and, therefore, two foreign reporter proteins can be expressed within the same virus (Figure 4A,B) [95,100,101,103,105]. Expression levels of RG are dependent on the location in the S segment [74,95,103,105]. Expression of an RG from the NP locus is greater than from the GPC locus, similar to the situation observed during viral infection (Figure 4A,B) [74,95,103,105]. Several r3 arenaviruses have been generated that express one or two additional RG [74,95,100,101,103–105]. Depending on the RG expressed, r3 arenaviruses showed little or no attenuation in cultured cells, and they exhibited long-term genetic stability as reflected by unaltered expression levels during serial virus passages [74,95,103,104]. In vivo, however, r3 arenaviruses exhibit significant attenuation compared to wild-type (WT) arenaviruses [103,104]. Since r3 arenaviruses are not drastically attenuated in vitro (ideal for vaccine production) but are attenuated in vivo (ideal for vaccine implementation), these r3 arenaviruses represent a great approach for arenavirus vaccine and vaccine vector development [95,100,101,104,105]. Importantly, the use of r3 arenaviruses expressing appropriate RG could be used to facilitate the identification of antiviral compounds or drugs amenable to HTS or siRNA-based library screens to identify host cell genes involved in the arenavirus life cycle [124].

To generate r3 reporter-expressing arenaviruses (Figure 3A), susceptible cells (e.g., murine or human cells using the appropriate murine or the human Pol-I promoter) [100] are co-transfected with the pCAGGS protein expression plasmids encoding NP and L, which are required to initiate viral gene transcription and genome replication (Figure 3A, left) [100,101]. The co-transfection also includes the pPol-I L segment, and two pPol-I S segments, where the GPC ORF is replaced with a RG (pPol-I S1 NP/RG 1) and the NP ORF is replaced by another RG in the second S segment (pPol-I S2 RG 2/GPC) (Figure 3A, right) [74,95,100,101,103,105]. Alternatively, the viral NP can be replaced by reporter gene one and the viral GPC by reporter gene two (Figure 4A,B) [74,95,103]. Since arenaviruses do not display classic cytopathic effect (CPE) observed with other NS RNA viruses, successful rescue of WT arenaviruses must be evaluated by performing classical plaque assays or by immunofluorescence using arenavirus-specific antibodies. Reporter-expressing r3 arenaviruses typically encode for two RG, such as fluorescent or luminescent reporter proteins [74,95,100,101,103–105]. In such cases, successful viral rescue and viral titers can be evaluated using fluorescent microscopy [74,95,100,101,103–105]. Alternatively, a luciferase assay can be used to evaluate the presence of viruses [74,95,100,101,103].

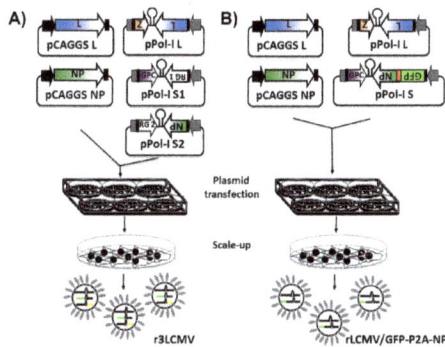

Figure 3. Generation of recombinant replicating competent reporter-expressing arenavirus: Arenavirus rescues are performed in rodent (using the mouse Pol-I promoter) or in human (using the human Pol-I promoter) cells in six-well plates. Alternatively, arenavirus rescues can be performed in T7-expressing cells using plasmids driving the expression of the arenaviral S and L segments under the T7 promoter. (**A**) Generation of r3 arenaviruses: Cells are transiently co-transfected, using LPF2000, with the pCAGGS protein expression plasmids encoding the viral NP and polymerase L (required to initiate viral gene transcription and genome replication) together with the pPol-I vRNA expression plasmids for the viral L segment and the two (pPol-I S1 and pPol-I S2) viral S segments. In the pPol-I S1 plasmid, the viral NP is replaced by a reporter gene 1 (RG 1), and, in the pPol-I S2 plasmid, the reporter gene 2 (RG 2) replaces the viral GPC. Alternatively, the viral NP can be replaced by RG 2 and the viral GPC by RG 1 (Figure 4A,B, respectively). At 72 h post-transfection, cells are trypsinized and scaled-up into 10 cm dishes. After an additional 72 h incubation period, presence of virus is determined by RG expression. R3 arenaviruses typically encode for two RGs, such as fluorescent or luminescent proteins (Table 1). In such cases, successful viral rescue and viral titers can be evaluated under a fluorescent microscope (i.e., fluorescent protein expression). Alternatively, a luciferase assay can be used to evaluate the presence of virus from tissue culture supernatant; (**B**) Generation of recombinant bicistronic arenaviruses: to rescue rLCMV/GFP-P2A-NP, susceptible cells are transiently co-transfected with the pCAGGS NP and L plasmids, together with the pPol-I vRNA expression plasmids encoding the L segment and the modified S segment encoding GFP-P2A-GFP (Figure 4C). At 72 h post-transfection, cells are trypsinized and scaled-up into 10 cm dishes. After an additional 72 h, presence of rLCMV/GFP-P2A-NP is determined by green fluorescent protein (GFP) expression. The chicken beta-actin promoter (black arrow) and the rabbit beta-globin polyadenylation (pA) signal (black boxes) are indicated in the pCAGGS protein expression plasmids. Viral untranslated regions (UTR, black boxes) and intergenic regions (IGR) in the pPol-I vRNA expression plasmid are indicated. The Pol-I promoter and terminator sequences in the pPol-I plasmids are indicated by gray arrows and boxes, respectively. The red box (**B**) indicates the porcine teschovirus (PTV1) 2A peptide sequence. For more details, see text.

285

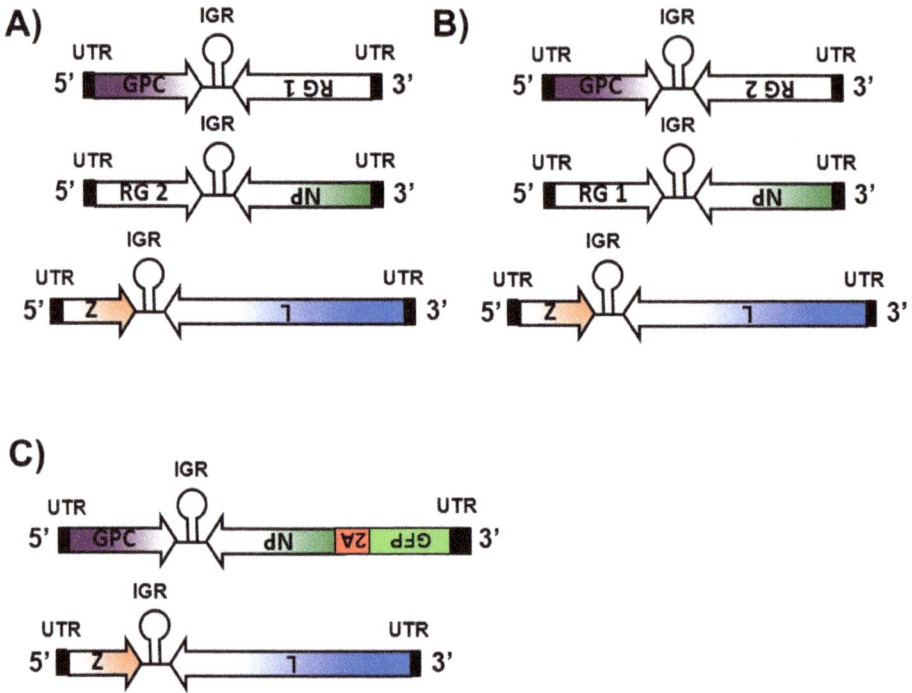

Figure 4. Reporter-expressing recombinant arenaviruses. (**A,B**) R3 arenaviruses: For the generation of r3 arenavirus, the pPol-I plasmid expressing the S vRNA segment is separated in two S plasmids. In the pPol-I S1 plasmid, the viral NP is replaced by RG 1 and in the pPol-I S2 plasmid, the viral GPC is replaced by RG 2 (**A**); alternatively, RG 1 can be expressed instead of the viral GPC in the pPol-I S2 plasmid and RG 2 from the pPol-I S1 plasmid instead of the viral NP (**B**). Regulation of RG expression depends on their location in the S segment. Expression of an RG in the NP locus is higher than that observed when the RG is located in the GPC locus. The physical separation of the GP and NP proteins into two different S segments (S1 and S2) represents a strong selective pressure to maintain a virus capable of packaging one L segment and two S segments; (**C**) recombinant bicistronic arenaviruses: In the bicistronic rLCMV/GFP-P2A-NP, the NP open reading frame in the pPol-I S plasmid is replaced by the GFP-P2A-NP sequence that contains GFP tagged to the N terminus of NP, separated by the PTV1 2A peptide sequence (P2A). The P2A sequence allows for production of both GFP and NP from the same bicistronic mRNA transcribed from the NP locus of the S genome segment. Untranslated regions (UTR, black boxes) and intergenic regions (IGR) in each of the vRNA L and S segments are indicated.

The generation of a recombinant bicistronic, reporter-expressing LCMV (rLCMV/GFP-P2A-NP) has been recently described (Figure 3B) [91]. In the rLCMV/GFP-P2A-NP, the NP ORF in the pPol-I S plasmid was replaced by the bicistronic ORF GFP-P2A-NP, which contained the ORF of green fluorescent protein (GFP) tagged to the N terminus of NP and was separated by the 2A peptide (P2A) sequence derived from the porcine teschovirus (PTV1) (Figure 4C) [91]. The P2A sequence allows for production of both GFP and NP proteins from the same mRNA transcribed from the NP locus of the S genome segments (Figure 4C) [91]. Thus, GFP expression levels serve as an accurate surrogate of virus multiplication levels in infected cells [91]. To generate rLCMV/GFP-P2A-NP, rodent BHK-21 cells were co-transfected with the pCAGGS plasmids encoding the minimal components of viral replication and transcription (NP and L) (Figure 3B, left) together with the pPol-I L and the pPol-I S GFP-P2A-NP plasmids (Figure 3B, right) [91]. Because rLCMV/GFP-P2A-NP expresses GFP upon viral infection, successful virus rescue can be monitored by the presence of GFP-expressing cells [91]. Characterization of rLCMV/GFP-P2A-NP indicates that GFP expression levels were higher than those expressed from r3LCMV viruses that express GFP from either the GPC or the NP loci [91]. Importantly, the growth kinetics of rLCMV/GFP-P2A-NP in BHK-21, A549 and Vero cells, were slower early in infection but reached similar peak titers as WT rLCMV (rLCMV/WT) [91], similar to the situation observed with r3 arenaviruses [74,95,100,101,103,104]. Notably, the early growth kinetic differences between rLCMV/GFP-2A-NP and rLCMV/WT were more noticeable in human A549 cells when using a low multiplicity of infection (MOI 0.01) [91]. This probably reflects that rLCMV/GFP-P2A-NP has a modest fitness decrease that was magnified in A549 cells due to the presence of a fully active IFN-I pathway. Thus, it is possible that incomplete processing during viral infection results in a GFP-P2A-NP polyprotein that cannot counteract the IFN-I response to levels comparable to WT NP [52–58]. As with the r3 arenaviruses, recombinant bicistronic LCMVs expressing the GPC of LASV [48] or JUNV [125] could be used for large HTS aimed at identifying inhibitors of LASV and JUNV GPC-mediated cell entry, respectively. Whether rLCMV/GFP-P2A-NP could be used for in vivo studies remains to be determined.

6. Applications of Reporter-Expressing Recombinant Arenavirus

6.1. Identification of Anti-Arenavirus Drugs

The development of HTS to screen a broad class of compounds that can target functions involved in different steps of the arenavirus infectious cycle would be of great value to identify potential novel anti-arenaviral drugs. This task would be facilitated by the generation of recombinant arenaviruses expressing appropriate

RG, leading to the development of assays amenable to HTS. For this aim, the r3 arenavirus platform has opened the possibility of rescuing recombinant arenaviruses containing two S and one L segment(s), where each of the two S segments contains a RG instead of either GPC or NP [74,95,100,101,103–105]. This strategy has been used to generate a variety of r3 arenaviruses (LCMV, JUNV and PICV) expressing different reporter genes including chloramphenicol acetyltransferase (CAT), GFP and luciferases (Table 1) [74,95,100,101,103–105]. Notably, these r3 arenaviruses have been shown to be phenotypically and genetically very stable, providing investigators with a fantastic tool for the development of HTS to globally identify inhibitors of arenavirus multiplication [74,95,100,101,103–105] (Table 1). This was illustrated by the use of r3LCMV CAT/FLuc that expressed CAT and the firefly luciferase gene (FLuc) in lieu of GPC and NP, respectively, to evaluate the effect of the nucleoside analog ribavirin and DL-2-hydroxymyristic acid (2-OHM), an inhibitor of arenavirus Z myristoylation that is required for efficient viral budding, on FLuc expression and virus production [103]. Levels of FLuc activity paralleled closely to titers of infectious progeny, demonstrating that RG expression can be used as a valid surrogate of inhibition of viral infection [103]. Similarly, r3Candid#1 reporter-expressing viruses grew to high titers in cultured cells and stably expressed both reporter genes [74]. Thus, RG-expressing r3Candid#1 viruses could allow the assessment of the antiviral activity of the compounds against NW arenaviruses [74]. Accordingly, the antiviral activity of the pyrimidine biosynthesis inhibitor A3 against LCMV and JUNV (Candid#1) using r3, reporter-expressing, arenaviruses has been evaluated [86] (Table 1). Results showed that the half maximal inhibitory concentration (IC_{50}) for A3 was about 100-fold lower than for ribavirin [86]. The anti-arenaviral activity of A3 was also observed in different cell types and species, including human A549 cells, and at drug concentrations that showed minimal effects on cell viability [86]. Moreover, readouts based on RG expression levels and viral titers gave similar IC_{50} values for each compound, similar to findings obtained with rLCMV/WT and rCandid#1/WT, further validating the use of r3 arenaviruses to screen for antiviral compounds using RG expression as readouts [86]. The antiviral activity of A3 on arenavirus was reverted by the exogenous addition of orotic acid, suggesting the involvement of the de novo pyrimidine biosynthesis pathway as the primary target of A3 [86]. Ribavirin and A3 appear to target different processes involved in arenavirus multiplication, and accordingly their use in combination therapy exhibited more potent arenavirus inhibitory activity than either single-drug treatment [86]. This combinatory therapy would allow circumventing some of the limitations of the current ribavirin monotherapy used for the treatment of arenavirus infection [1]. The recombinant bicistronic rLCMV/GFP-P2A-GFP has been also used for the identification of inhibitors of LCMV infection in the context of a cell-based HTS format (Table 1) [91]. Altogether, these results demonstrate the feasibility of using reporter-expressing r3 or bicistronic arenaviruses in cell-based assays to identify compounds with antiviral activity. Moreover, since RG expression can be used as a valid surrogate for viral replication, the replicating competent, RG-expressing arenaviruses can be used to screen large libraries of compounds for the rapid identification of compounds with anti-arenaviral activity [91].

Table 1. Reporter-expressing recombinant arenaviruses.

Virus	NP Loci	GPC Loci	Reference
r3LCMV GFP/CAT	CAT	GFP	[103]
r3LCMV CAT/GFP	GFP	CAT	[86,103]
r3LCMV GFP/GFP	GFP	GFP	[103,105]
r3LCMV CAT/FLuc	FLuc	CAT	[103]
r3LCMV FLuc/FLuc	FLuc	FLuc	[103]
r3LCMV GFP/Gluc	GFP	Gluc	[86,95,100,101]
r3LCMV Gluc/GFP	NP	GFP	[95]
r3LCMV/TransS GFP/Gluc	Gluc	GFP	[95]
r3LCMV/TransS Gluc/GFP	GFP	Gluc	[95]
r3LCMV/TransS GFP/Gluc	GFP	Gluc	[95]
r3LCMV GFP/IL-10	IL-10	GFP	[105]
r3LCMV GFP/Cre	Cre	GFP	[105]
rLCMV/GFP-P2A-NP	GFP	GPC	[91]
r3Candid#1 GFP/CAT	GFP	CAT	[74,86]
r3Candid#1 CAT/GFP	CAT	GFP	[74]
r3Candid#1 GFP/Gluc	GFP	Gluc	[86,101]
r3PICV GFP (rP18tri-G)	No RG	GFP	[104]
r3PICV GFP (rP18tri-G/H)	Influenza HA	GFP	[104]
r3PICV GFP (rP18tri-G/P)	Influenza NP	GFP	[104]

GFP: green fluorescent protein; r3: recombinant tri-segmented; LCMV: lymphocytic choriomeningitis virus; CAT: chloramphenicol acetyltransferase; FLuc: firefly luciferase; Gluc: Gaussia luciferase; PICV: Pinchinde virus; RG: reporter gene; HA: hemagglutinin; NP: nucleoprotein; IL-10: interleukin-10.

6.2. Studying the Biology of Arenaviruses In Vitro

Evidence indicated that r3LCMVs expressing different RG were stable both genetically and phenotypically and exhibited rLCMV/WT-like growth properties in culture cells [103]. Accordingly, RG expression from r3LCMV-infected cells provided an accurate surrogate of virus multiplication levels [86,103]. These results demonstrate the feasibility of using r3LCMVs to study several aspects in the biology of arenavirus in cell culture [86,103]. Thus, r3LCMV CAT/GFP (where CAT substituted for the viral GPC and GFP substituted for the viral NP) and r3LCMV GFP/CAT (where GFP substitute for the viral GPC and CAT substituted for the viral NP) (Table 1) (Figure 4A,B) [103] had similar growth properties in cultured cells, but normalized levels of CAT activity or GFP expression revealed higher levels of expression when the reporter gene was in the NP loci rather than in the GPC loci [103]. These results were further confirmed by results with rLCMV containing rearranged ORFs as a new approach to develop live attenuated vaccines [95], as well as with different r3LCMV expressing GFP or Gaussia luciferase (Gluc) from the NP or GPC loci (Table 1) [95]. Importantly, differences in RG expression by the different r3LCMV were not due to differences in numbers of infected cells or viral growth kinetics [95]. Consistent with these results, Cheng et al. generated a rLCMV containing a translocated viral S segment (rLCMV/TransS) where the viral NP and GPC ORF replaced one another [95]. The rLCMV/TransS showed slower growth kinetics in cell culture and was completely attenuated in a mouse model of lethal LCMV infection [95]. Notably, a single immunization dose of rLCMV/TransS conferred complete protection against a lethal challenge with rLCMV/WT [95]. To gain insights and to demonstrate that the mechanism of attenuation for LCMV/TransS was associated with reduced NP expression levels, Cheng et al. generated r3LCMVs containing both the NP and GPC genes under the regulation of either the antigenome (r3LCMV/TransNP) or the genome (r3LCMV/TransGPC) promoters [95]. To prevent discrepancies in measurements of RG expression under conditions of different UTRs, Cheng et al. used for comparison r3LCMVs encoding GFP and Gluc in positions corresponding to those found in r3LCMV/TransNP and r3LCMV/TransGPC (Table 1) [95]. They observed that r3LCMV/TransNP replicated similarly to r3LCMV/TransS. Moreover, viral growth kinetics corroborated that r3LCMV/TransS and r3LCMV/TransNP exhibited similar degrees of attenuation in cell culture [95]. In contrast, r3LCMV/TransGPC-infected and r3LCMV/WT-infected cells expressed similar GFP levels and viral growth kinetics, demonstrating that the two viruses also had similar replication capabilities [95]. Together, these results demonstrated that reduced NP expression levels, rather than increased GPC expression levels, were the major contributor of the impaired growth properties, and probably in vivo attenuation, observed with rLCMV/TransS [95].

In addition to demonstrating regulation of NP and GPC expression, results using r3LCMVs suggested that arenavirus coding regions do not seem to play a role in viral packaging and, therefore, both GPC and NP could be entirely replaced by a foreign gene without any attenuation in cell culture [74,86,95,100,101,103,104]. It remains to be evaluated the total length of a foreign RG to be incorporated and rescued in a r3 arenavirus. To date, FLuc (1.6 kb) is the largest foreign gene rescued in a r3LCMV (r3LCMV FLuc/FLuc) with a total genome size of 14.1 kb, compared to the 10.6 kb of r3LCMV/WT (Table 1) [103]. Notably, r3LCMV FLuc/FLuc was attenuated both in vitro and in vivo, suggesting a limit in the length of foreign genes that can be inserted in r3LCMVs [103]. Thus, the r3 arenavirus approach not only offers an opportunity to study the mechanisms responsible for controlling virus gene expression but also genome packaging for efficient viral fitness [103]. Moreover, the confirmation that arenavirus virions can package two S and one L genome segments in the r3 arenaviruses suggest this could be a very common event during the replication cycle of arenaviruses [121–123]. Likewise, it remains to be determined the size limit of the RG that can be expressed using the bicistronic approach without significantly affecting viral fitness [91].

6.3. Reporter-Expressing Arenaviruses for In Vivo Studies

Tri-segmented arenaviruses expressing RG provide investigators with a novel tool for the investigation of virus–host interactions in vivo. However, in contrast to results observed in cell culture, compared to rLCMV/WT, the r3LCMV GFP/GFP was attenuated in a mouse model of fatal lymphocytic choriomeningitis (LCM) following intracranial (i.c.) virus inoculation, as reflected by lower and delayed mortality [103]. By day 8 post-infection, only 37.5% of r3LCMV GFP/GFP-infected mice died and the remaining symptomatic mice recovered and did not show noticeable clinical symptoms by day 9 post-infection [103]. Nevertheless, the use of a higher i.c. dose (10^5 focus forming units, FFU) of r3LCMV GFP/GFP resulted in 100% lethality by day 7 post-infection [103]. Thus, although attenuated in mice, r3LCMVs can still be used for in vivo studies by increasing the amount of virus needed to obtain a phenotype similar to that of rLCMV/WT [103]. This outcome is similar to the situation observed with other replicating competent, RG-expressing NS segmented viruses, such as influenza [126,127]. To evaluate if a protective immune response was established in mice that survived the i.c. inoculation with 10^3 FFU of r3LCMV GFP/GFP, animals were challenged at day 21 post-infection with 10^3 FFU of rLCMV/WT [103]. All r3LCMV GFP/GFP inoculated mice exhibited complete protection [103]. The r3LCMV GFP/GFP and rLCMV/WT exhibited the same tropism within the brain of infected mice, although r3LCMV GFP/GFP displayed a more restricted distribution [103]. Cells in the ependymal, choroid plexus, and olfactory bulb were clearly infected by r3LCMV GFP/GFP, whereas infection of the meninges was patchy compared with rLCMV/WT [103]. GFP expression in r3LCMV GFP/GFP-infected mice was of sufficient intensity to be detected without the need for amplification or secondary quantification approaches, demonstrating the feasibility of using r3LCMV GFP/GFP for in vivo studies to identify the presence of infected cells [103].

The r3LCMV technology has also been used to express interleukin-10 (IL-10) and Cre recombinase genes in vivo (Table 1) [105]. Mice infected (i.c., 10^4 FFU) with r3LCMV/GFP-IL-10, but not r3LCMV GFP/GFP, were protected from lethal LCM, probably because IL-10 expression during viral infection leads to a decrease in immunopathology due to reduced CTL activity and modulation of macrophages and neutrophils pro-inflammatory activities [105]. These results support the potential of using r3LCMVs to determine the biological effects of candidate immune molecules, or other host genes, during the natural course of LCMV infection in vivo [105]. Infection of IL-10-deficient (IL-10−/−) mice with r3LCMV GFP/IL-10 resulted in a large number of hybridomas producing antibodies to IL-10 after a single immunization that after boosting resulted in several high-affinity clones specific to IL-10 [105]. These results also suggest the potential of using r3LCM viruses to generate antibodies, where LCMV infection can act as a natural adjuvant [105].

The characterization of a recombinant Pichinde virus, PICV (strain P18) with a trisegmented RNA genome (rP18tri-G) that expresses GFP from the GPC locus (Table 1) [104], showed that, similarly to other documented tri-segmented arenaviruses, the rP18tri was attenuated compared to rPICV/WT, but exhibited stability during serial passages in cultured cells [104]. rP18tri viruses expressing GFP from the GPC locus and either the hemagglutinin (HA, rP18tri-G/H) or the nucleoprotein (NP, rP18tri-G/P) of influenza A/Puerto Rico/8/34 (PR8) H1N1 from the NP locus have been also generated (Table 1) [104]. Mice immunized with a single low dose of rP18tri-G/H were protected against a lethal challenge with influenza PR8 [104]. Moreover, rP18tri-G/H was able to efficiently induce high levels of neutralizing antibodies against PR8 HA using intramuscular (i.m.), intranasal (i.n.) or intraperitoneal (i.p.) routes of infection [104]. Furthermore, the antibody neutralization titers were comparable to those induced by a formalin-inactivated influenza PR8 virus [104]. Likewise, mice immunized with rP18tri-G/P generated virus-specific CD8 and CD4 T cell responses above the background level seen in the rP18tri-G control group [104]. These data demonstrate that the rP18tri viruses can efficiently induce both humoral and cell-mediated immune responses via different immunization routes, leading to efficient protection. These results further demonstrate the feasibility of using the r3 arenavirus approach as a novel vaccine platform [104].

7. Conclusions

The development of arenavirus reverse genetics systems has provided investigators with a novel and powerful experimental approach to study basic aspects of arenavirus biology, including the identification of viral determinants, and their mechanisms of action, which contribute to arenaviral human diseases. The ability to manipulate the genome of arenaviruses has proven to be a superb model system to study virus–host interactions and associated disease. Moreover, generating recombinant arenaviruses with predetermined mutations allows investigators to gain a detailed understanding of arenavirus–host interactions and phenotypic outcomes of virus infection. Furthermore, the generation of r3 or bicistronic arenaviruses expressing appropriate RG, together with the development of specific cell-based assays for each of the different steps of the arenavirus life cycle, is facilitating novel approaches to discover and characterize antiviral drugs against these important human pathogens.

Acknowledgments: We want to thank past and current members in our laboratories and arenavirus virologists whose work has contributed to the generation of recombinant replicating-competent, reporter-expressing arenaviruses. Current arenavirus research in the Luis Martínez-Sobrido laboratory is funded by the National Institutes of Health (NIH) grants R21 AI119775-01 and R43 AI119775-01 and 1R21AI121550-01A1. Research in the Juan Carlos de la Torre laboratory is supported by grants RO1 AI047140, RO1 AI077719, and RO1 AI079665.

Author Contributions: Luis Martínez-Sobrido and Juan Carlos de la Torre wrote the manuscript.

Conflicts of Interest: The authors declare no conflict of interest.

References

1. Buchmeier, M.J.; Peter, C.J.; de la Torre, J.C. *Arenaviridae: The Viruses and Their Replication*; Lippincott William and Wilkins: Philadelphia, PA, USA, 2007; Volume 2.
2. Radoshitzky, S.R.; Bao, Y.; Buchmeier, M.J.; Charrel, R.N.; Clawson, A.N.; Clegg, C.S.; DeRisi, J.L.; Emonet, S.; Gonzalez, J.P.; Kuhn, J.H.; et al. Past, present, and future of arenavirus taxonomy. *Arch. Virol.* **2015**, *160*, 1851–1874.
3. Stenglein, M.D.; Jacobson, E.R.; Chang, L.W.; Sanders, C.; Hawkins, M.G.; Guzman, D.S.; Drazenovich, T.; Dunker, F.; Kamaka, E.K.; Fisher, D.; et al. Widespread recombination, reassortment, and transmission of unbalanced compound viral genotypes in natural arenavirus infections. *PLoS Pathog.* **2015**, *11*, e1004900.
4. Stenglein, M.D.; Leavitt, E.B.; Abramovitch, M.A.; McGuire, J.A.; DeRisi, J.L. Genome sequence of a bornavirus recovered from an African garter snake (*Elapsoidea loveridgei*). *Genome Announc.* **2014**, *2*.
5. Stenglein, M.D.; Sanders, C.; Kistler, A.L.; Ruby, J.G.; Franco, J.Y.; Reavill, D.R.; Dunker, F.; Derisi, J.L. Identification, characterization, and in vitro culture of highly divergent arenaviruses from boa constrictors and annulated tree boas: Candidate etiological agents for snake inclusion body disease. *mBio* **2012**, *3*.
6. Enria, D.A.; Briggiler, A.M.; Sanchez, Z. Treatment of argentine hemorrhagic fever. *Antivir. Res.* **2008**, *78*, 132–139.
7. Geisbert, T.W.; Jahrling, P.B. Exotic emerging viral diseases: Progress and challenges. *Nat. Med.* **2004**, *10*, S110–S121.
8. Khan, S.H.; Goba, A.; Chu, M.; Roth, C.; Healing, T.; Marx, A.; Fair, J.; Guttieri, M.C.; Ferro, P.; Imes, T.; et al. New opportunities for field research on the pathogenesis and treatment of Lassa fever. *Antivir. Res.* **2008**, *78*, 103–115.
9. McCormick, J.B.; Fisher-Hoch, S.P. Lassa fever. In *Arenaviruses i*; Oldstone, M.B., Ed.; Springer-Verlag: Berlin/Heidelberg, Germany, 2002; Volume 262, pp. 75–110.
10. Peters, C.J. Human infection with arenaviruses in the Americas. In *Arenaviruses I*; Oldstone, M.B., Ed.; Springer-Verlag: Berlin/Heidelberg, Germany, 2002; Volume 262, pp. 65–74.

11. Freedman, D.O.; Woodall, J. Emerging infectious diseases and risk to the traveler. *Med. Clin. N. Am.* **1999**, *83*, 865–883.

12. Holmes, G.P.; McCormick, J.B.; Trock, S.C.; Chase, R.A.; Lewis, S.M.; Mason, C.A.; Hall, P.A.; Brammer, L.S.; Perez-Oronoz, G.I.; McDonnell, M.K.; et al. Lassa fever in the United States. Investigation of a case and new guidelines for management. *N. Engl. J. Med.* **1990**, *323*, 1120–1123.

13. Isaacson, M. Viral hemorrhagic fever hazards for travelers in Africa. *Clin. Infect. Dis.* **2001**, *33*, 1707–1712.

14. Richmond, J.K.; Baglole, D.J. Lassa fever: Epidemiology, clinical features, and social consequences. *BMJ* **2003**, *327*, 1271–1275.

15. Briese, T.; Paweska, J.T.; McMullan, L.K.; Hutchison, S.K.; Street, C.; Palacios, G.; Khristova, M.L.; Weyer, J.; Swanepoel, R.; Egholm, M.; et al. Genetic detection and characterization of Lujo virus, a new hemorrhagic fever-associated arenavirus from Southern Africa. *PLoS Pathog.* **2009**, *5*, e1000455.

16. Kuns, M.L. Epidemiology of Machupo virus infection. II. Ecological and control studies of hemorrhagic fever. *Am. J. Trop. Med. Hyg.* **1965**, *14*, 813–816.

17. Webb, P.A.; Johnson, K.M.; Mackenzie, R.B.; Kuns, M.L. Some characteristics of Machupo virus, causative agent of Bolivian hemorrhagic fever. *Am. J. Trop. Med. Hyg.* **1967**, *16*, 531–538.

18. Delgado, S.; Erickson, B.R.; Agudo, R.; Blair, P.J.; Vallejo, E.; Albarino, C.G.; Vargas, J.; Comer, J.A.; Rollin, P.E.; Ksiazek, T.G.; et al. Chapare virus, a newly discovered arenavirus isolated from a fatal hemorrhagic fever case in Bolivia. *PLoS Pathog.* **2008**, *4*, e1000047.

19. Gonzalez, J.P.; Bowen, M.D.; Nichol, S.T.; Rico-Hesse, R. Genetic characterization and phylogeny of Sabia virus, an emergent pathogen in Brazil. *Virology* **1996**, *221*, 318–324.

20. Armstrong, L.R.; Dembry, L.M.; Rainey, P.M.; Russi, M.B.; Khan, A.S.; Fischer, S.H.; Edberg, S.C.; Ksiazek, T.G.; Rollin, P.E.; Peters, C.J. Management of a Sabia virus-infected patients in a US hospital. *Infect. Control Hosp. Epidemiol.* **1999**, *20*, 176–182.

21. Tesh, R.B.; Jahrling, P.B.; Salas, R.; Shope, R.E. Description of Guanarito virus (Arenaviridae: Arenavirus), the etiologic agent of Venezuelan hemorrhagic fever. *Am. J. Trop. Med. Hyg.* **1994**, *50*, 452–459.

22. Weaver, S.C.; Salas, R.A.; de Manzione, N.; Fulhorst, C.F.; Duno, G.; Utrera, A.; Mills, J.N.; Ksiazek, T.G.; Tovar, D.; Tesh, R.B. Guanarito virus (Arenaviridae) isolates from endemic and outlying localities in Venezuela: Sequence comparisons among and within strains isolated from Venezuelan hemorrhagic fever patients and rodents. *Virology* **2000**, *266*, 189–195.

23. Gonzalez, J.P.; Sanchez, A.; Rico-Hesse, R. Molecular phylogeny of Guanarito virus, an emerging arenavirus affecting humans. *Am. J. Trop. Med. Hyg.* **1995**, *53*, 1–6.

24. Fulhorst, C.F.; Bowen, M.D.; Ksiazek, T.G.; Rollin, P.E.; Nichol, S.T.; Kosoy, M.Y.; Peters, C.J. Isolation and characterization of whitewater arroyo virus, a novel North American arenavirus. *Virology* **1996**, *224*, 114–120.

25. Charrel, R.N.; de Lamballerie, X.; Fulhorst, C.F. The Whitewater Arroyo virus: Natural evidence for genetic recombination among Tacaribe serocomplex viruses (family Arenaviridae). *Virology* **2001**, *283*, 161–166.

26. Cajimat, M.N.; Milazzo, M.L.; Bradley, R.D.; Fulhorst, C.F. Ocozocoautla de espinosa virus and hemorrhagic fever, Mexico. *Emerg. Infect. Dis.* **2012**, *18*, 401–405.

27. Barton, L.L.; Mets, M.B. Lymphocytic choriomeningitis virus: Pediatric pathogen and fetal teratogen. *Pediatr. Infect. Dis. J.* **1999**, *18*, 540–541.

28. Barton, L.L.; Mets, M.B. Congenital lymphocytic choriomeningitis virus infection: Decade of rediscovery. *Clin. Infect. Dis.* **2001**, *33*, 370–374.

29. Barton, L.L.; Mets, M.B.; Beauchamp, C.L. Lymphocytic choriomeningitis virus: Emerging fetal teratogen. *Am. J. Obstet. Gynecol.* **2002**, *187*, 1715–1716.

30. Jahrling, P.B.; Peters, C.J. Lymphocytic choriomeningitis virus. A neglected pathogen of man. *Arch. Pathol. Lab. Med.* **1992**, *116*, 486–488.

31. Mets, M.B.; Barton, L.L.; Khan, A.S.; Ksiazek, T.G. Lymphocytic choriomeningitis virus: An underdiagnosed cause of congenital chorioretinitis. *Am. J. Ophthalmol.* **2000**, *130*, 209–215.

32. Fischer, S.A.; Graham, M.B.; Kuehnert, M.J.; Kotton, C.N.; Srinivasan, A.; Marty, F.M.; Comer, J.A.; Guarner, J.; Paddock, C.D.; DeMeo, D.L.; et al. Transmission of lymphocytic choriomeningitis virus by organ transplantation. *N. Engl. J. Med.* **2006**, *354*, 2235–2249.

33. Palacios, G.; Druce, J.; Du, L.; Tran, T.; Birch, C.; Briese, T.; Conlan, S.; Quan, P.L.; Hui, J.; Marshall, J.; et al. A new arenavirus in a cluster of fatal transplant-associated diseases. *N. Engl. J. Med.* **2008**, *358*, 991–998.

34. Peters, C.J. Lymphocytic choriomeningitis virus—An old enemy up to new tricks. *N. Engl. J. Med.* **2006**, *354*, 2208–2211.

35. Borio, L.; Inglesby, T.; Peters, C.J.; Schmaljohn, A.L.; Hughes, J.M.; Jahrling, P.B.; Ksiazek, T.; Johnson, K.M.; Meyerhoff, A.; O'Toole, T.; et al. Hemorrhagic fever viruses as biological weapons: Medical and public health management. *JAMA* **2002**, *287*, 2391–2405.

36. Damonte, E.B.; Coto, C.E. Treatment of arenavirus infections: From basic studies to the challenge of antiviral therapy. *Adv. Virus Res.* **2002**, *58*, 125–155.

37. Harvie, P.; Omar, R.F.; Dusserre, N.; Desormeaux, A.; Gourde, P.; Tremblay, M.; Beauchamp, D.; Bergeron, M.G. Antiviral efficacy and toxicity of ribavirin in murine acquired immunodeficiency syndrome model. *J. Acquir. Immune Defic. Syndr. Hum. Retrovirol.* **1996**, *12*, 451–461.

38. Omar, R.F.; Harvie, P.; Gourde, P.; Desormeaux, A.; Tremblay, M.; Beauchamp, D.; Bergeron, M.G. Antiviral efficacy and toxicity of ribavirin and foscarnet each given alone or in combination in the murine aids model. *Toxicol. Appl. Pharmacol.* **1997**, *143*, 140–151.

39. Snell, N.J. Ribavirin—Current status of a broad spectrum antiviral agent. *Expert Opin. Pharmacother.* **2001**, *2*, 1317–1324.

40. Oldstone, M.B. Biology and pathogenesis of lymphocytic choriomeningitis virus infection. In *Arenaviruses*; Berlin/Heidelberg, Germany; New York, NY, USA, 2002; Volume 263, pp. 83–118.

41. Zinkernagel, R.M. Lymphocytic choriomeningitis virus and immunology. *Curr. Top. Microbiol. Immunol.* **2002**, *263*, 1–5.

42. Lee, K.J.; Novella, I.S.; Teng, M.N.; Oldstone, M.B.; de La Torre, J.C. NP and L proteins of lymphocytic choriomeningitis virus (LCMV) are sufficient for efficient transcription and replication of LCMV genomic RNA analogs. *J. Virol.* **2000**, *74*, 3470–3477.

43. Strecker, T.; Eichler, R.; Meulen, J.; Weissenhorn, W.; Dieter Klenk, H.; Garten, W.; Lenz, O. Lassa virus Z protein is a matrix protein and sufficient for the release of virus-like particles [corrected]. *J. Virol.* **2003**, *77*, 10700–10705.

44. Perez, M.; Craven, R.C.; de la Torre, J.C. The small RING finger protein Z drives arenavirus budding: Implications for antiviral strategies. *Proc. Natl. Acad. Sci. USA* **2003**, *100*, 12978–12983.

45. Urata, S.; Noda, T.; Kawaoka, Y.; Yokosawa, H.; Yasuda, J. Cellular factors required for Lassa virus budding. *J. Virol.* **2006**, *80*, 4191–4195.

46. Pinschewer, D.D.; Perez, M.; Sanchez, A.B.; de la Torre, J.C. Recombinant lymphocytic choriomeningitis virus expressing vesicular stomatitis virus glycoprotein. *Proc. Natl. Acad. Sci. USA* **2003**, *100*, 7895–7900.

47. Beyer, W.R.; Popplau, D.; Garten, W.; von Laer, D.; Lenz, O. Endoproteolytic processing of the lymphocytic choriomeningitis virus glycoprotein by the subtilase SKI-1/S1P. *J. Virol.* **2003**, *77*, 2866–2872.

48. Rojek, J.M.; Sanchez, A.B.; Nguyen, N.T.; de la Torre, J.C.; Kunz, S. Different mechanisms of cell entry by human-pathogenic Old World and New World arenaviruses. *J. Virol.* **2008**, *82*, 7677–7687.

49. Lee, K.J.; Perez, M.; Pinschewer, D.D.; de la Torre, J.C. Identification of the lymphocytic choriomeningitis virus (LCMV) proteins required to rescue LCMV RNA analogs into LCMV-like particles. *J. Virol.* **2002**, *76*, 6393–6397.

50. Ortiz-Riano, E.; Cheng, B.Y.; de la Torre, J.C.; Martinez-Sobrido, L. Self-association of lymphocytic choriomeningitis virus nucleoprotein is mediated by its N-terminal region and is not required for its anti-interferon function. *J. Virol.* **2012**, *86*, 3307–3317.

51. Ortiz-Riano, E.; Cheng, B.Y.; de la Torre, J.C.; Martinez-Sobrido, L. The C-terminal region of lymphocytic choriomeningitis virus nucleoprotein contains distinct and segregable functional domains involved in NP-Z interaction and counteraction of the type I interferon response. *J. Virol.* **2011**, *85*, 13038–13048.

52. Pythoud, C.; Rodrigo, W.W.; Pasqual, G.; Rothenberger, S.; Martinez-Sobrido, L.; de la Torre, J.C.; Kunz, S. Arenavirus nucleoprotein targets interferon regulatory factor-activating kinase IKKepsilon. *J. Virol.* **2012**, *86*, 7728–7738.

53. Martinez-Sobrido, L.; Emonet, S.; Giannakas, P.; Cubitt, B.; Garcia-Sastre, A.; de la Torre, J.C. Identification of amino acid residues critical for the anti-interferon activity of the nucleoprotein of the prototypic arenavirus lymphocytic choriomeningitis virus. *J. Virol.* **2009**, *83*, 11330–11340.

54. Martinez-Sobrido, L.; Giannakas, P.; Cubitt, B.; Garcia-Sastre, A.; de la Torre, J.C. Differential inhibition of type I interferon induction by arenavirus nucleoproteins. *J. Virol.* **2007**, *81*, 12696–12703.

55. Martinez-Sobrido, L.; Zuniga, E.I.; Rosario, D.; Garcia-Sastre, A.; de la Torre, J.C. Inhibition of the type I interferon response by the nucleoprotein of the prototypic arenavirus lymphocytic choriomeningitis virus. *J. Virol.* **2006**, *80*, 9192–9199.

56. Borrow, P.; Martinez-Sobrido, L.; de la Torre, J.C. Inhibition of the type I interferon antiviral response during arenavirus infection. *Viruses* **2010**, *2*, 2443–2480.

57. Pythoud, C.; Rothenberger, S.; Martinez-Sobrido, L.; de la Torre, J.C.; Kunz, S. Lymphocytic choriomeningitis virus differentially affects the virus-induced type I interferon response and mitochondrial apoptosis mediated by RIG-I/MAVS. *J. Virol.* **2015**, *89*, 6240–6250.

58. Rodrigo, W.W.; Ortiz-Riano, E.; Pythoud, C.; Kunz, S.; de la Torre, J.C.; Martinez-Sobrido, L. Arenavirus nucleoproteins prevent activation of nuclear factor Kappa B. *J. Virol.* **2012**, *86*, 8185–8197.

59. Igonet, S.; Vaney, M.C.; Vonrhein, C.; Bricogne, G.; Stura, E.A.; Hengartner, H.; Eschli, B.; Rey, F.A. X-ray structure of the arenavirus glycoprotein GP2 in its postfusion hairpin conformation. *Proc. Natl. Acad. Sci. USA* **2011**, *108*, 19967–19972.

60. Burri, D.J.; da Palma, J.R.; Kunz, S.; Pasquato, A. Envelope glycoprotein of arenaviruses. *Viruses* **2012**, *4*, 2162–2181.

61. Cao, W.; Henry, M.D.; Borrow, P.; Yamada, H.; Elder, J.H.; Ravkov, E.V.; Nichol, S.T.; Compans, R.W.; Campbell, K.P.; Oldstone, M.B. Identification of alpha-dystroglycan as a receptor for lymphocytic choriomeningitis virus and Lassa fever virus. *Science* **1998**, *282*, 2079–2081.

62. Kunz, S.; Borrow, P.; Oldstone, M.B. Receptor structure, binding, and cell entry of arenaviruses. *Curr. Top. Microbiol. Immunol.* **2002**, *262*, 111–137.

63. Kunz, S.; Sevilla, N.; McGavern, D.B.; Campbell, K.P.; Oldstone, M.B. Molecular analysis of the interaction of LCMV with its cellular receptor [alpha]-dystroglycan. *J. Cell Biol.* **2001**, *155*, 301–310.

64. Radoshitzky, S.R.; Abraham, J.; Spiropoulou, C.F.; Kuhn, J.H.; Nguyen, D.; Li, W.; Nagel, J.; Schmidt, P.J.; Nunberg, J.H.; Andrews, N.C.; et al. Transferrin receptor 1 is a cellular receptor for new world haemorrhagic fever arenaviruses. *Nature* **2007**, *446*, 92–96.

65. Pasqual, G.; Rojek, J.M.; Masin, M.; Chatton, J.Y.; Kunz, S. Old World arenaviruses enter the host cell via the multivesicular body and depend on the endosomal sorting complex required for transport. *PLoS Pathog.* **2011**, *7*, e1002232.

66. Pinschewer, D.D.; Perez, M.; de la Torre, J.C. Dual role of the lymphocytic choriomeningitis virus intergenic region in transcription termination and virus propagation. *J. Virol.* **2005**, *79*, 4519–4526.

67. Capul, A.A.; Perez, M.; Burke, E.; Kunz, S.; Buchmeier, M.J.; de la Torre, J.C. Arenavirus Z-glycoprotein association requires Z myristoylation but not functional RING or late domains. *J. Virol.* **2007**, *81*, 9451–9460.

68. Perez, M.; Greenwald, D.L.; de la Torre, J.C. Myristoylation of the RING finger Z protein is essential for arenavirus budding. *J. Virol.* **2004**, *78*, 11443–11448.

69. Strecker, T.; Maisa, A.; Daffis, S.; Eichler, R.; Lenz, O.; Garten, W. The role of myristoylation in the membrane association of the Lassa virus matrix protein Z. *Virol. J.* **2006**, *3*, 93.

70. Enria, D.A.; Barrera Oro, J.G. Junin virus vaccines. *Curr. Top. Microbiol. Immunol.* **2002**, *263*, 239–261.

71. Maiztegui, J.I.; McKee, K.T., Jr.; Barrera Oro, J.G.; Harrison, L.H.; Gibbs, P.H.; Feuillade, M.R.; Enria, D.A.; Briggiler, A.M.; Levis, S.C.; Ambrosio, A.M.; et al. Protective efficacy of a live attenuated vaccine against Argentine hemorrhagic fever. AHF Study Group. *J. Infect. Dis.* **1998**, *177*, 277–283.

72. Falzarano, D.; Feldmann, H. Vaccines for viral hemorrhagic fevers—Progress and shortcomings. *Curr. Opin. Virol.* **2013**, *3*, 343–351.

73. Albarino, C.G.; Bergeron, E.; Erickson, B.R.; Khristova, M.L.; Rollin, P.E.; Nichol, S.T. Efficient reverse genetics generation of infectious Junin viruses differing in glycoprotein processing. *J. Virol.* **2009**, *83*, 5606–5614.

74. Emonet, S.F.; Seregin, A.V.; Yun, N.E.; Poussard, A.L.; Walker, A.G.; de la Torre, J.C.; Paessler, S. Rescue from cloned cDNAs and in vivo characterization of recombinant pathogenic Romero and live-attenuated Candid#1 strains of Junin virus, the causative agent of Argentine hemorrhagic fever disease. *J. Virol.* **2011**, *85*, 1473–1483.

75. Albarino, C.G.; Bird, B.H.; Chakrabarti, A.K.; Dodd, K.A.; Erickson, B.R.; Nichol, S.T. Efficient rescue of recombinant Lassa virus reveals the influence of S segment noncoding regions on virus replication and virulence. *J. Virol.* **2011**, *85*, 4020–4024.

76. Yun, N.E.; Seregin, A.V.; Walker, D.H.; Popov, V.L.; Walker, A.G.; Smith, J.N.; Miller, M.; de la Torre, J.C.; Smith, J.K.; Borisevich, V.; et al. Mice lacking functional STAT1 are highly susceptible to lethal infection with Lassa virus. *J. Virol.* **2013**, *87*, 10908–10911.

77. McCormick, J.B.; King, I.J.; Webb, P.A.; Scribner, C.L.; Craven, R.B.; Johnson, K.M.; Elliott, L.H.; Belmont-Williams, R. Lassa fever. Effective therapy with ribavirin. *N. Engl. J. Med.* **1986**, *314*, 20–26.

78. Kilgore, P.E.; Ksiazek, T.G.; Rollin, P.E.; Mills, J.N.; Villagra, M.R.; Montenegro, M.J.; Costales, M.A.; Paredes, L.C.; Peters, C.J. Treatment of Bolivian hemorrhagic fever with intravenous ribavirin. *Clin. Infect. Dis.* **1997**, *24*, 718–722.

79. McKee, K.T., Jr.; Huggins, J.W.; Trahan, C.J.; Mahlandt, B.G. Ribavirin prophylaxis and therapy for experimental Argentine hemorrhagic fever. *Antimicrob. Agents Chemother.* **1988**, *32*, 1304–1309.

80. Leyssen, P.; De Clercq, E.; Neyts, J. Molecular strategies to inhibit the replication of RNA viruses. *Antivir. Res.* **2008**, *78*, 9–25.

81. Parker, W.B. Metabolism and antiviral activity of ribavirin. *Virus Res.* **2005**, *107*, 165–171.

82. Cameron, C.E.; Castro, C. The mechanism of action of ribavirin: Lethal mutagenesis of RNA virus genomes mediated by the viral RNA-dependent RNA polymerase. *Curr. Opin. Infect. Dis.* **2001**, *14*, 757–764.

83. Crotty, S.; Maag, D.; Arnold, J.J.; Zhong, W.; Lau, J.Y.N.; Hong, Z.; Andino, R.; Cameron, C.E. The broad-spectrum antiviral ribonucleotide, ribavirin, is an RNA virus mutagen. *Nat. Med.* **2000**, *6*, 1375–1379.

84. Ruiz-Jarabo, C.M.; Ly, C.; Domingo, E.; de la Torre, J.C. Lethal mutagenesis of the prototypic arenavirus lymphocytic choriomeningitis virus (LCMV). *Virology* **2003**, *308*, 37–47.

85. Hoffmann, H.H.; Kunz, A.; Simon, V.A.; Palese, P.; Shaw, M.L. Broad-spectrum antiviral that interferes with de novo pyrimidine biosynthesis. *Proc. Natl. Acad. Sci. USA* **2011**, *108*, 5777–5782.

86. Ortiz-Riano, E.; Ngo, N.; Devito, S.; Eggink, D.; Munger, J.; Shaw, M.L.; de la Torre, J.C.; Martinez-Sobrido, L. Inhibition of arenavirus by A3, a pyrimidine biosynthesis inhibitor. *J. Virol.* **2014**, *88*, 878–889.

87. Gowen, B.B.; Juelich, T.L.; Sefing, E.J.; Brasel, T.; Smith, J.K.; Zhang, L.; Tigabu, B.; Hill, T.E.; Yun, T.; Pietzsch, C.; et al. Favipiravir (T-705) inhibits Junin virus infection and reduces mortality in a guinea pig model of Argentine hemorrhagic fever. *PLoS Negl. Trop. Dis.* **2013**, *7*, e2614.

88. Mendenhall, M.; Russell, A.; Juelich, T.; Messina, E.L.; Smee, D.F.; Freiberg, A.N.; Holbrook, M.R.; Furuta, Y.; de la Torre, J.C.; Nunberg, J.H.; et al. T-705 (favipiravir) inhibition of arenavirus replication in cell culture. *Antimicrob. Agents Chemother.* **2011**, *55*, 782–787.

89. Bolken, T.C.; Laquerre, S.; Zhang, Y.; Bailey, T.R.; Pevear, D.C.; Kickner, S.S.; Sperzel, L.E.; Jones, K.F.; Warren, T.K.; Amanda Lund, S.; et al. Identification and characterization of potent small molecule inhibitor of hemorrhagic fever New World arenaviruses. *Antivir. Res.* **2006**, *69*, 86–97.

90. Lee, A.M.; Rojek, J.M.; Spiropoulou, C.F.; Gundersen, A.T.; Jin, W.; Shaginian, A.; York, J.; Nunberg, J.H.; Boger, D.L.; Oldstone, M.B.; et al. Unique small molecule entry inhibitors of hemorrhagic fever arenaviruses. *J. Biol. Chem.* **2008**, *283*, 18734–18742.

91. Ngo, N.; Cubitt, B.; Iwasaki, M.; de la Torre, J.C. Identification and mechanism of action of a novel small-molecule inhibitor of arenavirus multiplication. *J. Virol.* **2015**, *89*, 10924–10933.

92. Loureiro, M.E.; D'Antuono, A.; Levingston Macleod, J.M.; Lopez, N. Uncovering viral protein-protein interactions and their role in arenavirus life cycle. *Viruses* **2012**, *4*, 1651–1667.

93. De la Torre, J.C. Reverse genetics approaches to combat pathogenic arenaviruses. *Antivir. Res.* **2008**, *80*, 239–250.

94. Emonet, S.E.; Urata, S.; de la Torre, J.C. Arenavirus reverse genetics: New approaches for the investigation of arenavirus biology and development of antiviral strategies. *Virology* **2011**, *411*, 416–425.

95. Cheng, B.Y.; Ortiz-Riano, E.; de la Torre, J.C.; Martinez-Sobrido, L. Arenavirus genome rearrangement for the development of live-attenuated vaccines. *J. Virol.* **2015**.

96. Ortiz-Riano, E.; Cheng, B.Y.; de la Torre, J.C.; Martinez-Sobrido, L. D471G mutation in LCMV-NP affects its ability to self-associate and results in a dominant negative effect in viral RNA synthesis. *Viruses* **2012**, *4*, 2137–2161.

97. Russier, M.; Reynard, S.; Carnec, X.; Baize, S. The exonuclease domain of Lassa virus nucleoprotein is involved in antigen-presenting-cell-mediated NK cell responses. *J. Virol.* **2014**, *88*, 13811–13820.

98. Reynard, S.; Russier, M.; Fizet, A.; Carnec, X.; Baize, S. Exonuclease domain of the Lassa virus nucleoprotein is critical to avoid RIG-I signaling and to inhibit the innate immune response. *J. Virol.* **2014**, *88*, 13923–13927.

99. Seregin, A.V.; Yun, N.E.; Miller, M.; Aronson, J.; Smith, J.K.; Walker, A.G.; Smith, J.N.; Huang, C.; Manning, J.T.; de la Torre, J.C.; et al. The glycoprotein precursor gene of Junin virus determines the virulence of Romero strain and attenuation of Candid#1 strain in a representative animal model of Argentine hemorrhagic fever. *J. Virol.* **2015**, *89*, 5949–5956.

100. Ortiz-Riano, E.; Cheng, B.Y.; Carlos de la Torre, J.; Martinez-Sobrido, L. Arenavirus reverse genetics for vaccine development. *J. Gen. Virol.* **2013**, *94*, 1175–1188.

101. Cheng, B.Y.; Ortiz-Riano, E.; de la Torre, J.C.; Martinez-Sobrido, L. Generation of recombinant arenavirus for vaccine development in FDA-approved Vero cells. *J. Vis. Exp.* **2013**.

102. Cheng, B.Y.; Ortiz-Riano, E.; Nogales, A.; de la Torre, J.C.; Martinez-Sobrido, L. Development of live-attenuated arenavirus vaccines based on codon deoptimization. *J. Virol.* **2015**.

103. Emonet, S.F.; Garidou, L.; McGavern, D.B.; de la Torre, J.C. Generation of recombinant lymphocytic choriomeningitis viruses with trisegmented genomes stably expressing two additional genes of interest. *Proc. Natl. Acad. Sci. USA* **2009**, *106*, 3473–3478.

104. Dhanwani, R.; Zhou, Y.; Huang, Q.; Verma, V.; Dileepan, M.; Ly, H.; Liang, Y. A novel live pichinde virus-based vaccine vector induces enhanced humoral and cellular immunity after a booster dose. *J. Virol.* **2015**, *90*, 2551–2560.

105. Popkin, D.L.; Teijaro, J.R.; Lee, A.M.; Lewicki, H.; Emonet, S.; de la Torre, J.C.; Oldstone, M. Expanded potential for recombinant trisegmented lymphocytic choriomeningitis viruses: Protein production, antibody production, and in vivo assessment of biological function of genes of interest. *J. Virol.* **2011**, *85*, 7928–7932.

106. Rodrigo, W.W.; de la Torre, J.C.; Martinez-Sobrido, L. Use of single-cycle infectious lymphocytic choriomeningitis virus to study hemorrhagic fever arenaviruses. *J. Virol.* **2011**, *85*, 1684–1695.

107. Flatz, L.; Hegazy, A.N.; Bergthaler, A.; Verschoor, A.; Claus, C.; Fernandez, M.; Gattinoni, L.; Johnson, S.; Kreppel, F.; Kochanek, S.; et al. Development of replication-defective lymphocytic choriomeningitis virus vectors for the induction of potent CD^{8+} T cell immunity. *Nat. Med.* **2010**, *16*, 339–345.

108. Nisii, C.; Castilletti, C.; Raoul, H.; Hewson, R.; Brown, D.; Gopal, R.; Eickmann, M.; Gunther, S.; Mirazimi, A.; Koivula, T.; et al. Biosafety level-4 laboratories in europe: Opportunities for public health, diagnostics, and research. *PLoS Pathog.* **2013**, *9*, e1003105.

109. Sanchez, A.B.; de la Torre, J.C. Rescue of the prototypic arenavirus LCMV entirely from plasmid. *Virology* **2006**, *350*, 370–380.

110. Flatz, L.; Bergthaler, A.; de la Torre, J.C.; Pinschewer, D.D. Recovery of an arenavirus entirely from RNA polymerase I/II-driven cDNA. *Proc. Natl. Acad. Sci. USA* **2006**, *103*, 4663–4668.

111. Hass, M.; Golnitz, U.; Muller, S.; Becker-Ziaja, B.; Gunther, S. Replicon system for Lassa virus. *J. Virol.* **2004**, *78*, 13793–13803.

112. Bergeron, E.; Chakrabarti, A.K.; Bird, B.H.; Dodd, K.A.; McMullan, L.K.; Spiropoulou, C.F.; Nichol, S.T.; Albarino, C.G. Reverse genetics recovery of Lujo virus and role of virus RNA secondary structures in efficient virus growth. *J. Virol.* **2012**, *86*, 10759–10765.

113. Lan, S.; McLay Schelde, L.; Wang, J.; Kumar, N.; Ly, H.; Liang, Y. Development of infectious clones for virulent and avirulent pichinde viruses: A model virus to study arenavirus-induced hemorrhagic fevers. *J. Virol.* **2009**, *83*, 6357–6362.

114. Patterson, M.; Seregin, A.; Huang, C.; Kolokoltsova, O.; Smith, J.; Miller, M.; Smith, J.; Yun, N.; Poussard, A.; Grant, A.; et al. Rescue of a recombinant Machupo virus from cloned cDNAs and in vivo characterization in interferon (alphabeta/gamma) receptor double knockout mice. *J. Virol.* **2014**, *88*, 1914–1923.

115. Flick, R.; Hobom, G. Transient bicistronic vRNA segments for indirect selection of recombinant influenza viruses. *Virology* **1999**, *262*, 93–103.

116. Vieira Machado, A.; Naffakh, N.; Gerbaud, S.; van der Werf, S.; Escriou, N. Recombinant influenza a viruses harboring optimized dicistronic NA segment with an extended native 5′ terminal sequence: Induction of heterospecific B and T cell responses in mice. *Virology* **2006**, *345*, 73–87.

117. Marschalek, A.; Finke, S.; Schwemmle, M.; Mayer, D.; Heimrich, B.; Stitz, L.; Conzelmann, K.K. Attenuation of rabies virus replication and virulence by picornavirus internal ribosome entry site elements. *J. Virol.* **2009**, *83*, 1911–1919.

118. Garcia-Sastre, A.; Muster, T.; Barclay, W.S.; Percy, N.; Palese, P. Use of a mammalian internal ribosomal entry site element for expression of a foreign protein by a transfectant influenza virus. *J. Virol.* **1994**, *68*, 6254–6261.

119. Goto, H.; Muramoto, Y.; Noda, T.; Kawaoka, Y. The genome-packaging signal of the influenza a virus genome comprises a genome incorporation signal and a genome-bundling signal. *J. Virol.* **2013**, *87*, 11316–11322.

120. Liang, Y.; Hong, Y.; Parslow, T.G. Cis-acting packaging signals in the influenza virus PB1, PB2, and PA genomic RNA segments. *J. Virol.* **2005**, *79*, 10348–10355.

121. Meyer, B.J.; de la Torre, J.C.; Southern, P.J. Arenaviruses: Genomic RNAs, transcription, and replication. *Curr. Top. Microbiol. Immunol.* **2002**, *262*, 139–157.

122. Buchmeier, M.J. Arenaviruses: Protein structure and function. *Curr. Top. Microbiol. Immunol.* **2002**, *262*, 159–173.

123. Young, P.R.; Howard, C.R. Fine structure analysis of pichinde virus nucleocapsids. *J. Gen. Virol.* **1983**, *64*, 833–842.

124. Lavanya, M.; Cuevas, C.D.; Thomas, M.; Cherry, S.; Ross, S.R. SiRNA screen for genes that affect Junin virus entry uncovers voltage-gated calcium channels as a therapeutic target. *Sci. Transl. Med.* **2013**, *5*, 204ra131.

125. Iwasaki, M.; Urata, S.; Cho, Y.; Ngo, N.; de la Torre, J.C. Cell entry of lymphocytic choriomeningitis virus is restricted in myotubes. *Virology* **2014**, *458–459*, 22–32.

126. Nogales, A.; Baker, S.F.; Martinez-Sobrido, L. Replication-competent influenza a viruses expressing a red fluorescent protein. *Virology* **2015**, *476*, 206–216.

127. Breen, M.; Nogales, A.; Baker, S.F.; Perez, D.R.; Martinez-Sobrido, L. Replication-competent influenza A and B viruses expressing a fluorescent dynamic timer protein for in vitro and in vivo studies. *PLoS ONE* **2016**, *11*, e0147723.

Fluorescent and Bioluminescent Reporter Myxoviruses

Christina A. Rostad, Michael C. Currier and Martin L. Moore

Abstract: The advent of virus reverse genetics has enabled the incorporation of genetically encoded reporter proteins into replication-competent viruses. These reporters include fluorescent proteins which have intrinsic chromophores that absorb light and re-emit it at lower wavelengths, and bioluminescent proteins which are luciferase enzymes that react with substrates to produce visible light. The incorporation of these reporters into replication-competent viruses has revolutionized our understanding of molecular virology and aspects of viral tropism and transmission. Reporter viruses have also enabled the development of high-throughput assays to screen antiviral compounds and antibodies and to perform neutralization assays. However, there remain technical challenges with the design of replication-competent reporter viruses, and each reporter has unique advantages and disadvantages for specific applications. This review describes currently available reporters, design strategies for incorporating reporters into replication-competent paramyxoviruses and orthomyxoviruses, and the variety of applications for which these tools can be utilized both in vitro and in vivo.

Reprinted from *Viruses*. Cite as: Rostad, C.A.; Currier, M.C.; Moore, M.L. Fluorescent and Bioluminescent Reporter Myxoviruses. *Viruses* **2016**, *8*, 214.

1. Introduction

Fluorescent proteins (FPs) were first described in the 1960s when the green fluorescent protein (GFP) was isolated from the *Aequorea victoria* jellyfish [1]. GFP was first sequenced and cloned in 1992 [2], and it was first used as a reporter protein to mark gene expression in *Escherichia coli* and in *Caenorhabditis elegans* in 1994 [3]. Since that time, an array of naturally occurring fluorescent proteins and genetically engineered derivatives have been developed that encompass nearly the entire the spectrum of visible wavelengths. As reporters, these proteins have made the monitoring and visualization of numerous intracellular events with high temporal and spatial resolution possible, ranging from gene expression, to protein localization and interaction, to cellular signaling and trafficking. Using virus reverse genetics, fluorescent proteins have also been incorporated into an increasing number of replication-competent viruses, elucidating aspects of viral pathogenesis, tropism, and transmission. Bioluminescent proteins have similarly been incorporated into live viruses and have recently gained traction as tools for characterizing viral infections, especially in vivo [4]. The purpose of this review is to examine the properties of fluorescent and bioluminescent proteins, their applications in replication-competent paramyxoviruses and orthomyxoviruses, and considerations for their design and development.

2. Characteristics of Fluorescent and Bioluminescent Proteins

Fluorescent proteins of the GFP family are homologous genetically-encoded proteins with intrinsic fluorescence which is not dependent on exogenous substrates other than molecular oxygen [5]. The structure of GFP consists of 11 β-sheets which form a barrel around a centrally located chromophore [6]. The chromophore is derived from three amino acids within the polypeptide sequence, which undergo unique post-translational modifications at residues 65–67 (Ser-Tyr-Gly in *Aequorea victoria* GFP). The central location of the chromophore renders it relatively stable to changes in temperature, pH, and physical stress. Naturally occurring fluorescent proteins of the GFP family have been described which encompass a broad palette of colors, including red (DsRed) [7], yellow (zFP538) [7], and green-to-red (the photoconvertible protein Kaede) [8]. Mutagenesis of residues within naturally occurring chromophores has enabled the development of unique fluorescent proteins with broadened spectral diversity. An important example has been the development of the monomeric, far-red fluorescent proteins Katushka (mKate) [9] and its brighter variant mKate2 [10] via random and site-directed mutagenesis of eqFP586 from the sea anemone *Entacmaea quadricolor*, which have been used in tissue (Figure 1) and whole body imaging. The genetically engineered fluorescent proteins also have a range of brightness, maturation rate, photostability, pH stability, and tendency for oligomerization; each of these properties may play an important role in the selection of an appropriate protein for a specific application [11].

Bioluminescent proteins (BPs) are luciferase enzymes, which react with substrates to produce visible light. They are distinct from fluorescent proteins, which have intrinsic chromophores that absorb light and re-emit it at lower wavelengths. Although bioluminescent and fluorescent proteins are utilized in many analogous applications, BPs have established a particular niche in the live bioluminescent imaging of small animal models [12]. BPs are advantageous in live imaging because they are highly sensitive (can detect down to 10 [2] virus plaque forming units/mL) [13] and have high signal-to-noise ratios. One disadvantage of BPs is their requirement for an exogenous substrate, which must be delivered to cells or tissues at sufficient concentration to generate signal. The firefly luciferase has been the most commonly utilized bioluminescent protein for in vivo imaging because its substrate luciferin has a good pharmacokinetic profile, sufficient bioavailability, and a red-shifted emission spectrum [14]. Recently, a novel small luciferase enzyme called NanoLuc and its substrate furimazine were engineered from the deep-sea shrimp *Oplophorus gracilirostris* [15]. Purified NanoLuc produced 150-fold greater luminescence than either firefly or *Renilla* luciferases and exhibited prolonged enzyme stability and signal duration in vitro [15]. The utility of this promising new reporter has been demonstrated with in vivo imaging of viral infections in small animals, although the pharmacokinetic profile of the substrate furimazine has not yet been fully elucidated and the blue-shifted emission wavelength may be disadvantageous [16,17].

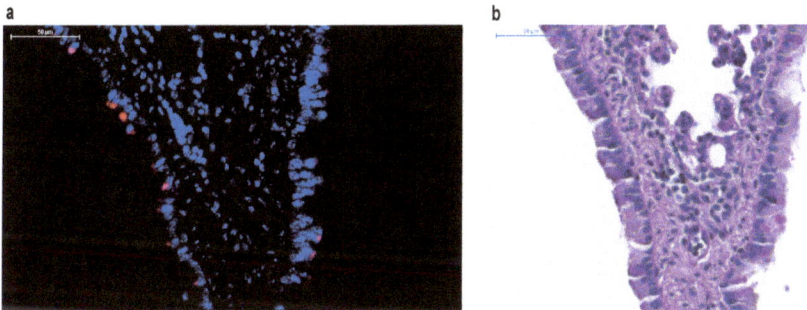

Figure 1. Respiratory syncytial virus (RSV) labeled with monomeric red fluorescent protein mKate2 in mouse lower respiratory tract epithelial cells. (**a**) RSV localizes to the apical surface of respiratory epithelial cells and can be tracked in vivo in paraffin-embedded tissues using mKate2 as demonstrated with fluorescent imaging using 4′,6-diamidino-2-phenylindole (DAPI) counterstaining; (**b**) Periodic Acid-Schiff (PAS) stain of the same lung specimen.

3. Reverse Genetics Systems for Paramyxoviruses and Orthomyxoviruses

The advent of virus reverse genetics has enabled the incorporation of genetically encoded reporter proteins into replication-competent viruses. The first negative-sense RNA virus to be rescued entirely from cloned cDNA was rabies virus in 1994 [18]. By 1996, vesicular stomatitis virus [19] and three paramyxoviruses, measles [20], Sendai virus (murine parainfluenza 1) [21], and respiratory syncytial virus [22] had been rescued by similar methods. Reverse genetics systems for all genera of the *Paramyxoviridae* family have now been successfully developed.

Paramyxoviridae are non-segmented, single-stranded, negative-sense RNA viruses which consist of two subfamilies: *Paramyxovirinae* and *Pneumovirinae*. Clinically significant paramyxoviruses which are pathogenic in humans include measles (MeV), mumps, the highly virulent Nipah and Hendra viruses, and respiratory viruses such as human metapneumovirus (HMPV), human respiratory syncytial virus (RSV), and the human parainfluenza viruses. Paramyxoviruses consist of 6–10 genes, which are transcribed by the viral ribonucleoprotein (RNP) complex down a transcriptional gradient, with the 3′ genes transcribed in higher quantities than the 5′ genes. Unlike positive-sense viruses, negative-sense viral RNA is not infectious, as it requires a functional RNP complex to perform transcription and replication. Reverse genetics approaches have therefore relied upon co-transfection of genes encoding proteins of the RNP complex (nucleoprotein, phosphoprotein, and large polymerase) with viral cDNA under the control of a promoter, often T7 polymerase. The T7 polymerase is typically supplied either by a constitutively expressing cell line (e.g., BSR-T7/5 cells), a co-transfected plasmid, or co-infection with a T7 polymerase-producing virus. Alternative approaches have implemented inherent cellular RNA polymerase I or II under control of the CMV promoter to transcribe viral genes [23,24].

Orthomyxoviridae are segmented, single-stranded, negative-sense RNA viruses. The most important human pathogens in the *Orthomyxoviridae* family are the influenza viruses, which cause annual seasonal pandemics worldwide. Early attempts to rescue recombinant influenza virus via reverse genetics utilized helper viruses to supply viral transcription and replication machinery [25,26]; however, these early approaches required selection against the helper viruses, which was often associated with technical challenges. In 1999, Neumann et al. successfully generated influenza virus entirely from cloned cDNA by co-transfecting eight plasmids encoding the viral gene segments under the control of RNA polymerase I along with four plasmids encoding the components of viral RNP (N, PB1, PB2, and PA) under the control of RNA polymerase II [27]. In 2000, Hoffmann et al. then developed an eight-plasmid cDNA transfection system for the rescue of influenza A by inserting viral RNA flanked by RNA polymerase 1 promoter and terminator sequences between RNA polymerase 2 promoter and polyadenylation sites oriented in the reverse direction. This schema enabled the synthesis of both negative- and positive-sense RNA from the same cDNA template and reduced the number of plasmids needed for successful viral rescue to eight [28]. Because eukaryotic cell lines approved for vaccine production (e.g., Vero cells) have low transfection efficiencies, further efforts were made to decrease the number of plasmids required for influenza virus production. In 2005, Neumann et al. successfully rescued influenza virus from a single plasmid encoding all eight viral RNA segments under the control of RNA polymerase I. Virus yield was further increased when this plasmid was co-transfected with helper plasmids encoding the components of the RNP complex [29].

4. Design of Reporter Paramyxoviruses and Orthomyxoviruses

4.1. Selection of Fluorescent Proteins

There is not a universal approach to the design of fluorescent and bioluminescent reporter viruses using reverse genetics. Key design considerations include the selection of the appropriate reporter proteins and expression strategies for specific applications.

4.2. Brightness

The brightness of a chromophore is proportional to the product of the extinction coefficient (ε)—which is the capacity for light absorption at a specific wavelength—and the quantum yield (QY), which is the number of fluorescent photons emitted per photon absorbed. A bright fluorophore can be advantageous in that it can increase the signal-to-noise ratio and reduce the amount of excitation light needed, thereby decreasing phototoxic effects. The brightest fluorophores tend to have yellow-green emission spectra, although many naturally occurring fluorophores have been genetically modified to increase brightness. One commonly used example is the enhanced green fluorescent protein (eGFP), which contains a single S65T mutation and generates a five-fold brighter fluorophore with a faster maturation rate than native GFP [30].

For in vivo imaging, fluorophores with red and far-red shifted excitation and emission spectra are preferable because whole tissue absorbs light up to 600 nm in wavelength. Investigators have therefore utilized site-directed mutagenesis and directed evolution approaches to generate brighter red and far-red fluorescent proteins such as mKate2 for utilization in whole animal imaging [9,31,32].

4.3. Photostability

All fluorescent proteins eventually undergo irreversible photobleaching after prolonged or high-intensity illumination. Thus, photostability is of particular importance when selecting a fluorescent protein for experiments with long time series or multiple sequential images. Shaner and colleagues [33] developed an assay for screening libraries of fluorescent proteins for enhanced photostability. Using this method, they generated TagRFP-T and mOrange2, which were 9-fold and 25-fold more photostable than their parental molecules, respectively. A uniform methodology to quantify fluorophore photostability is lacking. Some experts therefore recommend direct comparison of fluorophore photostability for a particular application prior to embarking on a large-scale experiment [11].

4.4. Maturation Time

Once a fluorescent protein has been translated, it must undergo folding and multiple post-translational modifications in order to become fluorescent. This process is called maturation, and the time it requires is the maturation time. Most fluorescent proteins have maturation times of several minutes to 1–2 h. The maturation rate may become important in experiments with short time courses relative to the maturation time of the fluorophore. In contrast, fluorescent proteins with prolonged maturation times [34] or with maturation processes that progress through intermediate chromophores [35] known as "Timers" have been exploited to study the dynamics of intracellular processes on larger time scales [36].

4.5. pH Stability

The pH stability of a fluorescent protein is expressed as the pK_a, which is the pH at which the FP fluorescence is half its maximal brightness. Because pH varies among intracellular compartments, fluorophore pH stability should be considered when labeling proteins that traverse these organelles. The pK_a of wild-type GFP is approximately 4.8, but GFP has a broad range of pH stability from 6 to 10 [37]. Interestingly, some GFP variants, which have been mutated to optimize spectral properties have increased acid sensitivity. This is true of eGFP, which has a pK_a of approximately 6 [30]. Fluorescent proteins with enhanced pH stability—such as mTagBFP2, with a pK_a of 2.7—have been developed [38]. However, the imaging of lysosomes and other acidic organelles is still fraught with challenges [39]. As with other properties, the pH sensitivity of fluorescent proteins has been exploited to develop intracellular pH sensors [40].

4.6. Oligomerization

The tendency of oligomerization is inherent in many fluorescent proteins, which can lead to protein aggregation and attenuation of signal. When these FPs are fused to viral proteins, oligomerization may also render the protein of interest non-functional. While this may be of lesser concern for in vitro assays such as high throughput screenings, it may substantially affect the monitoring of intracellular processes. Wild-type GFP has only a weak tendency to dimerize at high concentrations. In contrast, the first red fluorescent protein discovered was DsRed from *Discosoma* sp., which forms obligate tetramers. The derivation of a monomeric red fluorescent protein from DsRed was elusive until 2002 when Campbell et al. used a directed evolution approach to impede the interaction of each subunit interface to generate mRFP1 [41]. From mRFP1, several monomeric fluorescent proteins have since been generated known collectively as the "mFruits" with improved brightness and photostability and broader excitation and emission spectra compared to the first generation mRFP1 [42]. Monomeric versions of fluorescent proteins are now available across the spectrum of visible light.

4.7. Selection of Luciferase Proteins

As mentioned above, bioluminescent reporter proteins generally have higher sensitivity and signal-to-noise ratio than fluorescent proteins, primarily due to the intrinsically low background bioluminescence observed in cells and tissues. This is useful for in vivo imaging applications, although the luciferase substrate must be administered and must achieve sufficient biodistribution to generate signal.

Important determinants in selecting a bioluminescent label include the size of the luciferase, the intensity and wavelength of luminescence produced, and the bioavailability and ease of administration of the substrate.

Many different luciferases have been identified in nature, but three have been studied extensively and implemented for biomedical research: the firefly *Photinus pyralis* luciferase, the sea pansy *Renilla reniformis* luciferase, and the marine copepod *Gaussia princeps* luciferase. Of these, the firefly luciferase is by far the most commonly utilized bioluminescent reporter. Firefly luciferase is 61 kDa in size and emits light at a wavelength of 562 nm (yellow to green), which is less-readily absorbed by tissues than the blue light (wavelength 480 nm) emitted by the smaller *Renilla* (36 kDa) and *Gaussia* (19.9 kDa) luciferases. Firefly and *Renilla* luciferases also remain intracellular, whereas *Gaussia* luciferase is secreted, making it more difficult to localize intracellular processes. The substrate of firefly luciferase, D-luciferin, also has a good pharmacokinetic profile and sufficient bioavailability. In mice, D-luciferin can be injected intraperitoneally and achieve a peak concentration in 10 min that remains stable for 30 min. Intranasal administration of D-luciferin also confers an increase in bioluminescence of one to two orders of magnitude in the nose and lungs of mice [43,44]. In contrast, the substrate of *Renilla* and *Gaussia* luciferases—coelenterazine—is not soluble in aqueous solutions and commonly precipitates when diluted in 100% alcohol. In animals, coelenterazine must be administered intravenously; it rapidly degrades in vivo, and it lacks the biodistribution profile of D-luciferin.

More recently, a smaller luciferase NanoLuc (19.1 kDa) was engineered based on the luciferase from the deep-sea shrimp *Oplophorus gracilirostris* [15]. This enzyme catalyzes its substrate furimazine to generate high-intensity bioluminescence with greater than two-hour half-life and 150-fold higher specificity than the firefly and *Renilla* luciferases [15]. NanoLuc also exhibits improved physical stability and prolonged signal duration in vitro compared to firefly and *Renilla* luciferases. Because of its increased specificity and unique physical properties, NanoLuc bioluminescent reporter viruses hold promise for many in vivo imaging applications. However, like other marine luciferases, NanoLuc emits blue-shifted light (460 nm), which is more readily absorbed by tissues than the yellow-green light emitted by firefly luciferase. The NanoLuc substrate furimazine must also be administered intravenously, and its pharmacokinetic profile and toxicities have not been fully elucidated. For these reasons, the firefly luciferase is still often considered the preferred luciferase for live imaging applications.

4.8. Expression Strategies for Fluorescent and Bioluminescent Labels

An additional consideration in the design of reporter viruses is the location of the reporter sequence within the viral genome. In general, a reporter protein can be encoded as an additional transcriptional unit (ATU) or as fusion protein with a viral gene product (Figure 2). Because paramyxoviruses are transcribed down a gradient according to proximity to the viral promoter, insertion of ATUs can decrease the expression of downstream genes and result in virus attenuation [45]. One strategy to address this has been to replace endogenous transcription initiation (gene start) signals downstream of the ATU with gene start signals known to be more efficient, to offset the inevitable disassociation of RNPs at gene junctions [46–48]. In paramyxoviruses, ATUs are most commonly placed upstream of the first gene position to preserve the transcriptional balance. Several groups have found that insertion of ATUs at this site was relatively well tolerated and resulted in little to no viral attenuation [49,50]. However, ATUs have been inserted into several other locations within the paramyxovirus genome, including between genes N and P [51]; P and M [44,52–55]; M and F [52]; and H and L [56,57], which have resulted in varying levels of attenuation. Van Remmerden et al. noted that insertion of an ATU encoding eGFP in the first gene position was attenuating, whereas insertion between the SH and G genes was not [58]. Unfortunately, few other studies have directly compared different expression strategies for fluorescent and bioluminescent reporter proteins, and many studies do not report attenuation levels in vitro or in vivo.

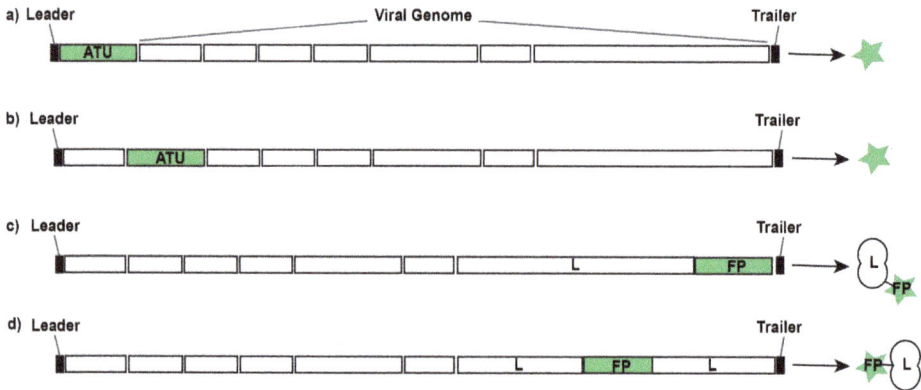

Figure 2. Design strategies for replication-competent fluorescent or bioluminescent reporter viruses. Additional transcription units (ATUs) can be inserted 3′ of the first gene in the viral genome (**a**) or 5′ of the final gene in the viral genome; between viral genes (**b**); or their sequences can be fused with viral genes to generate fusion proteins (FPs) either at the N-terminus or C-terminus (**c**) of a viral gene; or at inert hinge regions within a viral protein (**d**).

Regarding influenza, the gene segments are relatively small (890 to 2341 nucleotides), which can make it difficult for them to accommodate large reporter gene sequences. One site which has been demonstrated to tolerate insertion of reporters is the C-terminus of the PB2 segment, where reporters have been incorporated as either PB2-fusion proteins [59] or as ATUs [60]. Other tolerant sites have included the C-termini of segments PB1 [61], NA [62], and PA [17]. Rearrangements of the influenza genome—e.g., to move NS2 downstream of PB1—also permitted insertion of GFP or *Gaussia* luciferase in the NS gene segment [63]. Reporters have also commonly been incorporated as fusion proteins with NS1 after Manicassamy et al. demonstrated this strategy in 2010 [64].

Fluorescent fusion proteins are commonly designed to study the actions of specific viral proteins. To minimize the steric hindrance of the reporter on the native protein, the fusion protein sequence is often inserted at the COOH- or NH-terminus of the gene [17,64], or within hinge regions [65] which may be biologically inert. Genetic linkers can then be added between the sequences [64] to further diminish steric hindrance. However, the reporter gene may have various effects on protein functions and interactions, which must be considered in the interpretation of results. One strategy recently employed to counteract the attenuating effects of reporter genes was the serial passage of reporter viruses in mice [66]. This strategy generated fluorescent viruses that grew to wild-type titers in mice and broadened the repertoire of fluorescent replication-competent reporter viruses for in vivo use.

5. Applications of Reporter Paramyxoviruses and Orthomyxoviruses

5.1. In Vitro Applications

Fluorescent and bioluminescent reporter viruses have been developed for a variety of uses, ranging from basic science applications to anti-viral screening tools. The first fluorescent reporter paramyxovirus was a GFP-expressing measles virus used to track in vitro MeV spread in human astrocytoma cells [46]. Since then, fluorescent reporter paramyxoviruses have elucidated many aspects of viral pathogenesis, ranging from functional analysis of viral glycoproteins [67], to receptors for viral entry [49], the role of matrix protein [47], the trafficking of viral RNP [57], the role of surface glycoproteins on infection [68], and cellular tropism [69]. Influenza reporter viruses have likewise been utilized to elucidate various aspects of influenza pathogenesis [70,71]. Replacement of a non-essential protein with a fluorescent protein has been one strategy to study its effects on viral pathogenesis. Alternatively, the design of fusion proteins has enabled monitoring of specific viral protein localization and interaction [57,65]. Time-lapse confocal microscopy has also enabled the real-time visualization of intracellular processes with the help of fluorescent labels.

As research tools, fluorescent and bioluminescent reporter viruses have also eased the process of viral detection and quantification. Titration of fluorescent viral foci or quantification of luminescence is often faster than standard plaque assays, which require several days for plaque formation. The correlation between titers obtained by standard methods and fluorescent or bioluminescent methods have been demonstrated in several models [16]. This characteristic has been exploited to generate new virus-neutralization assays which depend on the reduction of fluorescent foci rather than plaque or tissue culture infectious dose reduction [72]. Neutralization assays have also been developed using bioluminescent reporter proteins, which can be automated to measure the reduction of virally-expressed luciferase [73]. Such assays have been developed for influenza viruses [61,74], in addition to several paramyxoviruses including measles virus [75], mumps virus [55], HMPV [54], and RSV [72,73].

Another application of fluorescent reporter viruses as research tools has been in the generation of high-throughput assays to screen antiviral drugs and to measure susceptibility or acquisition of resistance. Such assays have been developed to screen for antivirals against single viruses [76,77] or multiple distinct viruses to identify broad-spectrum inhibitors. In 2013, Yan and colleagues developed a high-throughput assay to screen a library of antiviral compounds against multiple myxoviruses using firefly and *Renilla* luciferase reporter constructs. Their strategy identifiedbroad-spectrum anti-myxovirus compounds in addition to several paramyxovirus- and orthomyxovirus-specific hits [78]. They built upon this technology in 2015, when they synchronized influenza and RSV luciferase reporter expression kinetics to generate a more robust, high-throughput screen [79]. Such assays hold promise to accelerate the discovery of novel antiviral therapies.

5.2. In Vivo Applications

Even in the early 2000s, fluorescently-labeled reporter paramxyoviruses were utilized for in vivo imaging of infection in small animal models. Using an eGFP-expressing measles virus, Duprex et al. imaged measles virus infection of the murine central nervous system [80]. This application has been extended to elucidate MeV target cells and tropism in vivo [56], particularly of oncolytic MeV [81], and to evaluate strategies to target and de-target oncolytic viruses for purposes of gene therapy [82]. Other paramyxoviruses that have been imaged in vivo using fluorescent reporters include Hendra virus [53] and RSV [44]. For RSV, one important application of this technology is to evaluate infectivity and attenuation levels of vaccine strains in animal models [72].

The first fluorescent influenza reporter virus was developed by Manicassamy et al. in 2010, who inserted a GFP reporter in the NS segment to generate an NS1-GFP fusion protein separated by a Gly-Ser-Gly-Gly (GSGG) linker [64]. This group tracked the dynamics of influenza virus infection in mice, identified target cells within the immune system, and measured the effects of antiviral compounds on in vivo infection dynamics. However, the GFP reporter virus did lack genetic stability and was attenuated in cell culture and in mice. To address these obstacles, Fukuyama et al. used a similar expression strategy to generate a reporter influenza expressing Venus, but subsequently serially passaged the virus in mice and identified deattenuated variants that had high genetic stability [66]. This group then generated a panel of mouse-adapted reporter viruses expressing one of four FPs: Venous, eGFP, eCFP, and mCherry. They called this panel of viruses "Color-Flu," which they demonstrated could be co-imaged and utilized to track viral co-infections in animals. However, the application of these viruses to live, real-time imaging is limited by high background autofluorescence and attenuation of signal in vivo.

Bioluminescent reporter viruses are advantageous in that they generate bright signals that can be detected by relatively inexpensive imaging technology. Because of the high signal-to-noise ratio, viruses can be imaged in live animals in real time. The first bioluminescent reporter virus to be utilized for in vivo imaging was herpes simplex virus 1 in 2002. Using firefly luciferase-labeled HSV-1, Luker et al. imaged viral spread in living mice and assessed response to antiviral therapy [13]. Among the paramyxoviruses, this technology has subsequently been applied to the live imaging of Sendai virus (murine parainfluenza type 1) [48] and RSV [44] in mouse models. Burke et al. engineered a firefly luciferase-labeled Sendai virus to image the spatial and temporal progression of infection and transmission following intranasal inoculation of mice. They found that transmissibility to naïve mice was associated with high viral replication in the upper respiratory tract and trachea of donors, but was independent of replication in the donor lower respiratory tracts. This study revealed a previously unknown dichotomy between parainfluenza virus transmissibility and pathogenesis in the mouse model. Bioluminescent imaging has similarly been used to image influenza virus infection dynamics and transmission in mice [17,60,62]. In 2015, Karlsson and colleagues utilized bioluminescence to image a NanoLuc-labeled influenza virus in ferrets, demonstrating the potential of this approach to assess tissue distribution and transmissibility of infection in larger animal models [16].

6. Future Directions

Fluorescent and bioluminescent reporter viruses have revolutionized our understanding of molecular virology and aspects of viral tropism and transmission. These reporters have also enabled the development of high-throughput assays, which have greatly reduced the time and effort necessary to quantify viral infection, to screen antiviral compounds and antibodies, and to perform neutralization assays. Although there is still no one-size-fits-all approach to the design of reporter viruses, new fluorescent and bioluminescent labels continue to be developed which have more optimal characteristics for labeling and detection. The development of new far-red and near infrared fluorescent proteins holds promise for the future of live fluorescent imaging. In addition, improvements in bioluminescent imaging technology could help hone spatial resolution of this already powerful modality.

Author Contributions: C.A.R. and M.L.M. coauthored the manuscript. M.C.C. performed fluorescent imaging shown in Figure 1.

Conflicts of Interest: This work was supported by NIH grants 1R01AI087798, 1U19AI095227, 5K12HD072245, and by Emory University and Children's Healthcare of Atlanta funds.

References

1. Shimomura, O.; Johnson, F.H.; Saiga, Y. Extraction, purification and properties of aequorin, a bioluminescent protein from the luminous hydromedusan, Aequorea. *J. Cell. Comp. Physiol.* **1962**, *59*, 223–239.

2. Prasher, D.C.; Eckenrode, V.K.; Ward, W.W.; Prendergast, F.G.; Cormier, M.J. Primary structure of the Aequorea victoria green-fluorescent protein. *Gene* **1992**, *111*, 229–233.

3. Chalfie, M.; Tu, Y.; Euskirchen, G.; Ward, W.W.; Prasher, D.C. Green fluorescent protein as a marker for gene expression. *Science* **1994**, *263*, 802–805.

4. Luker, K.E.; Luker, G.D. Bioluminescence imaging of reporter mice for studies of infection and inflammation. *Antivir. Res.* **2010**, *86*, 93–100.

5. Cody, C.W.; Prasher, D.C.; Westler, W.M.; Prendergast, F.G.; Ward, W.W. Chemical structure of the hexapeptide chromophore of the Aequorea green-fluorescent protein. *Biochemistry* **1993**, *32*, 1212–1218.

6. Ormo, M.; Cubitt, A.B.; Kallio, K.; Gross, L.A.; Tsien, R.Y.; Reminqton, S.J. Crystal structure of the Aequorea victoria green fluorescent protein. *Science* **1996**, *273*, 1392–1395.

7. Matz, M.V.; Fradkov, A.F.; Labas, Y.A.; Savitsky, A.P.; Zaraisky, A.G.; Markelov, M.L.; Lukyanov, S.A. Fluorescent proteins from nonbioluminescent Anthozoa species. *Nat. Biotechnol.* **1999**, *17*, 969–973.

8. Ando, R.; Hama, H.; Yamamoto-Hino, M.; Mizuno, H.; Miyawaki, A. An optical marker based on the UV-induced green-to-red photoconversion of a fluorescent protein. *Proc. Natl. Acad. Sci. USA* **2002**, *99*, 12651–12656.

9. Shcherbo, D.; Merzlyak, E.M.; Chepurnykh, T.V.; Fradkov, A.F.; Ermakova, G.V.; Solovieva, E.A.; Lukyanov, K.A.; Bogdanova, E.A.; Zaraisky, A.G.; Lukyanov, S.; et al. Bright far-red fluorescent protein for whole-body imaging. *Nat. Methods* **2007**, *4*, 741–746.

10. Shcherbo, D.; Murphy, C.S.; Ermakova, G.V.; Solovievs, E.A.; Chepurnykh, T.V.; Shcheqlov, A.S.; Verkhusha, W.; Pletnev, V.Z.; Hazelwood, K.L.; Roche, P.M.; et al. Far-red fluorescent tags for protein imaging in living tissues. *Biochem. J.* **2009**, *418*, 567–574.

11. Chudakov, D.M.; Matz, M.V.; Lukyanov, S.; Lukyanov, K.A. Fluorescent proteins and their applications in imaging living cells and tissues. *Physiol. Rev.* **2010**, *90*, 1103–1163.

12. Badr, C.E.; Tannous, B.A. Bioluminescence imaging: Progress and applications. *Trends Biotechnol.* **2011**, *29*, 624–633.

13. Luker, G.D.; Bardill, J.P.; Prior, J.L.; Pica, C.M.; Piwnica-Worms, D.; Leib, D.A. Noninvasive bioluminescence imaging of herpes simplex virus type 1 infection and therapy in living mice. *J. Virol.* **2002**, *76*, 12149–12161.

14. Berger, F.; Paulmurugan, R.; Bhaumik, S.; Gambhir, S.S. Uptake kinetics and biodistribution of 14C-D-luciferin—A radiolabeled substrate for the firefly luciferase catalyzed bioluminescence reaction: Impact on bioluminescence based reporter gene imaging. *Eur. J. Nucl. Med. Mol. Imaging* **2008**, *35*, 2275–2285.

15. Hall, M.P.; Unch, J.; Binkowski, B.F.; Vally, M.P.; Butler, B.L.; Wood, M.G.; Otto, P.; Zimmerman, K.; Vidugiris, G.; Machleidt, T.; et al. Engineered luciferase reporter from a deep sea shrimp utilizing a novel imidazopyrazinone substrate. *ACS Chem. Biol.* **2012**, *7*, 1848–1857.

16. Karlsson, E.A.; Meliopoulos, V.A.; Savaqe, C.; Livinqston, B.; Mehle, A.; Schultz-Cherry, S. Visualizing real-time influenza virus infection, transmission and protection in ferrets. *Nat. Commun.* **2015**, *6*, 6378.

17. Tran, V.; Moser, L.A.; Poole, D.S.; Mehle, A. Highly sensitive real-time in vivo imaging of an influenza reporter virus reveals dynamics of replication and spread. *J. Virol.* **2013**, *87*, 13321–13329.

18. Schnell, M.J.; Mebatsion, T.; Conzelmann, K.K. Infectious rabies viruses from cloned cDNA. *EMBO J.* **1994**, *13*, 4195–4203.

19. Lawson, N.D.; Stillman, E.A.; Whitt, M.A.; Rose, J.K. Recombinant vesicular stomatitis viruses from DNA. *Proc. Natl. Acad. Sci. USA* **1995**, *92*, 4477–4481.

20. Radecke, F.; Spieihofer, P.; Schneider, H.; Kaelin, K.; Huber, M.; Dotsch, C.; Christiansen, G.; Billeter, M.A. Rescue of measles viruses from cloned DNA. *EMBO J.* **1995**, *14*, 5773–5784.

21. Garcin, D.; PELET, T.; Calain, P.; Roux, L.; Curran, J.; Kolakofsky, D. A highly recombinogenic system for the recovery of infectious Sendai paramyxovirus from cDNA: Generation of a novel copy-back nondefective interfering virus. *EMBO J.* **1995**, *14*, 6087–6094.

22. Collins, P.L.; HILL, M.G.; Camargo, E.; Grosfeld, H.; Chanock, R.M.; Murphy, B.R. Production of infectious human respiratory syncytial virus from cloned cDNA confirms an essential role for the transcription elongation factor from the 5′ proximal open reading frame of the M2 mRNA in gene expression and provides a capability for vaccine development. *Proc. Natl. Acad. Sci. USA* **1995**, *92*, 11563–11567.

23. Wang, J.; Wang, C.; Feng, N.; Wang, H.; Zheng, X.; Yang, S.; Gao, Y.; Xia, X.; Yi, R.; Liu, X.; et al. Development of a reverse genetics system based on RNA polymerase II for Newcastle disease virus genotype VII. *Virus Genes* **2015**, *50*, 152–155.

24. Martin, A.; Staeheli, P.; Schneider, U. RNA polymerase II-controlled expression of antigenomic RNA enhances the rescue efficacies of two different members of the Mononegavirales independently of the site of viral genome replication. *J. Virol.* **2006**, *80*, 5708–5715.

25. Luytjes, W.; Krystal, M.; Enami, M.; Parvin, J.D.; Palese, P. Amplification, expression, and packaging of foreign gene by influenza virus. *Cell* **1989**, *59*, 1107–1113.

26. Enami, M.; Luytjes, W.; Krystal, M.; Palese, P. Introduction of site-specific mutations into the genome of influenza virus. *Proc. Natl. Acad. Sci. USA* **1990**, *87*, 3802–3805.

27. Neumann, G.; Watanabe, T.; Ito, H.; Watanabe, S.; Goto, H.; Gao, P.; Huqhes, M.; Perez, D.R.; Donis, R.; Hoffmann, E.; et al. Generation of influenza A viruses entirely from cloned cDNAs. *Proc. Natl. Acad. Sci. USA* **1999**, *96*, 9345–9350.

28. Hoffmann, E.; Neumann, G.; Kawaoka, Y.; Hobom, G.; Webster, R.G. A DNA transfection system for generation of influenza A virus from eight plasmids. *Proc. Natl. Acad. Sci. USA* **2000**, *97*, 6108–6113.

29. Neumann, G.; Fujii, K.; Kino, Y.; Kawaoka, Y. An improved reverse genetics system for influenza A virus generation and its implications for vaccine production. *Proc. Natl. Acad. Sci. USA* **2005**, *102*, 16825–16829.

30. Heim, R.; Cubitt, A.B.; Tsien, R.Y. Improved green fluorescence. *Nature* **1995**, *373*, 663–664.

31. Merzlyak, E.M.; Goedhart, J.; Shcherbo, D.; Bulina, M.E.; Shcheglov, A.S.; Fradkov, A.F.; Gaintzeva, A.; Lukyanov, K.A.; Lukyanov, S.; Gadella, T.W.J.; et al. Bright monomeric red fluorescent protein with an extended fluorescence lifetime. *Nat. Methods* **2007**, *4*, 555–557.

32. Lin, M.Z.; Mckeown, M.R.; Ng, H.L.; Aquilera, T.A.; Sjhaner, N.C.; Campbell, R.E.; Adams, S.R.; Gross, L.A.; Ma, W.; Alber, T.; et al. Autofluorescent proteins with excitation in the optical window for intravital imaging in mammals. *Chem. Biol.* **2009**, *16*, 1169–1179.

33. Shaner, N.C.; Lin, M.Z.; Mckeown, M.R.; Steinbach, P.A.; Hazelwood, K.L.; Davidson, M.W.; Tsien, R.Y. Improving the photostability of bright monomeric orange and red fluorescent proteins. *Nat. Methods* **2008**, *5*, 545–551.

34. Verkhusha, V.V.; Otsuna, H.; Awasaki, T.; Oda, H.; Tsukita, S.; Ito, K. An enhanced mutant of red fluorescent protein DsRed for double labeling and developmental timer of neural fiber bundle formation. *J. Biol. Chem.* **2001**, *276*, 29621–29624.

35. Terskikh, A.; Fradkov, A.; Ermakova, G.; Zaraisky, A.; Tan, P.; Kajava, A.V.; Zhao, X.; Lukyanov, S.; Matz, M.; Kim, S.; et al. "Fluorescent timer": Protein that changes color with time. *Science* **2000**, *290*, 1585–1588.

36. Breen, M.; Nogales, A.; Baker, S.F.; Perez, D.R.; Martinez-Sobrido, L. Replication-competent influenza a and b viruses expressing a fluorescent dynamic timer protein for in vitro and in vivo studies. *PLoS ONE* **2016**, *11*, e0147723.

37. Ward, W.W. *Bioluminescence and Chemiluminescence*; Academic Press: New York, USA, 1981; pp. 235–242.

38. Subach, O.M.; Cranfill, P.J.; Davidson, M.W.; Verkhusha, V.V. An enhanced monomeric blue fluorescent protein with the high chemical stability of the chromophore. *PLoS ONE* **2011**, *6*, e28674.

39. Huang, L.; Pike, D.; Sleat, D.E.; Nanda, V.; Lobel, P. Potential pitfalls and solutions for use of fluorescent fusion proteins to study the lysosome. *PLoS ONE* **2014**, *9*, e88893.

40. Kneen, M.; Farinas, J.; Li, Y.X.; Verkman, A.S. Green fluorescent protein as a noninvasive intracellular pH indicator. *Biophys. J.* **1998**, *74*, 1591–1599.

41. Campbell, R.E.; Tour, O.; Plamer, A.E.; Stienbach, P.A.; Baird, G.S.; Zacharias, D.A.; Taien, R.Y. A monomeric red fluorescent protein. *Proc. Natl. Acad. Sci. USA* **2002**, *99*, 7877–7882.

42. Shaner, N.C.; Campbell, R.E.; Steinbach, P.A.; Giepmans, B.N.; Plamer, A.E.; Tsien, R.Y. Improved monomeric red, orange and yellow fluorescent proteins derived from Discosoma sp. red fluorescent protein. *Nat. Biotechnol.* **2004**, *22*, 1567–1572.

43. Buckley, S.M.; Howe, S.J.; Rahim, A.A.; Buning, H.; McIntosh, J.; Wong, S.P.; Baker, A.H.; Nathwani, A.; Thrasher, A.J.; Coutelle, C.; et al. Luciferin detection after intranasal vector delivery is improved by intranasal rather than intraperitoneal luciferin administration. *Hum. Gene Ther.* **2008**, *19*, 1050–1056.

44. Rameix-Welti, M.A.; Le Goffic, R.; Herve, P.L.; Sourimant, J.; Remot, A.; Riffault, S.; Yu, Q.; Galloux, M.; Gault, E.; Eleouet, J.F. Visualizing the replication of respiratory syncytial virus in cells and in living mice. *Nat. Commun.* **2014**, *5*, 5104.

45. Hasan, M.K.; Kato, A.; Shioda, T.; Sakai, Y.; Yu, D.; Nagai, Y. Creation of an infectious recombinant Sendai virus expressing the firefly luciferase gene from the 3′ proximal first locus. *J. Gen. Virol.* **1997**, *78*, 2813–2820.

46. Duprex, W.P.; McQuaid, S.; Hangartner, L.; Billeter, M.A.; Rima, B.K. Observation of measles virus cell-to-cell spread in astrocytoma cells by using a green fluorescent protein-expressing recombinant virus. *J. Virol.* **1999**, *73*, 9568–9575.

47. Naim, H.Y.; Ehler, E.; Billeter, M.A. Measles virus matrix protein specifies apical virus release and glycoprotein sorting in epithelial cells. *EMBO J.* **2000**, *19*, 3576–3585.

48. Burke, C.W.; Mason, J.N.; Surman, S.L.; Jones, B.G.; Dalloneau, E.; Hurwitz, J.L.; Russell, C.J. Illumination of parainfluenza virus infection and transmission in living animals reveals a tissue-specific dichotomy. *PLoS Pathog.* **2011**, *7*, e1002134.

49. Hashimoto, K.; Ono, N.; Tatsuo, H.; Minagawa, H.; Takeda, M.; Takeuchi, K.; Yanaqi, Y. SLAM (CD150)-independent measles virus entry as revealed by recombinant virus expressing green fluorescent protein. *J. Virol.* **2002**, *76*, 6743–6749.

50. Hotard, A.L.; Shaikh, F.Y.; Lee, S.; Yan, D.; Teng, M.N.; Plemper, R.K.; Crowe, J.E., Jr.; Moore, M.L. A stabilized respiratory syncytial virus reverse genetics system amenable to recombination-mediated mutagenesis. *Virology* **2012**, *434*, 129–136.

51. Yoneda, M.; Guillauma, V.; Ikeda, F.; Sakuma, Y.; Sato, H.; Wild, T.F.; Kai, C. Establishment of a Nipah virus rescue system. *Proc. Natl. Acad. Sci. USA* **2006**, *103*, 16508–16513.

52. Burke, C.W.; Bridges, O.; Brown, S.; Rahija, R.; Russell, C.J. Mode of parainfluenza virus transmission determines the dynamics of primary infection and protection from reinfection. *PLoS Pathog.* **2013**, *9*, e1003786.

53. Marsh, G.A.; Virtue, E.R.; Smith, I.S.; Todd, S.; Arinstall, R.; Frazer, L.; Monaghan, P.; Smith, G.A.; Broder, C.C.; Middleton, D.; et al. Recombinant Hendra viruses expressing a reporter gene retain pathogenicity in ferrets. *Virol. J.* **2013**, *10*, 95.

54. De Graaf, M.; Herfst, S.; Schrauwen, E.J.A.; van den Hoohen, B.G.; Osterhaus, A.D.M.E.; Fouchier, R.A.M. An improved plaque reduction virus neutralization assay for human metapneumovirus. *J. Virol. Methods* **2007**, *143*, 169–174.

55. Matsubara, K.; Fujino, M.; Takeuchi, K.; Iwata, S.; Nakayama, T. A new method for the detection of neutralizing antibodies against mumps virus. *PLoS ONE* **2013**, *8*, e65281.

56. Rennick, L.J.; de Vries, R.D.; Carsillo, T.J.; Lemon, K.; van Amerongen, G.; Ludlow, M.; Nquyen, D.T.; Yuksel, S.; Verburqh, R.J.; Haddock, P.; et al. Live-attenuated measles virus vaccine targets dendritic cells and macrophages in muscle of nonhuman primates. *J. Virol.* **2015**, *89*, 2192–2200.

57. Nakatsu, Y.; Ma, X.; Seki, F.; Suzuki, T.; Iwasaki, M.; Yanagi, Y.; Komase, K.; Takeda, M. Intracellular transport of the measles virus ribonucleoprotein complex is mediated by Rab11A-positive recycling endosomes and drives virus release from the apical membrane of polarized epithelial cells. *J. Virol.* **2013**, *87*, 4683–4693.

58. Van Remmerden, Y.; Xu, F.; van Eldik, M.; Heldens, J.G.; Huyisman, W.; Widjojoatmodjo, M.N. An improved respiratory syncytial virus neutralization assay based on the detection of green fluorescent protein expression and automated plaque counting. *Virol. J.* **2012**, *9*, 253.

59. Avilov, S.V.; Mosiy, D.; Munier, S.; Schraidt, O.; Naffakh, N.; Cusack, S. Replication-competent influenza A virus that encodes a split-green fluorescent protein-tagged PB2 polymerase subunit allows live-cell imaging of the virus life cycle. *J. Virol.* **2012**, *86*, 1433–1448.

60. Heaton, N.S.; Leyva-Grado, V.H.; Tan, G.S.; Eqqink, D.; Hai, R.; Palese, P. In vivo bioluminescent imaging of influenza a virus infection and characterization of novel cross-protective monoclonal antibodies. *J. Virol.* **2013**, *87*, 8272–8281.

61. Fulton, B.O.; Palese, P.; Heaton, N.S. Replication-Competent Influenza B Reporter Viruses as Tools for Screening Antivirals and Antibodies. *J. Virol.* **2015**, *89*, 12226–12231.

62. Pan, W.; Dong, Z.; Li, F.; Meng, W.; Feng, L.; Niu, X.; Li, C.; Luo, Q.; Li, Z.; Sun, C.; et al. Visualizing influenza virus infection in living mice. *Nat. Commun.* **2013**, *4*, 2369.

63. Sutton, T.C.; Obadan, A.; Laviqne, J.; Chan, H.; Li, W.; Perez, D.R. Genome rearrangement of influenza virus for anti-viral drug screening. *Virus Res.* **2014**, *189*, 14–23.

64. Manicassamy, B.; Manicassamy, S.; Belicha-Villanueva, A.; Pisanelli, G.; Pulendran, B.; Garcia-Sastre, A. Analysis of in vivo dynamics of influenza virus infection in mice using a GFP reporter virus. *Proc. Natl. Acad. Sci. USA* **2010**, *107*, 11531–11536.

65. Duprex, W.P.; Collins, F.M.; Rima, B.K. Modulating the function of the measles virus RNA-dependent RNA polymerase by insertion of green fluorescent protein into the open reading frame. *J. Virol.* **2002**, *76*, 7322–7328.

66. Fukuyama, S.; Kastura, H.; Zhao, D.; Ozawa, M.; Ando, T.; Shoemaker, J.E.; Ishikawa, I.; Yamada, S.; Neumann, G.; Watanabe, S.; et al. Multi-spectral fluorescent reporter influenza viruses (Color-flu) as powerful tools for in vivo studies. *Nat. Commun.* **2015**, *6*, 6600.

67. Techaarpornkul, S.; Barretto, N.; Peeples, M.E. Functional analysis of recombinant respiratory syncytial virus deletion mutants lacking the small hydrophobic and/or attachment glycoprotein gene. *J. Virol.* **2001**, *75*, 6825–6834.

68. Batonick, M.; Oomens, A.G.; Wertz, G.W. Human respiratory syncytial virus glycoproteins are not required for apical targeting and release from polarized epithelial cells. *J. Virol.* **2008**, *82*, 8664–8672.

69. Von Messling, V.; Milosevic, D.; Cattaneo, R. Tropism illuminated: lymphocyte-based pathways blazed by lethal morbillivirus through the host immune system. *Proc. Natl. Acad. Sci. USA* **2004**, *101*, 14216–14221.

70. Baker, S.F.; Noqales, A.; Finch, C.; Tuffy, K.M.; Domm, W.; Perez, D.R.; Topham, D.J.; Martinez-Sobrido, L. Influenza A and B virus intertypic reassortment through compatible viral packaging signals. *J. Virol.* **2014**, *88*, 10778–10791.

71. Roberts, K.L.; Manicassamy, B.; Lamb, R.A. Influenza A virus uses intercellular connections to spread to neighboring cells. *J. Virol.* **2015**, *89*, 1537–1549.

72. Meng, J.; Lee, S.; Hotard, A.L.; Moore, M.L. Refining the balance of attenuation and immunogenicity of respiratory syncytial virus by targeted codon deoptimization of virulence genes. *MBio* **2014**, *5*, e01704–e01714.

73. Fuentes, S.; Crim, R.L.; Beeler, J.; Teng, M.N.; Golding, H.; Khurana, S. Development of a simple, rapid, sensitive, high-throughput luciferase reporter based microneutralization test for measurement of virus neutralizing antibodies following Respiratory Syncytial Virus vaccination and infection. *Vaccine* **2013**, *31*, 3987–3994.

74. Baker, S.F.; Nogales, A.; Santiago, F.W.; Topham, D.J.; Martinez-Sobrido, L. Competitive detection of influenza neutralizing antibodies using a novel bivalent fluorescence-based microneutralization assay (BiFMA). *Vaccine* **2015**, *33*, 3562–3570.

75. Fujino, M.; Yoshida, N.; Kimura, K.; Zhou, J.; Motegi, Y.; Komase, K.; Nakayama, T. Development of a new neutralization test for measles virus. *J. Virol. Methods* **2007**, *142*, 15–20.

76. Kwanten, L.; De Clerck, B.; Roymans, D. A fluorescence-based high-throughput antiviral compound screening assay against respiratory syncytial virus. *Methods Mol. Biol.* **2013**, *1030*, 337–344.

77. Lo, M.K.; Nichol, S.T.; Spiropoulou, C.F. Evaluation of luciferase and GFP-expressing Nipah viruses for rapid quantitative antiviral screening. *Antivir. Res.* **2014**, *106*, 53–60.

78. Yan, D.; Krumm, S.A.; Sun, A.; Steinhuaer, D.A.; Luo, M.; Moore, M.L.; Plemper, R.K. Dual myxovirus screen identifies a small-molecule agonist of the host antiviral response. *J. Virol.* **2013**, *87*, 11076–11087.

79. Yan, D.; Weisshaar, M.; Lamb, K.; Chung, H.K.; Lin, M.Z.; Plemper, R.K. Replication-Competent Influenza Virus and Respiratory Syncytial Virus Luciferase Reporter Strains Engineered for Co-Infections Identify Antiviral Compounds in Combination Screens. *Biochemistry* **2015**, *54*, 5589–5604.

80. Duprex, W.P.; McQuaid, S.; Rima, B.K. Measles virus-induced disruption of the glial-fibrillary-acidic protein cytoskeleton in an astrocytoma cell line (U-251). *J. Virol.* **2000**, *74*, 3874–3880.

81. Peng, K.W.; Frenzke, M.; Myers, R.; Soeffker, D.; Harvery, M.; Greiner, S.; Galanis, E.; Cattaneo, R.; Federspiel, M.J.; Russell, S.J. Biodistribution of oncolytic measles virus after intraperitoneal administration into Ifnar-CD46Ge transgenic mice. *Hum. Gene Ther.* **2003**, *14*, 1565–1577.

82. Baertsch, M.A.; Leber, M.F.; Bossow, S.; Singh, M.; Engeland, C.E.; Alber, J.; Grossardt, C.; Japer, D.; von Kalle, C.; Ungerechts, G. MicroRNA-mediated multi-tissue detargeting of oncolytic measles virus. *Cancer Gene Ther.* **2014**, *21*, 373–380.

MDPI AG

St. Alban-Anlage 66

4052 Basel, Switzerland

Tel. +41 61 683 77 34

Fax +41 61 302 89 18

http://www.mdpi.com

Viruses Editorial Office

E-mail: viruses@mdpi.com

http://www.mdpi.com/journal/viruses